Polyamine Metabolism in Disease and Polyamine-Targeted Therapies

Polyamine Metabolism in Disease and Polyamine-Targeted Therapies

Special Issue Editor

Dr. Tracy Murray-Stewart

MDPI • Basel • Beijing • Wuhan • Barcelona • Belgrade

MDPI

Special Issue Editor
Dr. Tracy Murray-Stewart
Johns Hopkins University
USA

Editorial Office
MDPI
St. Alban-Anlage 66
4052 Basel, Switzerland

This is a reprint of articles from the Special Issue published online in the open access journal *Medical Sciences* (ISSN 2076-3271) from 2017 to 2018 (available at: https://www.mdpi.com/journal/medsci/special_issues/Polyamine_Metabolism_Disease)

For citation purposes, cite each article independently as indicated on the article page online and as indicated below:

LastName, A.A.; LastName, B.B.; LastName, C.C. Article Title. *Journal Name* **Year**, *Article Number*, Page Range.

ISBN 978-3-03921-152-4 (Pbk)
ISBN 978-3-03921-153-1 (PDF)

Contents

About the Special Issue Editor

Tracy Murray-Stewart, Ph.D., is a Research Associate on the faculty of the Johns Hopkins University School of Medicine. As a member of the Division of Cancer Biology at the Sidney Kimmel Comprehensive Cancer Center at Johns Hopkins for more than 20 years, Dr. Murray-Stewart studies the contributions of polyamines to the carcinogenic process as well as potential of the polyamine metabolic pathway as a target for therapeutic intervention. She received her Ph.D. and M.A. degrees from the Johns Hopkins University School of Medicine and her B.S. from Salisbury University. Along with her mentor, Dr. Robert A. Casero, Jr., she has authored numerous articles, reviews, and book chapters on polyamines as they relate to cancer and other pathologies.

Preface to "Polyamine Metabolism in Disease and Polyamine-Targeted Therapies"

Polyamines are ubiquitous polycations that are essential for all cellular life. The most common mammalian polyamines—spermine, spermidine, and putrescine—exist in millimolar intracellular concentrations that are tightly regulated through biosynthesis, catabolism, and transport [1]. Polyamines interact with and have regulatory roles involving including nucleic acids, proteins, and ion channels [2–4]. Accordingly, alterations in polyamine metabolism affect cellular proliferation and survival through changes in gene expression and transcription, translation, autophagy, oxidative stress, and apoptosis [5–8]. These multifaceted effects of polyamine dysregulation contribute to multiple disease processes, but also implicate polyamine metabolism and function as targets for preventive or therapeutic intervention. The correlation between elevated polyamine levels and cancer is well established, and ornithine decarboxylase (ODC), the rate-limiting polyamine biosynthetic enzyme, is considered a MYC-driven oncogene [9]. Furthermore, induced polyamine catabolism contributes to carcinogenesis, which is associated with certain forms of chronic infection and/or inflammation through the production of reactive oxygen species [5,10]. These and other characteristics specific to cancer cells have led to the development of polyamine-based agents and inhibitors targeting the polyamine metabolic pathway for chemotherapeutic and chemopreventive benefits. In addition to cancer, polyamines are involved in the pathologies of neurodegenerative diseases, parasitic and infectious diseases, wound healing, ischemia/reperfusion injuries, and certain age-related conditions, as polyamines are known to decrease with age [11]. As in cancer, polyamine-based therapies for these conditions are an area of active investigation. With recent advances in immunotherapy, interest has increased regarding the polyamine-associated modulation of immune responses as well as potential immunoregulation related to polyamine metabolism, the results of which could have relevance to multiple disease processes. The goal of this Special Issue of *Medical Sciences* is to present the most recent advances in polyamine research as it relates to health, disease, and/or therapy.

As polyamines are upregulated in hyperproliferative cells, a significant portion of this Special Issue focuses on polyamine regulation in cancer and ways in which the polyamine metabolic pathway may be targeted for therapy. Pancreatic ductal adenocarcinoma (PDAC), the most common type of pancreatic cancer, has a five-year survival rate of less than 9% [12]. In this Issue, a study by Massaro and colleagues investigates the antiproliferative effects of potential treatment combinations using a set of four inhibitors targeting different components of polyamine biosynthesis and transport, based on the fact that most PDAC cases are associated with oncogenic activation capable of stimulating polyamine biosynthesis and uptake [13]. They find that adding an inhibitor of spermine synthase to the combination of an ODC inhibitor and a polyamine uptake inhibitor most effectively decreases intracellular polyamine pools and inhibits cellular proliferation. A study by Peters *et al.* describes a novel arylmethyl-polyamine (AP) analogue with antitumor effects on melanoma cells with mutated serine/threonine protein kinase B-Raf (*BRAF*) [14]. BRAF is overexpressed in more than 50% of all melanomas, and although BRAF inhibitors are effective first-line treatments, essentially all patients relapse in less than a year [15,16]. The authors identify increased polyamine transport activity in constitutively active *BRAF*-mutant cells that correlates to increased sensitivity to the cytotoxic effects of AP, which was further increased by adding an inhibitor of ODC, difluoromethylornithine (DFMO). Furthermore, treatment of BRAF-inhibitor-resistant melanoma spheroid cultures with AP restored the sensitivity of these cultures toward BRAF inhibitors. As activating *BRAF* mutations are common in many solid tumor types, the implications of this study may extend beyond melanoma. Likewise, additional polyamine analogues known to use the polyamine transport system exist that have successfully completed early clinical trials. Thus, the knowledge of upregulated polyamine transport in *BRAF*-mutant tumors suggests a method for patient stratification in future clinical trial design. The use of a *Drosophila* epithelial model as an initial in vivo screen for novel polyamine transport inhibitors (PTIs) is described by Wang *et al.* [17]. The authors investigate the abilities of potential PTIs to interfere with the uptake of the natural polyamines or toxic polyamine analogues by observing the development of imaginal discs into legs. Their results suggest similarities between the polyamine transport systems of *Drosophila* and mammals and provide further evidence for multiple modes of selective polyamine uptake. Additionally, with regard to targeting polyamine metabolism as a chemotherapeutic strategy, Thomas and Thomas have provided a thorough review of animal and cellular studies investigating polyamine analogues as inhibitors of breast cancer growth [18].

The increased requirement for polyamines in oncogene-driven cancer types suggests that dysregulated polyamine metabolism is a downstream effect of oncogene activation [9]. In a comprehensive review by Flynn and Hogarty, activation of the *MYC* family of oncogenes and its downstream effects on polyamine homeostasis are presented in the context of their role in protein translation and synthesis to fulfill the increased biomass needs of tumor cells [19]. Targeting this dependency of tumor cells on polyamines to sustain tumor growth is discussed as a therapeutic strategy that may circumvent the challenges associated with direct pharmacological inhibition of the MYC proteins. As MYC-driven upregulation of ODC is a rate-limiting step in polyamine biosynthesis, its inhibition by DFMO has demonstrated utility in the chemopreventive setting, particularly in patients with elevated risk of developing or relapsing to certain cancers. This Issue features a review of these studies in colon cancer and neuroblastoma [20], and a novel study by Weicht and colleagues investigates the potential use of DFMO in preventing recurrence of osteosarcoma, the most common bone tumor in pediatric patients [21]. DFMO treatment not only decreases proliferation of osteosarcoma cells, it also results in increased expression of late osteogenic differentiation markers indicative of terminal differentiation, and these effects persist following drug removal. Thus, inhibiting polyamine biosynthesis might serve as an adjunctive therapy for osteosarcoma patients at high risk of relapse. Finally, like DFMO, the plant polyphenol curcumin has shown promise as both a chemopreventive and chemotherapeutic agent by targeting components of polyamine metabolism. A collection of studies investigating the antitumor effects of curcumin in animal and cell culture models and the molecular mechanisms associated with these effects reviews the potential of this natural plant derivative and dietary component as a modulator of the polyamine metabolic pathway [22].

Beyond its antitumor effects, DFMO targets hyperproliferating cells in general, and thus has proven useful for the treatment of other pathologies associated with rapid cell growth. Along with the outcomes of DFMO studies in cancer mentioned above, LoGiudice and colleagues review the pharmacokinetic and pharmacodynamic properties of DFMO as well as the formulations and conditions for which DFMO is approved for use, including female hirsutism and African sleeping sickness [20]. Polyamines are ubiquitous components of both prokaryotes and eukaryotes, and a study in this Issue by Nakamya and colleagues suggests a role for polyamines in capsule formation of *Streptococcus pneumoniae* [23]. Deleting a gene required for cadaverine synthesis resulted in decreased capsule synthesis and protein expression, thereby attenuating pneumococcal virulence. These data suggest a way through which targeting the polyamine pathway of an infectious agent may have antimicrobial benefits. Furthermore, as increasing attention is being directed toward the importance of the human immune system and its interactions and influences within the cellular environment, a timely review by Hesterberg and colleagues assesses the current knowledge pertaining to polyamine metabolic requirements for normal immune cell function as well as tumor-associated immunity and autoimmune conditions [24]. Immune and inflammatory cells also play major roles in asthma pathophysiology, and increasing evidence suggests that modulation of polyamine levels is a contributing factor affecting the inflammatory cell response as well as airway hyperresponsiveness. The details of these contributions and potential for targeting polyamines as a treatment for severe asthma are discussed in a review written by Vaibhav Jain [25].

With the emergence of advanced techniques and new discoveries, aspects of polyamine metabolism and its regulation continue to be elucidated at the molecular level. In this Issue, Nowotarski and Shantz further investigate the post-transcriptional regulation of ODC. Specifically, they identify the location of a negative regulatory element within the 3'-untranslated region of the ODC mRNA transcript [26]. A research article contributed by Abaandou and Shiloach used CRISPR gene editing to knock out the OAZ1 (ODC antizyme 1) gene in human embryonic kidney (HEK293) cells [27]. In addition to increased intracellular polyamine levels, the resultant cells gained an enhanced ability to produce recombinant proteins following either stable or transient transfection, without negative growth or metabolic effects. These data suggest the utility of OAZ1-deficient human cells in recombinant protein production as an alternative to commonly used rodent cell lines, which can have post-translational modification limitations. Further upstream, the regulation of polyamine biosynthesis and uptake is maintained by a family of antizyme inhibitors (AZINs), which are the subject of a review written by Ramos-Molina *et al.* [28]. Although both AZIN1 and AZIN2 interact with antizyme to positively effect intracellular polyamine concentration, their tissue distribution and physiological roles differ. While AZIN1 is associated with proliferation, survival, and oncogenic potential, AZIN2 tends to associate with more differentiated cells, including those of the brain and testes. The distribution of AZIN2 expression in the brain is complex and suggests important roles for AZIN2 in central nervous system physiology. Interestingly, evidence is also reviewed suggesting polyamine-independent functions for AZIN2, including a role in secretory processes.

In addition to AZIN2, other regulators of polyamines have drawn recent attention in the context of neurological pathology. Snyder–Robinson Syndrome (SRS) is a rare X-linked intellectual disability syndrome that results from a loss-of-function mutation in the spermine synthase gene [29]. The biochemical outcome is an accumulation of intracellular spermidine with very little detectable spermine [30]. Males affected with SRS display a broad range of symptoms, the most common of which include osteoporosis, hypotonia, intellectual disability, and seizures. In this Special Issue, the first comprehensive examination of polyamine homeostasis in SRS patient-derived tissue is described [31]. Using lymphoblastoid cell lines from SRS or unaffected males, the major enzymes and proteins responsible for maintaining polyamine homeostasis are assayed, and the previous assumption that SRS cells would have down-regulated polyamine transport systems was shown to be erroneous. Importantly, this study provided evidence that supplying exogenous spermine could normalize the intracellular spermidine and spermine ratios, thus providing hope for a treatment for SRS, of which there is currently none. Diminished muscle bulk is a common manifestation in SRS males, and studies have suggested that spermine oxidase (SMOX), which can serve as a source of spermidine, plays a role in skeletal muscle differentiation and physiology. A review by Cervelli and colleagues explores these studies and the current evidence associating changes in polyamine levels with neuromuscular diseases and conditions, such as Duchenne's muscular dystrophy, amyotrophic lateral sclerosis (ALS), and muscle atrophy [32].

In summary, this Issue has hopefully provided a collection of research articles and reviews of value to those new to the polyamine field as well as enhanced the awareness of those already seasoned in the field to recent advances outside of their own laboratory's interests. With new molecular tools and systems available, it is an exciting time in the field of polyamines, as investigators drill down into the critical molecular mechanisms in which polyamines are involved.

Tracy Murray-Stewart, Ph.D

Guest Editor

References

1. Pegg, A.E.; McCann, P.P. Polyamine metabolism and function. *Am. J. Physiol.* **1982**, *243*, C212–C221.
2. Nichols, C.G.; Lee, S.-j. Polyamines and potassium channels: A 25-year romance. *J. Biol. Chem.* **2018**, *293*, 18779–18788, doi:10.1074/jbc.TM118.003344.
3. Bowie, D. Polyamine-mediated channel block of ionotropic glutamate receptors and its regulation by auxiliary proteins. *J. Biol. Chem.* **2018**, *293*, 18789–18802, doi:10.1074/jbc.TM118.003794.
4. Pasini, A.; Caldarera, C.M.; Giordano, E. Chromatin remodeling by polyamines and polyamine analogs. *Amino Acids* **2014**, *46*, 595–603, doi:10.1007/s00726-013-1550-9.
5. Murray Stewart, T.; Dunston, T.T.; Woster, P.M.; Casero, R.A., Jr. Polyamine catabolism and oxidative damage. *J. Biol. Chem.* **2018**, doi:10.1074/jbc.TM118.003337.
6. Pegg, A.E. Mammalian polyamine metabolism and function. *IUBMB Life* **2009**, *61*, 880–894, doi:10.1002/iub.230.
7. Igarashi, K.; Kashiwagi, K. Modulation of cellular function by polyamines. *Int. J. Biochem. Cell Biol.* **2010**, *42*, 39–51, doi:10.1016/j.biocel.2009.07.009.
8. Dever, T.E.; Ivanov, I.P. Roles of polyamines in translation. *J. Biol. Chem.* **2018**, *293*, 18719–18729, doi:10.1074/jbc.TM118.003338.
9. Casero, R.A., Jr.; Murray Stewart, T.; Pegg, A.E. Polyamine metabolism and cancer: Treatments, challenges and opportunities. *Nat. Rev. Cancer* **2018**, *18*, 681, doi:10.1038/s41568-018-0050-3.
10. Murray-Stewart, T.; Casero, R., Jr. Mammalian Polyamine Catabolism. In *Polyamines*; Kusano, T., Suzuki, H., Eds.; Springer: Tokyo, Japan, 2015; pp. 61–75, doi:10.1007/978-4-431-55212-3_5.
11. Minois, N.; Carmona-Gutierrez, D.; Madeo, F. Polyamines in aging and disease. *Aging* **2011**, *3*, 716–732, doi:10.18632/aging.100361.
12. Ilic, M.; Ilic, I. Epidemiology of pancreatic cancer. *World J. Gastroenterol.* **2016**, *22*, 9694–9705, doi:10.3748/wjg.v22.i44.9694.
13. Massaro, C.; Thomas, J.; Phanstiel, O. Investigation of Polyamine Metabolism and Homeostasis in Pancreatic Cancers. *Med. Sci.* **2017**, *5*, 32.
14. Peters, M.C.; Minton, A.; Phanstiel IV, O.; Gilmour, S.K. A Novel Polyamine-Targeted Therapy for BRAF Mutant Melanoma Tumors. *Med. Sci.* **2018**, *6*, 3.

15. Davies, H.; Bignell, G.R.; Cox, C.; Stephens, P.; Edkins, S.; Clegg, S.; Teague, J.; Woffendin, H.; Garnett, M.J.; Bottomley, W.; et al. Mutations of the BRAF gene in human cancer. *Nature* **2002**, *417*, 949–954, doi:10.1038/nature00766.

16. Flaherty, K.T.; Puzanov, I.; Kim, K.B.; Ribas, A.; McArthur, G.A.; Sosman, J.A.; O'Dwyer, P.J.; Lee, R.J.; Grippo, J.F.; Nolop, K.; et al. Inhibition of mutated, activated BRAF in metastatic melanoma. *N. Engl. J. Med.* **2010**, *363*, 809–819, doi:10.1056/NEJMoa1002011.

17. Wang, M.; Phanstiel, O.; Von Kalm, L. Evaluation of Polyamine Transport Inhibitors in a Drosophila Epithelial Model Suggests the Existence of Multiple Transport Systems. *Med. Sci.* **2017**, *5*, 27.

18. Thomas, T.J.; Thomas, T. Cellular and Animal Model Studies on the Growth Inhibitory Effects of Polyamine Analogues on Breast Cancer. *Med. Sci.* **2018**, *6*, 24.

19. Flynn, A.T.; Hogarty, M.D. Myc, Oncogenic Protein Translation, and the Role of Polyamines. *Med. Sci. (Basel)* **2018**, *6*, 41, doi:10.3390/medsci6020041.

20. LoGiudice, N.; Le, L.; Abuan, I.; Leizorek, Y.; Roberts, S.C. Alpha-Difluoromethylornithine, an Irreversible Inhibitor of Polyamine Biosynthesis, as a Therapeutic Strategy against Hyperproliferative and Infectious Diseases. *Med. Sci. (Basel)* **2018**, *6*, 12, doi:10.3390/medsci6010012.

21. Weicht, R.R.; Schultz, C.R.; Geerts, D.; Uhl, K.L.; Bachmann, A.S. Polyamine Biosynthetic Pathway as a Drug Target for Osteosarcoma Therapy. *Med. Sci.* **2018**, *6*, 65.

22. Murray-Stewart, T.; Casero, R.A. Regulation of Polyamine Metabolism by Curcumin for Cancer Prevention and Therapy. *Med. Sci.* **2017**, *5*, 38.

23. Nakamya, M.F.; Ayoola, M.B.; Park, S.; Shack, L.A.; Swiatlo, E.; Nanduri, B. The Role of Cadaverine Synthesis on Pneumococcal Capsule and Protein Expression. *Med. Sci.* **2018**, *6*, 8.

24. Hesterberg, R.S.; Cleveland, J.L.; Epling-Burnette, P.K. Role of Polyamines in Immune Cell Functions. *Med. Sci.* **2018**, *6*, 22.

25. Jain, V. Role of Polyamines in Asthma Pathophysiology. *Med. Sci.* **2018**, *6*, 4.

26. Nowotarski, S.L.; Shantz, L.M. The ODC 3'-Untranslated Region and 5'-Untranslated Region Contain cis-Regulatory Elements: Implications for Carcinogenesis. *Med. Sci.* **2018**, *6*, 2.

27. Abaandou, L.; Shiloach, J. Knocking out Ornithine Decarboxylase Antizyme 1 (OAZ1) Improves Recombinant Protein Expression in the HEK293 Cell Line. *Med. Sci.* **2018**, *6*, 48.

28. Ramos-Molina, B.; Lambertos, A.; Peñafiel, R. Antizyme Inhibitors in Polyamine Metabolism and Beyond: Physiopathological Implications. *Med. Sci.* **2018**, *6*, 89.

29. Arena, J.F.; Schwartz, C.; Ouzts, L.; Stevenson, R.; Miller, M.; Garza, J.; Nance, M.; Lubs, H. X-linked mental retardation with thin habitus, osteoporosis, and kyphoscoliosis: linkage to Xp21.3-p22.12. *Am. J. Med. Genet.* **1996**, *64*, 50–58, doi:10.1002/(SICI)1096-8628(19960712)64:1<50::AID-AJMG7>3.0.CO;2-V.

30. Schwartz, C.E.; Wang, X.; Stevenson, R.E.; Pegg, A.E. Spermine synthase deficiency resulting in X-linked intellectual disability (Snyder-Robinson syndrome). *Methods Mol. Biol.* **2011**, *720*, 437–445, doi:10.1007/978-1-61779-034-8_28.

31. Murray-Stewart, T.; Dunworth, M.; Foley, J.R.; Schwartz, C.E.; Casero, R.A. Polyamine Homeostasis in Snyder-Robinson Syndrome. *Med. Sci.* **2018**, *6*, 112.

32. Cervelli, M.; Leonetti, A.; Duranti, G.; Sabatini, S.; Ceci, R.; Mariottini, P. Skeletal Muscle Pathophysiology: The Emerging Role of Spermine Oxidase and Spermidine. *Med. Sci. (Basel)* **2018**, *6*, 14, doi:10.3390/medsci6010014.

medical
sciences

MDPI

Article

Polyamine Homeostasis in Snyder-Robinson Syndrome

Tracy Murray-Stewart [1], Matthew Dunworth [1], Jackson R. Foley [1], Charles E. Schwartz [2] and Robert A. Casero Jr. [1,*]

[1] Sidney Kimmel Comprehensive Cancer Center, Johns Hopkins University, Baltimore, MD 21287, USA; tmurray2@jhmi.edu (T.M.-S.); matthewdunworth@jhmi.edu (M.D.); jfoley13@jhmi.edu (J.R.F.)
[2] The Greenwood Genetic Center, Greenwood, SC 29646, USA; ceschwartz@ggc.org
* Correspondence: rcasero@jhmi.edu; Tel.: +1-410-955-8580

Received: 15 November 2018; Accepted: 3 December 2018; Published: 7 December 2018

Abstract: Loss-of-function mutations of the spermine synthase gene (*SMS*) result in Snyder-Robinson Syndrome (SRS), a recessive X-linked syndrome characterized by intellectual disability, osteoporosis, hypotonia, speech abnormalities, kyphoscoliosis, and seizures. As SMS catalyzes the biosynthesis of the polyamine spermine from its precursor spermidine, SMS deficiency causes a lack of spermine with an accumulation of spermidine. As polyamines, spermine, and spermidine play essential cellular roles that require tight homeostatic control to ensure normal cell growth, differentiation, and survival. Using patient-derived lymphoblast cell lines, we sought to comprehensively investigate the effects of SMS deficiency on polyamine homeostatic mechanisms including polyamine biosynthetic and catabolic enzymes, derivatives of the natural polyamines, and polyamine transport activity. In addition to decreased spermine and increased spermidine in SRS cells, ornithine decarboxylase activity and its product putrescine were significantly decreased. Treatment of SRS cells with exogenous spermine revealed that polyamine transport was active, as the cells accumulated spermine, decreased their spermidine level, and established a spermidine-to-spermine ratio within the range of wildtype cells. SRS cells also demonstrated elevated levels of tissue transglutaminase, a change associated with certain neurodegenerative diseases. These studies form a basis for further investigations into the leading biochemical changes and properties of *SMS*-mutant cells that potentially represent therapeutic targets for the treatment of Snyder-Robinson Syndrome.

Keywords: Snyder-Robinson Syndrome; spermine synthase; X-linked intellectual disability; polyamine transport; spermidine; spermine; transglutaminase

1. Introduction

First described in 1969 [1], Snyder-Robinson Syndrome (SRS) is an X-linked intellectual disability syndrome resulting from mutation of the spermine synthase (*SMS*) gene, located at chromosome Xp22.11 [2]. Active only as a homodimer [3], SMS catalyzes the production of spermine (SPM) from its precursor, spermidine (SPD), via the transfer of an aminopropyl group, which is derived from decarboxylated S-adenosylmethionine (dcAdoMet) through the action of S-adenosylmethionine decarboxylase (AdoMetDC; Figure 1). SRS males with the most severe phenotypes lack functional SMS protein, biochemically resulting in elevated levels of intracellular spermidine and near complete depletion of spermine. Spermidine and spermine, along with their precursor putrescine (PUT), constitute the mammalian polyamines, organic polycations that are absolutely essential for growth and proliferation. As their amine groups are protonated at physiological pH, polyamines interact with negatively charged intracellular moieties, including nucleic acids, chromatin, ion channels, certain proteins, and phospholipids [4–7]. Thus, alterations in intracellular polyamine concentrations can elicit potentially detrimental effects, and polyamine homeostasis must be tightly regulated through biosynthesis, catabolism, uptake, and excretion. Additionally, the primary amino groups of polyamines are natural substrates for transglutaminase-catalyzed reactions that result in protein cross-linking that has been associated with a number of pathologies [8,9]. As polyamines have essential roles in growth, differentiation, and development, the imbalance that occurs in SRS results in a combination of clinical manifestations including moderate-to-severe cognitive impairment, osteoporosis, asthenic build, low muscle mass, facial asymmetry, speech abnormalities, and seizures [10].

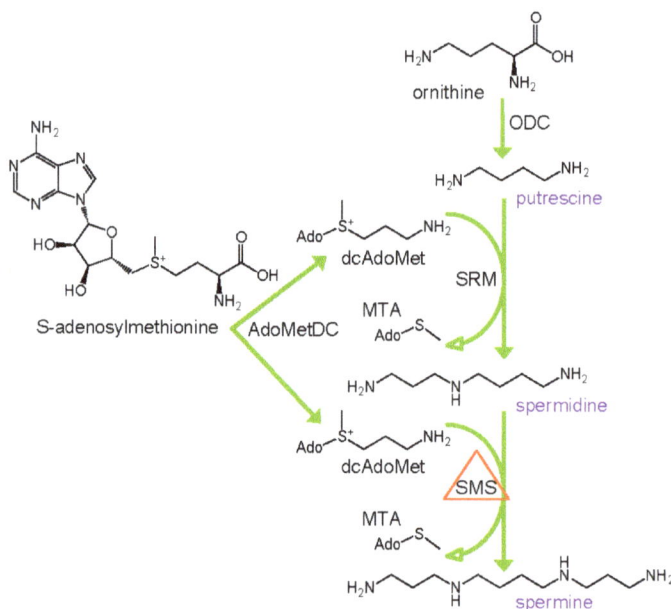

Figure 1. Mammalian polyamine biosynthesis. Polyamines are indicated in purple. Putrescine is created from ornithine via ornithine decarboxylase (ODC). Conversion of putrescine to spermidine and spermidine to spermine occurs through spermidine synthase (SRM) or spermine synthase (SMS), respectively. Both enzymes require the activity of S-adenosylmethionine decarboxylase (AdoMetDC) for the provision of the aminopropyl group donor (decarboxylated AdoMet, dcAdoMet). Snyder-Robinson Syndrome (SRS) patients are deficient in SMS activity, resulting in decreased spermine and accumulation of spermidine. MTA = methylthioadenosine.

The current study investigates the biochemical effects of decreased SMS activity on the individual enzymes in polyamine metabolism as well as its effect on polyamine uptake from the extracellular environment and transglutaminase (TG) expression. SRS patient-derived lymphoblastoid cell lines are used that range in severity of *SMS* loss-of-function and spermine pool depletion, in comparison with those from healthy donors, to ascertain compensatory changes that might occur in an attempt to regulate polyamine homeostasis. Results of these studies provide useful background knowledge towards the goal of developing treatment strategies for these patients, of which there are currently none.

2. Materials and Methods

2.1. Cell Lines and Culture Conditions

The lymphoblastoid cell lines were generated by transformation with Epstein–Barr virus as previously described [11–13]. The lines were derived from three SRS patients and two healthy male donors. Cells were grown in RPMI-1640 supplemented with 15% fetal bovine serum (Gemini Bio-Products, Sacramento, CA, USA), 2 mM glutamine, non-essential amino acids, sodium pyruvate, and penicillin/streptomycin in a humidified 5% CO_2 atmosphere at 37 °C. Uptake experiments were conducted in the presence of 1 mM aminoguanidine (AG) to inhibit extracellular oxidation of spermine by bovine serum amine oxidase present in the culture medium. For these experiments, cells were incubated with either exogenous SPM (5 μM) or the polyamine analog bis(ethyl)norspermine (BENSpm) (10 μM) for 24 h prior to collection and preparation for HPLC analysis. BENSpm was synthesized as previously reported [14].

2.2. Assay of Polyamine Concentrations and Enzyme Activities

Cell lysates were acid extracted and labeled with dansyl chloride, followed by determination of intracellular polyamine concentrations via HPLC, as previously described [15]. Diaminoheptane, PUT, SPD, SPM, and acetylated derivatives of SPD and SPM used for HPLC standards were purchased from Sigma Chemical Co. (St. Louis, MO, USA). For HPLC analysis of culture medium, after 24 h of growth, each culture was pelleted and 2 mLs (1/5 of the total volume) of medium were removed, dried in a speed-vac, and resuspended in perchloric acid for acid extraction and labeling.

Enzyme activity assays were performed for spermidine/spermine N^1-acetyltransferase (SSAT/*SAT1*), ornithine decarboxylase (ODC), and S-adenosylmethionine decarboxylase (AdoMetDC/*AMD1*) using radiolabeled substrates, as previously described [16–18]. Oxidation via spermine oxidase (SMOX) and N^1-acetylpolyamine oxidase (PAOX) was measured using luminol-based detection of H_2O_2 in the presence of either SPM or N^1-acetylated spermine (N^1-AcSPM) as a substrate [19]. Enzyme activities and intracellular polyamine concentrations are presented relative to total protein in the lysate, which was determined by the method of Bradford [20], with interpolation on a bovine serum albumin standard curve.

2.3. Protein Isolation and Western Blots

For Western blot analyses of proteins, cells were lysed in 4% SDS containing a protease inhibitor cocktail and homogenized using column-based centrifugation (Omega Bio-Tek, Norcross, GA, USA). The BioRad DC assay (Bio-Rad Laboratories, Hercules, CA, USA) was used for protein quantification. Equal amounts of reduced protein samples were separated on 4–12% Bis-Tris BOLT gels (Invitrogen, Carlsbad, CA, USA), transferred onto Immun-Blot PVDF (BioRad), and blocked in Odyssey blocking buffer (LI-COR, Lincoln, NE, USA). Primary antibodies were used targeting the following, ODC antizyme 1 (OAZ1) [21], spermidine synthase (SRM) (#19858-1-AP; Proteintech, Rosemont, IL, USA), histone deacetylase 10 (HDAC10) (#H3413; Sigma), and transglutaminase 2 (TGM2) (#ab421; Abcam, Cambridge, MA, USA), with pan histone H3 (#05-928; Upstate Cell Signaling Solutions, Lake Placid, NY, USA) as a normalization control. Secondary, species-specific, fluorophore-conjugated antibodies allowed visualization and quantification of bands via near-infrared imaging on an Odyssey detection system (LI-COR). Blot images were analyzed using Image Studio software (LI-COR, Lincoln, NE, USA).

2.4. RNA Isolation and Quantitation of Gene Expression

Total RNA was extracted from the lymphoblastoid cell lines using Trizol reagent (Invitrogen) and used for cDNA synthesis with qScript cDNA SuperMix (Quanta Biosciences, Gaithersburg, MD, USA). The mRNA expression levels of polyamine-metabolism-associated genes in the SRS versus wildtype (WT) lymphoblastoid lines were measured by SYBR-green-mediated (BioRad) quantitative real-time PCR on a BioRad iQ2 detection system. Custom primers specific for human *ODC1*, *OAZ1*, *AMD1*, *SRM*, *SMS*, *SAT1*, *HDAC10*, *SMOX*, *PAOX*, *TGM2*, and *GAPDH* were synthesized by Integrated DNA Technologies (Coralville, IA, USA). Primers were optimized on annealing temperature gradients with melt curve analyses and agarose gel electrophoresis. Triplicate determinations were obtained for each gene in each patient and normalized to *GAPDH* expression. The fold-change in expression was determined using the $2^{-\Delta\Delta Ct}$ algorithm.

2.5. Statistical Analyses

Statistically significant differences were determined by two-tailed Student's *t*-tests with 95% confidence interval using GraphPad Prism software (La Jolla, CA, USA).

3. Results

3.1. Alterations in Intracellular Polyamine Distribution

It has been previously reported that spermine concentrations are reduced while spermidine concentrations are increased in SRS lymphoblast cell lines [11–13]. However, the overall effects on other enzymes in the polyamine pathway have not been thoroughly evaluated. Based on these previous studies, we chose three SRS cell lines with varying degrees of spermine synthase deficiency [10] (Table 1), which was confirmed by our HPLC analyses of intracellular polyamine concentrations (Figure 2). Along with SPM levels, PUT concentrations were also significantly decreased in the SRS lines relative to the WT lines, while the intracellular SPD pools significantly increased, as observed previously [11]. Consequently, the SPD/SPM ratio increased nearly 10-fold in the most affected lines (Table 1 and Figure 2). The total intracellular concentrations of polyamines did not significantly differ among the genotypes examined, regardless of the severity of spermine deficiency (Figure 2d), and none of the lysates contained detectable levels of acetylated SPD or SPM derivatives.

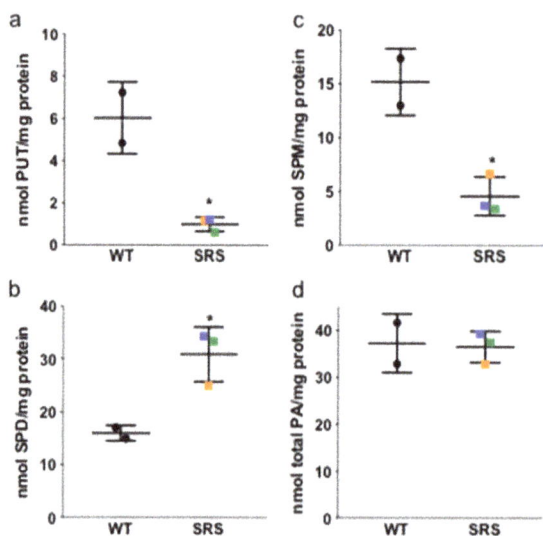

Figure 2. Alterations in basal intracellular concentrations of (**a**) putrescine (PUT), (**b**) spermidine (SPD), (**c**) spermine (SPM), and (**d**) total polyamines (PA) between *SMS* wildtype (WT) or mutant (SRS) lymphoblast cell lines (*n* = 5, each measured in duplicate). Concentrations are presented as nmol of polyamine per mg of cellular protein. The individual SRS cell line designations are orange for SRS1, blue for SRS2, and green for SRS3. Error bars indicate standard error of the mean (SEM). * $p < 0.05$.

Table 1. Characteristics of lymphoblastoid cell lines. SPD/SPM ratios represent means with (SEM) $n = 5$. Mutations, protein products, and spermine synthase (SMS) activity were as previously reported [10]. ND = none detected.

Cell Line	Mutation	Protein	SMS Activity	SPD/SPM
WT1	none	wildtype	yes	1.17 (0.04)
WT2	none	wildtype	yes	0.83 (0.07)
SRS1	c.329+5 G>A aberrant splice site	truncated; some functional SMS from read-through	reduced	3.76 (0.25)
SRS2	V132G	decreased dimerization	ND	9.56 (0.73)
SRS3	G56S	no dimerization	ND	9.85 (0.36)

3.2. Ornithine Decarboxylase Activity Is Decreased in Snyder-Robinson Syndrome

Ornithine decarboxylase activity is the first rate-limiting enzyme in polyamine biosynthesis and catalyzes the production of PUT from ornithine (Figure 1). ODC activity was significantly lower in each of the SRS lymphoblast lines compared to WT controls, consistent with the reduction in PUT levels observed in these cells (Figure 3a). As the reductions in ODC activity did not correspond with reductions in ODC1 mRNA expression (Figure S1a), we analyzed expression of ODC antizyme (OAZ1), a negative regulator of ODC protein that targets its degradation via the 26S proteasome. We consistently observed increased expression of OAZ1 protein only in SRS line 1, the least affected SRS line in terms of SMS activity and SPM depletion (Figure 3b). Although this increase might be responsible for the decreased ODC activity in these cells, it does not appear to contribute to that in SRS lines 2 or 3. Consequently, it is likely that product inhibition due to the increased levels of SPD plays a role in the reduced ODC activity. It is interesting that in spite of the obvious difference in SPD/SPM ratio between SRS1 (3.76) and the other 2 SRS lines (9.56 and 9.85), the decreases in ODC activity and PUT concentration among the three lines were quite similar, suggesting that the increased OAZ1 may serve to supplement the feedback regulation by SPD in SRS line 1. As with ODC, no apparent change in OAZ1 mRNA expression was observed to account for the change in protein (Figure S1a).

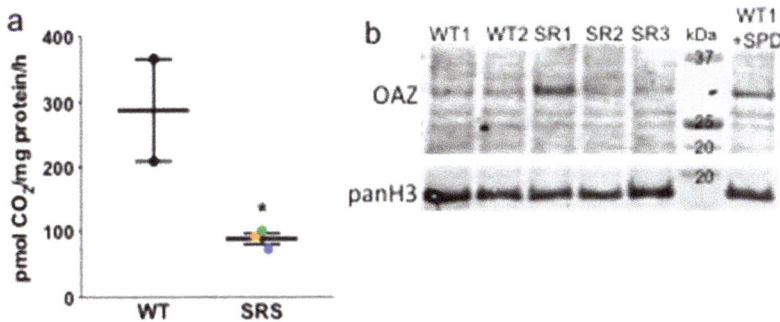

Figure 3. (**a**) ODC activity in donor or SRS lymphoblasts ($n = 2$, in triplicate; error bars = SEM), presented as pmol CO_2 produced per hour per mg of total protein. Color designations are orange (SRS1), blue (SRS2), and green (SRS3). (**b**) Representative Western blot of ODC antizyme 1 (OAZ1) with pan histone H3 as loading control. The WT1 cell line treated with SPD for 24 h was used as a positive control for OAZ1. * $p < 0.05$.

3.3. Effects of Spermine Synthase Mutations on Spermidine Biosynthesis

Our current study and others have observed that a common result of *SMS* loss-of-function includes SPD accumulation (Figure 2b) [11–13,22–24]. SPD biosynthesis is similar to that of SPM: SRM catalyzes an aminopropyl group transfer from dcAdoMet to PUT, producing SPD (Figure 1). Both SRM and

SMS aminopropyl transfer reactions are therefore limited by the availability of dcAdoMet and hence, the activity of AdoMetDC [25]. We examined the expression levels of these enzymes to determine the extent to which biosynthesis through these steps might be affected in SRS. We found that AdoMetDC activity and mRNA expression levels were similar among the five cell lines regardless of *SMS* status or intracellular SPM or SPD concentration (Figures 4a and S1b, respectively). Although SRM gene expression was consistently upregulated in SRS line 2 (Figure S1b), quantitative Western blots revealed SRM protein level in this line was similar to that of the WT lines (Figure 4b). As AdoMetDC activity levels are essentially equal among the cell lines, the possibility exists for increased availability of dcAdoMet for SPD biosynthesis in the absence of SMS activity, thus contributing to SPD accumulation in these patients as well as the observed PUT depletion. In regard to SMS, a severe reduction (>90% less than the average WT *SMS* expression level) in the expression of the full-length *SMS* transcript was noted in SRS1 cells, consistent with some read-through of the mutated splice site previously reported in these cells (Figure S1b) [11]. *SMS* transcript levels of the other two SRS lines were within the WT range.

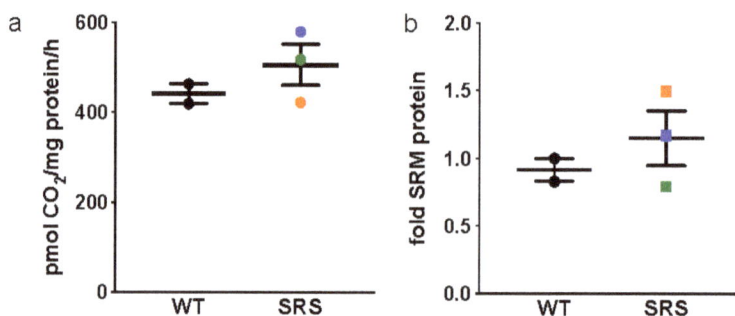

Figure 4. (**a**) Basal AdoMetDC activity (*n* = 2, in triplicate) of donor or SRS lymphoblast lines. Color designations are orange (SRS1), blue (SRS2), and green (SRS3). (**b**) Quantitative Western blots of SRM in lymphoblast cell lines (*n* = 2). All error bars indicate SEM.

3.4. Effects of Spermine Synthase Mutations on Polyamine Catabolism

One possible fate for excess SPD is its catabolism via SSAT (Figure 5a). A rate-limiting enzyme, SSAT catalyzes the transfer of an acetyl group from acetyl CoA to the N^1 position of spermidine or spermine. These N^1-acetylated polyamines are then either exported from the cell or oxidized by PAOX. This 2-step back-conversion via SSAT/PAOX thereby returns SPD or SPM to its precursor (PUT or SPD, respectively). As SPD and SPM induce SSAT expression at multiple levels [26,27], we investigated the possibility that SSAT mRNA and activity levels in the SRS lines might respond to the altered SPD/SPM ratios. Line SRS3 demonstrated elevated SSAT transcript levels as well as activity; however, a similar increase in activity was evident in WT line 2 (Figure 5b and Figure S1c). Additionally, PAOX activity, which typically is dependent upon substrate availability, was significantly decreased by approximately 50% in each of the SRS lines compared to WT (Figure 5c). Overall, these data suggest that loss of SMS has little effect on the basal catabolism of SPD by SSAT, with the decreased PAOX activity in the SRS lines potentially serving to increase export of any SPD that does become N^1-acetylated in lieu of its back-conversion to PUT. However, examination of culture medium removed from the lymphoblasts after 24 h of growth revealed a complete absence of polyamines, including the acetylated derivatives, suggesting that polyamine catabolism is not induced by excess SPD. SMOX, which directly oxidizes SPM back to SPD, was expressed at very low mRNA levels in nearly all of the lymphoblast lines (Figure S1d), with no protein or activity detected regardless of *SMS* status.

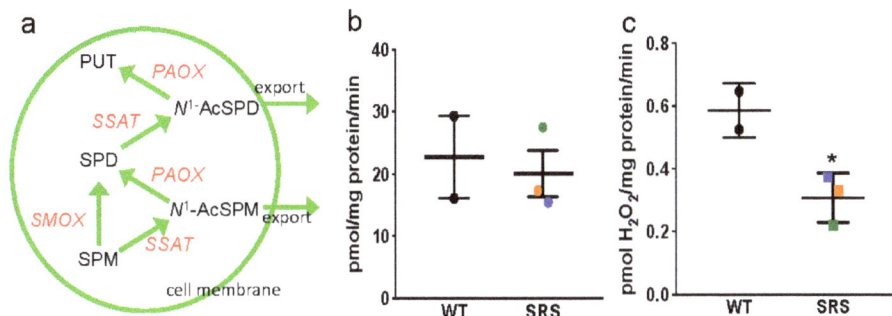

Figure 5. (**a**) The polyamine catabolic pathway. (**b**) Spermidine/spermine N^1-acetyltransferase (SSAT) and (**c**) N^1-acetylpolyamine oxidase (PAOX) activity assays. Color designations in (b,c) are orange for SRS1, blue for SRS2, and green for SRS3. Error bars indicate SEM ($n \geq 2$, in triplicate; * $p < 0.05$).

3.5. Snyder-Robinson Syndrome Lymphoblasts Maintain Active Polyamine Transport

In addition to downregulating their own biosynthesis, an accumulation of polyamines also suppresses the polyamine transport system, thereby inhibiting the uptake of additional polyamines from the extracellular environment [21]. The lymphoblast cell lines were incubated in the presence of exogenous SPM to determine if polyamine uptake/transport was altered in SRS lymphoblasts. Treatment with 5 μM SPM for 24 h not only increased SPM levels in the SRS lines, but simultaneously decreased SPD levels (Figure 6a). With the exception of PUT, polyamine levels were effectively restored to those similar to WT SMS cells, indicating that the polyamine transport system was active and the SRS lymphoblast lines could self-regulate their polyamine pools in the presence of exogenous SPM. The SPD/SPM ratios of the SRS lines decreased from their baseline values (3.07, 7.35, and 9.31) to 1.11, 1.35, and 1.4, respectively (Figure 6b). SSAT activity following treatment with SPM was unchanged (Figure 6c), and HPLC analyses revealed a lack of N^1-acetylated polyamines in both intracellular lysates and medium samples, indicating that SPD reduction in SRS cells is independent of SSAT induction.

To more accurately quantify the ability of the SRS lines to uptake polyamines, cells were incubated with the polyamine analog BENSpm, followed by HPLC analysis (Figure 6d). Overall, the WT and SRS lines were equally capable of accumulating BENSpm over 24 h. Line SRS1 accumulated the least BENSpm, but this was not significantly less than the WT line 1. Like biosynthesis, polyamine transport is negatively regulated by antizyme expression [21]; thus, the decreased uptake of BENSpm in SRS1 cells is likely associated with its increased antizyme level. Regardless of this decrease, the amount of SPM transported into each of the SRS cell lines was sufficient to reduce their intracellular SPD levels to within the range observed in the wildtype cells.

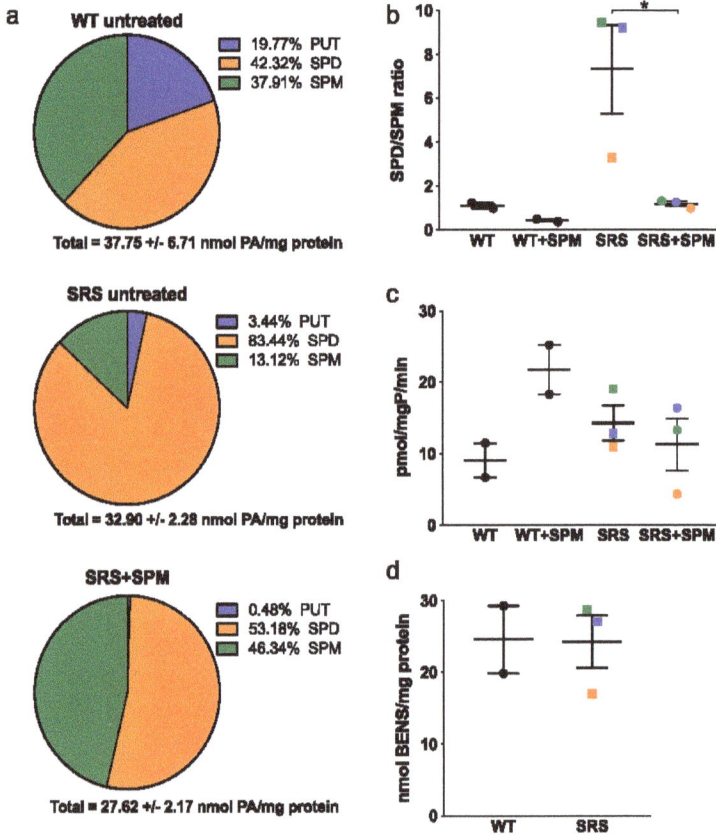

Figure 6. (**a**) Average polyamine levels of SRS lymphoblast lines before (middle) and after (bottom) 5 µM SPM treatment for 24 h, compared with untreated WT lymphoblast lines (top). (**b**) SPD/SPM ratios and (**c**) SSAT activity before and after SPM supplementation. (**d**) Intracellular accumulation of bis(ethyl)norspermine (BENSpm) following 10 µM treatment for 24 h. Color designations in (b–d) are orange for SRS1, blue for SRS2, and green for SRS3. All error bars indicate SEM ($n \geq 2$, in triplicate; * $p < 0.05$).

3.6. N^8-Acetylation of Spermidine

Spermidine localized in the nucleus can undergo acetylation of its N^8 position, which allows relocation of the spermidine moiety into the cytoplasm via charge neutralization. N^8-acetylated spermidine was recently reported as a potential biomarker for SRS due to its elevated levels in the plasma of three SRS patients [28]. While the acetyltransferase responsible for this acetylation has yet to be definitively determined [29,30], it was recently found that deacetylation of N^8-acetylspermidine to yield native spermidine is catalyzed by the cytoplasmic enzyme HDAC10 [31]. We therefore determined if HDAC10 expression levels were altered in the SRS lymphoblasts in our study. Although HDAC10 mRNA levels were significantly elevated in SRS line 2, it did not translate to an increase in protein observed via Western blot, and there was no overall difference between the wildtype and mutant SMS lines (Figure 7). Additionally, we failed to detect N^8-acetylated spermidine in either the cell lysates or excreted into the culture medium, suggesting that lymphocytes were not the likely source of the plasma metabolite detected by Abela and colleagues [28].

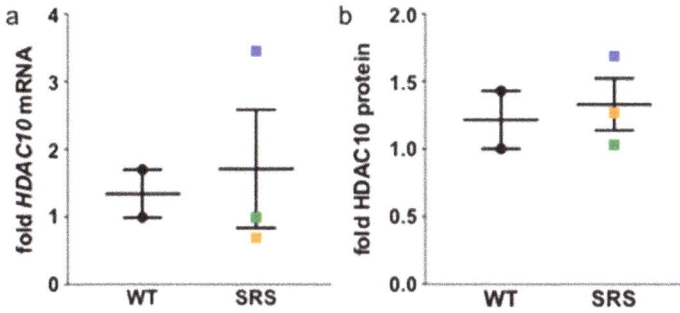

Figure 7. HDAC10 mRNA (**a**) and protein (**b**) levels in wildtype versus SRS lymphoblasts. Color designations are orange for SRS1, blue for SRS2, and green for SRS3. Error bars indicate SEM (*n* = 3).

3.7. Transglutaminase 2 Expression Is Upregulated in Snyder-Robinson Syndrome Patient Lymphoblasts

As low-molecular-weight amines, polyamines are natural acceptor substrates for the TG family of enzymes [8], which catalyze the calcium-dependent cross-linking of glutamine and lysine residues within or between proteins. TG activity also incorporates polyamines into certain glutamine residues of cellular proteins via one or both of their primary amino groups, thereby forming mono- or bis(γ-glutamyl)-PUT, SPD, or SPM, and potentially interfering with isopeptide bond cross-linking. To determine if the altered polyamine concentrations in the SRS lymphoblasts correspond to changes in TG, we analyzed the expression of transglutaminase 2 (TGM2), a ubiquitously expressed tissue TG (Figure 8). *TGM2* mRNA expression was significantly increased in all three SRS cell lines when compared to the WT cell lines, and TGM2 protein followed a similar trend, with the highest expression of both mRNA and protein in SRS line 2.

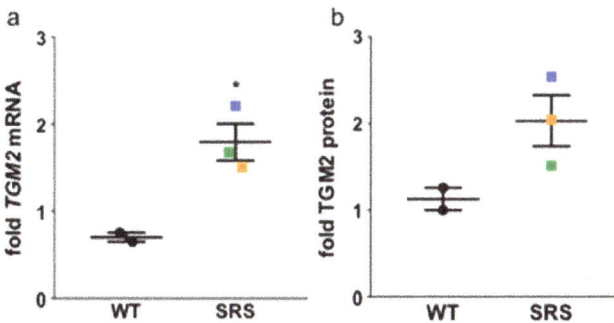

Figure 8. Transglutaminase 2 mRNA (**a**) and protein (**b**) expression levels are increased in SRS cell lines. Colors designate SRS line 1 (orange), SRS line 2 (blue), and SRS line 3 (green). All error bars indicate SEM (*n* = 3; * *p* < 0.05).

4. Discussion

We have investigated the differences in polyamine metabolism and transport in lymphoblastoid cell lines derived from three SRS patients and two wildtype donors. The SRS cell lines were chosen based on their extent of spermine deficiency, which ranged from approximately 50 to 25% of the WT lines. Both SPM and PUT levels were significantly decreased in the SRS lines relative to the WT lines, while SPD concentration was increased. The SPD/SPM ratio averaged 1.14 in the WT lines, while the SRS lines had elevated SPD/SPM ratios of approximately 3.76, 8.72, and 9.87, confirming previous observations [10–13].

The only significant change in gene expression between lymphoblasts derived from SRS patients versus controls was in *TGM2*. Although there was some variation in the levels of other genes investigated among the three SRS lines, the two WT cell lines also varied and there were no statistically significant differences between the groups. Analyses of enzyme activities and protein levels of polyamine-related genes did reveal changes between the SRS versus WT lymphoblasts. In particular, ODC activity was significantly decreased in all three SRS lines, which correlated with the decreased putrescine levels observed in these lines. Interestingly, an elevated level of ODC antizyme (OAZ1) was detected only in SRS line 1, which also displayed slightly decreased uptake of exogenous polyamines, consistent with the known ability of OAZ1 to downregulate the polyamine transport system. Thus, the decrease in ODC activity in the SRS cell lines is likely due, at least in part, to product inhibition of ODC by the high concentrations of SPD in the affected cells. The three SRS lines also had significantly reduced PAOX activity (~50%) compared to the WT lines, consistent with a lack of back-conversion to putrescine.

The ability of the SRS lymphoblasts to normalize their intracellular polyamine levels upon treatment with exogenous spermine was impressive and highlights the potential of administering spermine as a therapeutic strategy. As ODC antizyme is stimulated by excess polyamines to negatively regulate polyamine biosynthesis and uptake, we anticipated that elevated antizyme in response to the elevated spermidine concentrations in the SRS patients that would prevent uptake via the polyamine transport system. This assumption was based on observations in the Gy mouse model, in which *Sms*-deficient male mice fed a diet containing SPM still had no detectable levels of SPM in brain, liver, or heart tissue [32]. However, studies using embryonic fibroblasts from these mice did indicate that spermine could be acquired from the culture medium [33]. In our study, although antizyme was induced in SRS line 1, its effect on polyamine transport was not sufficient to limit the maintenance of homeostasis in response to excess spermine. Although cell types more relevant to the SRS phenotype might differ in this regard, when considering systemic delivery of a therapeutic agent, the avoidance of off-target effects is also necessary. As the accumulation of excess polyamines is associated with certain pathologies, including cancer [4], the fact that these cells reestablish polyamine homeostasis rather than accumulating additional polyamines following SPM administration may have important consequences over the long term.

The finding that the lymphoblast cell lines derived from SRS patients exhibit levels of TGM2 that exceed those of normal donors may have significant clinical implications if conserved or exaggerated in cell types that contribute more to the SRS phenotype. Stimulation of peripheral blood lymphocytes with mitogen to induce blastogenesis induces the influx of Ca2+, which increases TGM activity as well as polyamine concentrations. However, even with mitogen stimulation, only a small number of lymphoblast proteins became conjugated with spermidine, and this binding was subsequently determined to be independent of transglutaminase activity [8,34]. Very few spermine-conjugated proteins were detected in mitogen-stimulated lymphocytes, and attempts to identify polyamines released from hydrolysates of acid-insoluble protein from our SRS lymphoblast lines have been unsuccessful. However, in other systems, such as seminal secretions, spermine becomes highly conjugated, and both spermine and spermidine are capable of cross-linking proteins via the incorporation of both primary amines [8]. As high affinity substrates for TG reactions, polyamines in excess, such as SPD in SRS patients, may serve as competitive substrate inhibitors of the cross-linking transamidation reactions, or they may themselves become incorporated into the cross-link when bound through both primary amines. Importantly, transglutaminases play roles in several processes associated with the characteristic phenotype of SRS males, including osteoblast, neuronal, and myoblast differentiation [35]. As it has been shown that transglutaminase activity responds to changes in ODC activity and polyamine concentrations, further investigation into its role in the SRS phenotype is warranted and ongoing [36].

Snyder-Robinson Syndrome studies using patient-derived material are limited by small patient number and acquisition of suitable cell lines for study. Lymphoblasts were chosen for this initial

Med. Sci. **2018**, *6*, 112

characterization, as they are most easily obtained from the patient and had been previously established. However, they do not correspond to a tissue that displays an obvious phenotype in SRS. These studies do, however, form a basis and provide knowledge for leading biochemical changes to study in affected tissues as well as for development of therapeutic strategies, particularly those aimed at replenishing intracellular spermine. Recently, a novel neurodevelopmental disorder was described resulting from gene variants that increase ODC activity [37,38]. Although, unlike SRS, these patients have elevated levels of ODC and putrescine, several clinical manifestations are similar, including hypotonia and developmental delays. Of note, with the exception of red blood cells, intracellular polyamine levels have not been reported in these patients. In a mouse model where ODC is specifically overexpressed in the skin, increased intracellular spermidine levels with concurrent decreases in spermine, relative to wildtype littermates were observed [39]. Thus, patients with *ODC1* gene variants may have intracellular SPD:SPM ratios similar to those of SRS patients. Similarly altered polyamine metabolism was also recently described in a mouse model of tuberous sclerosis complex (TSC), a neurodevelopmental disorder associated with mechanistic Target of Rapamycin Complex 1 (mTORC1) dysregulation. TSC also shares clinical manifestations with SRS, including intellectual disability and epileptic seizures, and inhibition of ODC reduced astrogliosis, an indicator of TSC neuropathology [40]. The common neurological phenotypes resulting from mutations in *SMS*, *ODC1*, and *TSC* and their association with dysregulated polyamine metabolism suggest that the knowledge obtained from the current studies might serve to benefit a patient population beyond those with Snyder-Robinson Syndrome.

Supplementary Materials: The following are available online at http://www.mdpi.com/2076-3271/6/4/112/s1, Figure S1: mRNA expression levels of polyamine metabolic genes that do not significantly differ between wildtype and SMS mutant lymphoblasts.

Author Contributions: Conceptualization, T.M.-S., C.E.S., and R.A.C.J.; Data curation, T.M.-S., M.D., J.R.F., and R.A.C.J.; Formal analysis, T.M.-S., M.D., J.R.F., and R.A.C.J.; Funding acquisition, C.E.S. and R.A.C.J.; Investigation, T.M.-S., M.D., and J.R.F.; Methodology, T.M.-S., C.E.S., and R.A.C.J.; Project administration, T.M.-S. and R.A.C.J.; Resources, C.E.S. and R.A.C.J.; Software, R.A.C.J.; Supervision, R.A.C.J.; Validation, T.M.-S., M.D., and J.R.F.; Visualization, T.M.-S.; Writing—Original draft, T.M.-S.; Writing—Review & editing, T.M.-S., C.E.S., and R.A.C.J.

Funding: This research was funded, in part, by the Million Dollar Bike Ride, Orphan Disease Center at the University of Pennsylvania (#MDBR-18-127-SR to R.A.C.), the National Institutes of Health National Cancer Institute (#R01CA204345 to R.A.C.), the National Institutes of Health National Institute of Neurological Disorders and Stroke (Grant #R01NS073854 to C.E.S.), and a grant from the South Carolina Department of Disabilities and Special Needs (to C.E.S.).

Acknowledgments: The authors thank the Snyder-Robinson families for their cooperation. In memory of Ethan Francis Schwartz, 1996–1998.

Conflicts of Interest: The authors declare no conflicts of interest. The funders had no role in the design of the study; in the collection, analyses, or interpretation of data; in the writing of the manuscript, or in the decision to publish the results.

References

1. Snyder, R.D.; Robinson, A. Recessive sex-linked mental retardation in the absence of other recognizable abnormalities. Report of a family. *Clin. Pediatr.* **1969**, *8*, 669–674. [CrossRef] [PubMed]

2. Arena, J.F.; Schwartz, C.; Ouzts, L.; Stevenson, R.; Miller, M.; Garza, J.; Nance, M.; Lubs, H. X-linked mental retardation with thin habitus, osteoporosis, and kyphoscoliosis: Linkage to Xp21.3-p22.12. *Am. J. Med. Genet.* **1996**, *64*, 50–58. [CrossRef]

3. Wu, H.; Min, J.; Zeng, H.; McCloskey, D.E.; Ikeguchi, Y.; Loppnau, P.; Michael, A.J.; Pegg, A.E.; Plotnikov, A.N. Crystal structure of human spermine synthase: Implications of substrate binding and catalytic mechanism. *J. Biol. Chem.* **2008**, *283*, 16135–16146. [CrossRef]

4. Casero, R.A., Jr.; Murray Stewart, T.; Pegg, A.E. Polyamine metabolism and cancer: Treatments, challenges and opportunities. *Nat. Rev. Cancer* **2018**. [CrossRef]

5. Baronas, V.A.; Kurata, H.T. Inward rectifiers and their regulation by endogenous polyamines. *Front. Physiol.* **2014**, *5*, 325. [CrossRef]

6. Lightfoot, H.L.; Hall, J. Endogenous polyamine function—The RNA perspective. *Nucleic Acids Res.* **2014**, *42*, 11275–11290. [CrossRef] [PubMed]

7. Pasini, A.; Caldarera, C.M.; Giordano, E. Chromatin remodeling by polyamines and polyamine analogs. *Amino Acids* **2014**, *46*, 595–603. [CrossRef] [PubMed]

8. Folk, J.E.; Park, M.H.; Chung, S.I.; Schrode, J.; Lester, E.P.; Cooper, H.L. Polyamines as physiological substrates for transglutaminases. *J. Biol. Chem.* **1980**, *255*, 3695–3700. [PubMed]

9. Lai, T.S.; Lin, C.J.; Greenberg, C.S. Role of tissue transglutaminase-2 (TG2)-mediated aminylation in biological processes. *Amino Acids* **2017**, *49*, 501–515. [CrossRef] [PubMed]

10. Schwartz, C.E.; Wang, X.; Stevenson, R.E.; Pegg, A.E. Spermine synthase deficiency resulting in X-linked intellectual disability (Snyder-Robinson syndrome). *Methods Mol. Biol.* **2011**, *720*, 437–445. [PubMed]

11. Cason, A.L.; Ikeguchi, Y.; Skinner, C.; Wood, T.C.; Holden, K.R.; Lubs, H.A.; Martinez, F.; Simensen, R.J.; Stevenson, R.E.; Pegg, A.E.; et al. X-linked spermine synthase gene (SMS) defect: The first polyamine deficiency syndrome. *Eur. J. Hum. Genet.* **2003**, *11*, 937–944. [CrossRef]

12. de Alencastro, G.; McCloskey, D.E.; Kliemann, S.E.; Maranduba, C.M.; Pegg, A.E.; Wang, X.; Bertola, D.R.; Schwartz, C.E.; Passos-Bueno, M.R.; Sertie, A.L. New SMS mutation leads to a striking reduction in spermine synthase protein function and a severe form of Snyder-Robinson X-linked recessive mental retardation syndrome. *J. Med. Genet.* **2008**, *45*, 539–543. [CrossRef]

13. Becerra-Solano, L.E.; Butler, J.; Castaneda-Cisneros, G.; McCloskey, D.E.; Wang, X.; Pegg, A.E.; Schwartz, C.E.; Sanchez-Corona, J.; Garcia-Ortiz, J.E. A missense mutation, p.V132G, in the X-linked spermine synthase gene (SMS) causes Snyder-Robinson syndrome. *Am. J. Med. Genet. A* **2009**, *149A*, 328–335. [CrossRef]

14. Bergeron, R.J.; Neims, A.H.; McManis, J.S.; Hawthorne, T.R.; Vinson, J.R.; Bortell, R.; Ingeno, M.J. Synthetic polyamine analogues as antineoplastics. *J. Med. Chem.* **1988**, *31*, 1183–1190. [CrossRef]

15. Kabra, P.M.; Lee, H.K.; Lubich, W.P.; Marton, L.J. Solid-phase extraction and determination of dansyl derivatives of unconjugated and acetylated polyamines by reversed-phase liquid chromatography: Improved separation systems for polyamines in cerebrospinal fluid, urine and tissue. *J. Chromatogr.* **1986**, *380*, 19–32. [CrossRef]

16. Casero, R.A., Jr.; Celano, P.; Ervin, S.J.; Porter, C.W.; Bergeron, R.J.; Libby, P.R. Differential induction of spermidine/spermine N1-acetyltransferase in human lung cancer cells by the bis(ethyl)polyamine analogues. *Cancer Res.* **1989**, *49*, 3829–3833.

17. Seely, J.E.; Pegg, A.E. Ornithine decarboxylase (mouse kidney). *Methods Enzymol.* **1983**, *94*, 158–161.

18. Pegg, A.E. The role of polyamine depletion and accumulation of decarboxylated S-adenosylmethionine in the inhibition of growth of SV-3T3 cells treated with alpha-difluoromethylornithine. *Biochem. J.* **1984**, *224*, 29–38. [CrossRef]

19. Goodwin, A.C.; Murray-Stewart, T.R.; Casero, R.A., Jr. A simple assay for mammalian spermine oxidase: A polyamine catabolic enzyme implicated in drug response and disease. *Methods Mol. Biol.* **2011**, *720*, 173–181.

20. Bradford, M.M. A rapid and sensitive method for the quantitation of microgram quantities of protein utilizing the principle of protein-dye binding. *Anal. Biochem.* **1976**, *72*, 248–254. [CrossRef]

21. Mitchell, J.L.; Judd, G.G.; Bareyal-Leyser, A.; Ling, S.Y. Feedback repression of polyamine transport is mediated by antizyme in mammalian tissue-culture cells. *Biochem. J.* **1994**, *299*, 19–22. [CrossRef]

22. Albert, J.S.; Bhattacharyya, N.; Wolfe, L.A.; Bone, W.P.; Maduro, V.; Accardi, J.; Adams, D.R.; Schwartz, C.E.; Norris, J.; Wood, T.; et al. Impaired osteoblast and osteoclast function characterize the osteoporosis of Snyder—Robinson syndrome. *Orphanet. J. Rare Dis.* **2015**, *10*, 27. [CrossRef]

23. Peron, A.; Spaccini, L.; Norris, J.; Bova, S.M.; Selicorni, A.; Weber, G.; Wood, T.; Schwartz, C.E.; Mastrangelo, M. Snyder-Robinson syndrome: A novel nonsense mutation in spermine synthase and expansion of the phenotype. *Am. J. Med. Genet. A* **2013**, *161A*, 2316–2320. [CrossRef]

24. Korhonen, V.P.; Niiranen, K.; Halmekyto, M.; Pietila, M.; Diegelman, P.; Parkkinen, J.J.; Eloranta, T.; Porter, C.W.; Alhonen, L.; Janne, J. Spermine deficiency resulting from targeted disruption of the spermine synthase gene in embryonic stem cells leads to enhanced sensitivity to antiproliferative drugs. *Mol. Pharmacol.* **2001**, *59*, 231–238. [CrossRef]

25. Pegg, A.E. Mammalian polyamine metabolism and function. *IUBMB Life* **2009**, *61*, 880–894. [CrossRef]

26. Casero, R.A., Jr.; Pegg, A.E. Spermidine/spermine N1-acetyltransferase—The turning point in polyamine metabolism. *FASEB J.* **1993**, *7*, 653–661. [CrossRef]

27. Pegg, A.E. Recent advances in the biochemistry of polyamines in eukaryotes. *Biochem. J.* **1986**, *234*, 249–262. [CrossRef]

28. Abela, L.; Simmons, L.; Steindl, K.; Schmitt, B.; Mastrangelo, M.; Joset, P.; Papuc, M.; Sticht, H.; Baumer, A.; Crowther, L.M.; et al. N(8)-acetylspermidine as a potential plasma biomarker for Snyder-Robinson syndrome identified by clinical metabolomics. *J. Inherit. Metab. Dis.* **2016**, *39*, 131–137. [CrossRef]

29. Libby, P.R. Rat liver nuclear N-acetyltransferases: Separation of two enzymes with both histone and spermidine acetyltransferase activity. *Arch. Biochem. Biophys.* **1980**, *203*, 384–389. [CrossRef]

30. Burgio, G.; Corona, D.F.; Nicotra, C.M.; Carruba, G.; Taibi, G. P/CAF-mediated spermidine acetylation regulates histone acetyltransferase activity. *J. Enzyme Inhib. Med. Chem.* **2016**, *31*, 75–82. [CrossRef]

31. Hai, Y.; Shinsky, S.A.; Porter, N.J.; Christianson, D.W. Histone deacetylase 10 structure and molecular function as a polyamine deacetylase. *Nat. Commun.* **2017**, *8*, 15368. [CrossRef]

32. Mackintosh, C.A.; Pegg, A.E. Effect of spermine synthase deficiency on polyamine biosynthesis and content in mice and embryonic fibroblasts, and the sensitivity of fibroblasts to 1,3-bis-(2-chloroethyl)-N-nitrosourea. *Biochem. J.* **2000**, *351*, 439–447. [CrossRef]

33. Rider, J.E.; Hacker, A.; Mackintosh, C.A.; Pegg, A.E.; Woster, P.M.; Casero, R.A., Jr. Spermine and spermidine mediate protection against oxidative damage caused by hydrogen peroxide. *Amino Acids* **2007**, *33*, 231–240. [CrossRef]

34. Park, M.H.; Cooper, H.L.; Folk, J.E. Identification of hypusine, an unusual amino acid, in a protein from human lymphocytes and of spermidine as its biosynthetic precursor. *Proc. Natl. Acad. Sci. USA* **1981**, *78*, 2869–2873. [CrossRef]

35. Eckert, R.L.; Kaartinen, M.T.; Nurminskaya, M.; Belkin, A.M.; Colak, G.; Johnson, G.V.; Mehta, K. Transglutaminase regulation of cell function. *Physiol. Rev.* **2014**, *94*, 383–417. [CrossRef]

36. Wang, J.Y.; Viar, M.J.; Johnson, L.R. Regulation of transglutaminase activity by polyamines in the gastrointestinal mucosa of rats. *Proc. Soc. Exp. Biol. Med.* **1994**, *205*, 20–28. [CrossRef]

37. Bupp, C.P.; Schultz, C.R.; Uhl, K.L.; Rajasekaran, S.; Bachmann, A.S. Novel de novo pathogenic variant in the ODC1 gene in a girl with developmental delay, alopecia, and dysmorphic features. *Am. J. Med. Genet. A* **2018**. [CrossRef]

38. Rodan, L.H.; Anyane-Yeboa, K.; Chong, K.; Klein Wassink-Ruiter, J.S.; Wilson, A.; Smith, L.; Kothare, S.V.; Rajabi, F.; Blaser, S.; Ni, M.; et al. Gain-of-function variants in the ODC1 gene cause a syndromic neurodevelopmental disorder associated with macrocephaly, alopecia, dysmorphic features, and neuroimaging abnormalities. *Am. J. Med. Genet. A* **2018**. [CrossRef]

39. Megosh, L.; Gilmour, S.K.; Rosson, D.; Soler, A.P.; Blessing, M.; Sawicki, J.A.; O'Brien, T.G. Increased frequency of spontaneous skin tumors in transgenic mice which overexpress ornithine decarboxylase. *Cancer Res.* **1995**, *55*, 4205–4209.

40. McKenna, J.; Kapfhamer, D.; Kinchen, J.M.; Wasek, B.; Dunworth, M.; Murray-Stewart, T.; Bottiglieri, T.; Casero, R.A.; Gambello, M.J. Metabolomic studies identify changes in transmethylation and polyamine metabolism in a brain-specific mouse model of tuberous sclerosis complex. *Hum. Mol. Genet.* **2018**, *27*, 2113–2124. [CrossRef]

medical sciences

MDPI

Article

Polyamine Biosynthetic Pathway as a Drug Target for Osteosarcoma Therapy

Rebecca R. Weicht [1,2], Chad R. Schultz [1], Dirk Geerts [3], Katie L. Uhl [1] and André S. Bachmann [1,2,*]

[1]   Department of Pediatrics and Human Development, College of Human Medicine, Michigan State University, 400 Monroe Avenue, NW, Grand Rapids, MI 49503, USA; weichtgr@msu.edu (R.R.W.); Chad.Schultz@hc.msu.edu (C.R.S.); Katie.Uhl@hc.msu.edu (K.L.U.)

[2]   Helen DeVos Children's Hospital, Department of Pediatric Hematology Oncology, Grand Rapids, MI 49503, USA

[3]   Department of Medical Biology, Amsterdam University Medical Center, University of Amsterdam, 1105 AZ Amsterdam, The Netherlands; h.a.geerts@amc.uva.nl

*   Correspondence: andre.bachmann@hc.msu.edu; Tel.: +1-616-234-2841; Fax: +1-616-234-2838

Received: 13 July 2018; Accepted: 13 August 2018; Published: 16 August 2018

Abstract: Osteosarcoma (OS) is the most common bone tumor in children. Polyamines (PAs) are ubiquitous cations involved in many cell processes including tumor development, invasion and metastasis. In other pediatric cancer models, inhibition of the PA biosynthesis pathway with ornithine decarboxylase (ODC) inhibitor alpha-difluoromethylornithine (DFMO) results in decreased cell proliferation and differentiation. In OS, the PA pathway has not been evaluated. DFMO is an attractive, orally administered drug, is well tolerated, can be given for prolonged periods, and is already used in pediatric patients. Three OS cell lines were used to study the cellular effects of PA inhibition with DFMO: MG-63, U-2 OS and Saos-2. Effects on proliferation were analyzed by cell count, flow cytometry-based cell cycle analysis and RealTime-Glo™ MT Cell Viability assays. Intracellular PA levels were measured with high-performance liquid chromatography (HPLC). Western blot analysis was used to evaluate cell differentiation. DFMO exposure resulted in significantly decreased cell proliferation in all cell lines. After treatment, intracellular spermidine levels were drastically decreased. Cell cycle arrest at G_2/M was observed in U-2 OS and Saos-2. Cell differentiation was most prominent in MG-63 and U-2 OS as determined by increases in the terminal differentiation markers osteopontin and collagen 1a1. Cell proliferation continued to be suppressed for several days after removal of DFMO. Based on our findings, DFMO is a promising new adjunct to current osteosarcoma therapy in patients at high risk of relapse, such as those with poor necrosis at resection or those with metastatic or recurrent osteosarcoma. It is a well-tolerated oral drug that is currently in phase II clinical trials in pediatric neuroblastoma patients as a maintenance therapy. The same type of regimen may also improve outcomes in osteosarcoma patients in whom there have been essentially no medical advances in the last 30 years.

Keywords: cell differentiation; DFMO; ornithine decarboxylase; osteosarcoma; polyamines

1. Introduction

Osteosarcoma (OS) is the most common bone tumor in children, with approximately 400 children diagnosed annually in the United States. With our current treatments we have achieved approximately 70% cure rates for patients presenting with localized disease. However, OS is often metastatic at diagnosis and only about 30% of children survive in this scenario. Treatment of OS involves a combination of aggressive chemotherapy and surgery [1,2]. Unfortunately, there have not been any significant advances in OS treatment or outcomes since the 1980s [3].

Polyamines (PAs) are small molecules found in all cells [4–6]. They participate in many cell processes including angiogenesis, immune regulation, cell growth, cell signaling and apoptosis [7–9]. They are also known to be involved in tumor development, invasion and metastasis [10–16]. Polyamines are absorbed from the diet and intrinsically produced. Ornithine decarboxylase (ODC) is a rate-limiting enzyme in PA biosynthesis [10,17]. Alpha-difluoromethylornithine (DFMO) blocks PA synthesis by inhibiting ODC [10,11,15,18,19]. DFMO has been evaluated for both treatment and chemoprevention in a number of adult cancers with promising outcomes [15,19–24]. Its investigation in pediatric cancers has largely been limited to neuroblastoma, in which PA depletion resulted in G_1 cell cycle arrest and differentiation [10,14,16,25–27]. These findings led to several clinical trials which show promising results [28].

In OS, the role of PAs and the effect of DFMO have not been evaluated. However, PAs are known to be involved in osteogenic differentiation in a complex way. Some studies have shown that exogenous PAs stimulate osteogenic differentiation, while others have shown that polyamine depletion, via ODC inhibition with DFMO, promotes osteogenic differentiation of mesenchymal stem cells [29–31]. In this study we found that PA depletion with DFMO in OS cell lines resulted in decreased cell viability and differentiation. Remarkably, the effects of DFMO were persistent even after removal of the drug. Our results suggest that PA biosynthesis plays an important role in OS and that the targeting of this pathway may have clinically significant effects.

2. Materials and Methods

2.1. Chemicals, Reagents and Antibodies

The ODC inhibitor DFMO was provided by Dr. Patrick Woster (Medical University of South Carolina, Charleston, SC, USA). Dansylated spermidine and 1,7-diaminoheptane standards were provided by Dr. Otto Phanstiel (University of Central Florida, Orlando, FL, USA). High-performance liquid chromatography (HPLC)-grade methanol, HPLC-grade acetonitrile, and methylene chloride were obtained from Fisher Scientific (Hampton, NH, USA). Rabbit monoclonal glyceraldehyde 3-phosphate dehydrogenase (GAPDH) antibodies were obtained from GeneTex (Irvine, CA, USA). Mouse monoclonal antibodies against osteopontin (OPN), GAPDH, and collagen 1a1 (Col1a1) were obtained from Santa Cruz Biotechnology (Dallas, TX, USA). Rabbit monoclonal antibodies against alkaline phosphatase were obtained from Abcam (Cambridge, MA, USA). RealTime-Glo MT Cell Viability Assay was obtained from Promega (Madison, WI, USA). Goat anti-mouse or anti-rabbit secondary antibodies conjugated to IRDye® 680RD or IRDye® 800CW were obtained from Licor (Lincoln, NE, USA). Protein assay dye reagent was obtained from Bio-Rad Laboratories (Hercules, CA, USA).

2.2. Culturing of Osteosarcoma Cells

The human OS cell lines MG-63 (CRL-1427), U-2 OS (HTB-96), and Saos-2 (HTB-85) were cultured in Roswell Park Memorial Institute (RPMI) 1640 media (VWR, Radnor, PA, USA), supplemented with 10% heat-inactivated fetal bovine serum (FBS) (Invitrogen, Carlsbad, CA, USA), penicillin (100 IU/mL) and streptomycin (100 µg/mL) (Corning, Corning, NY, USA). The three OS cell lines which express c-MYC and ODC (Supplementary Figure S1) were purchased from the American Type Culture Collection (ATCC, Manassas, VA, USA) within the last two years. The cells were maintained at 37 °C in a humidified atmosphere containing 5% CO_2. For DFMO treatments, cells were plated, allowed to settle overnight, and then exposed to 5 mM DFMO for six days. DFMO and/or media of cells were replaced on day 3.

2.3. Cell Viability

Cell viability assays were performed using the RealTime-Glo™ MT Cell Viability Assay according to the manufacturer's protocol (Promega, Madison, WI, USA). After cells had been exposed to 5 mM DFMO for six days they were reseeded in standard media in white-walled 96-well plates at a density of 2000 cells/well (MG-63) or 4000 cells/well (U-2 OS and Saos-2). Cells were allowed to attach overnight. For time-zero measurements, cells were incubated with RealTime-Glo™ MT Cell Viability reagent for 20 min at 37 °C, and luminescence was measured on a Synergy (Biotek, Winooski, VT, USA) microplate reader. Luminescence was then measured at 24, 48, and 72 h after the addition of RealTime-Glo™ reagent.

2.4. Western Blot Analysis

Whole cell lysates were prepared using radioimmunoprecipitation assay (RIPA) buffer (20 mM Tris-HCl (pH 7.5), 135 mM NaCl, 2 mM ethylenediaminetetraacetic acid (EDTA), 0.1% (w/v) sodium lauryl sulfate, 10% (v/v) glycerol, 0.5% (w/v) sodium deoxycholate, and 1% (v/v) Triton X-100). The RIPA buffer was supplemented with cOmplete™ Protease Inhibitor Cocktail (Roche, Basel, Switzerland), and 0.27 mM Na_3VO_4 and 20 mM NaF as phosphatase inhibitors. Protein concentration was determined using the Bradford dye reagent protein assay (Bio-Rad, Hercules, CA, USA). Cell lysates in sodium dodecyl sulfate (SDS) sample buffer were boiled for 5 min and equal amounts of protein were resolved by 10% or 12% SDS–polyacrylamide gel electrophoresis (PAGE). Protein was electrotransferred onto 0.45 µM polyvinylidene difluoride Immobilon-P membrane (Millipore, Burlington, MA, USA). Primary antibodies were incubated overnight at 4 °C in 5% bovine serum albumin (BSA) in Tris-buffered saline containing 0.1% Tween-20. Secondary antibodies were incubated for up to 2 h at room temperature in Tris-buffered saline containing 0.1% Tween-20. Blots were imaged using an Odyssey Fc or Odyssey CLx (Licor, Lincoln, NE, USA) Western blot scanner. Western blot quantitation was performed using Image Studio Lite version 5.2 (Licor).

2.5. Measurement of Polyamines

Polyamines from treated OS cells were isolated, dansylated, and analyzed by HPLC as previously described [32,33]. Briefly, PAs were extracted and protonated in perchloric acid/sodium chloride buffer. To 100 µL of sample, 4.5 nmol of 1,7-diaminoheptane internal standard and 200 µL of 1 M sodium carbonate were added prior to dansylation with 400 µL of 5 mg/mL dansyl chloride (Sigma-Aldrich, St. Louis, MO, USA). Samples were analyzed using a Thermo Scientific/Dionex Ultimate 3000 HPLC (Thermo Scientific, Waltham, MA, USA) equipped with a Syncronis C18 column (250 × 4.6 mm, 5 µM pore size). The dansylated PA derivatives were visualized by excitation at 340 nm and emission at 515 nm. Using the relative molar response derived from N-dansylated PAs and 1,7-diaminoheptane standards, the amount of N-dansylated PAs was calculated and normalized to total sample protein.

2.6. Flow Cytometry Cell Cycle Analysis

Cells were collected, fixed in 70% cold ethanol overnight and stained with phosphate-buffered saline (PBS) containing 50 µg/mL propidium iodide and 100 ug/mL RNase A for 2 h at 37 °C. Cells were subjected to flow-cytometric analysis using a Cytoflex S flow cytometry instrument (Beckman Coulter, Miami, FL, USA). Cell cycle distribution was determined using the ModFit software (Verity Software House, Topsham, ME, USA).

2.7. Statistical Analyses

The statistical significance of DFMO treatment in cell viability experiments and polyamine analysis was determined using an unpaired Student's *t*-test assuming the null hypothesis. For all tests, a value of $p < 0.05$ was considered statistically significant.

Additional materials and methods are provided under Supplementary Information.

3. Results

3.1. Alpha-difluoromethylornithine Treatment Decreases Osteosarcoma Cell Proliferation

To study the effect of DFMO on OS cell morphology and viability, OS cells were plated in 10-cm dishes and exposed to DFMO for six days. By the end of the exposure, treated cells were significantly less confluent, which was most evident with the MG-63 and U-2 OS cell lines (Figure 1A). This was not associated with morphologic changes seen with apoptosis. When viable cells were counted with a hemocytometer and trypan blue, the cell numbers in the treated samples were significantly reduced compared to control samples (Figure 1B). The half maximal inhibitory concentration (IC-50) values at 72 h were determined at 4.43 ± 1.19 mM (MG-63), 4.78 ± 1.41 mM (U-2 OS), and 5.14 ± 1.12 mM (Supplementary Figure S2).

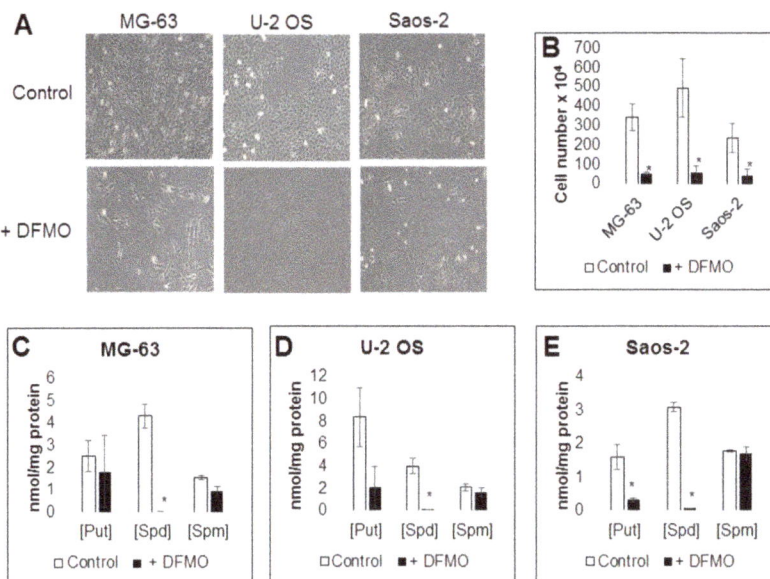

Figure 1. Effects of polyamine inhibitor alpha-difluoromethylornithine (DFMO) on tumor cell growth and polyamine profile of human osteosarcoma (OS) cells. (A) Representative light micrographs of MG-63, U-2 OS and Saos-2 cells grown in the presence or absence of 5 mM DFMO for six days. Light micrographs were routinely taken with a Leica (Wetzlar, Germany) DMi1 light microscope to document most of the experiments throughout this study and the pictures are representative of 5 separate experiments ($N = 5$); (B) viable cells were counted with a hemocytometer and trypan blue after 6 days of 5 mM DFMO exposure. DFMO treatment drastically reduced cell number in all cell lines; (C–E) levels of intracellular polyamines putrescine (Put), spermidine (Spd) and spermine (Spm) were measured by high-performance liquid chromatography (HPLC) in OS cells after exposure to 5 mM DFMO for six days. DFMO treatment resulted in a consistent decrease of putrescine and spermidine levels in all cell lines. Putrescine levels were more variable, however, trended toward decreased concentration in all cell lines and were significantly decreased in Saos-2 cells. Spermine levels were essentially unchanged, which is not uncommon and found also after DFMO treatment of neuroblastoma and other tumor cell lines [16]. Data represents three independent experiments ($N = 3$). * denotes statistically significant changes compared with control ($p < 0.05$).

3.2. Alpha-difluoromethylornithine Treatment Decreases Intracellular Polyamine Levels

To evaluate the effect of DFMO on intracellular polyamine levels, OS cells were exposed to DFMO for six days after which levels of putrescine, spermidine and spermine were measured by HPLC. In all cell lines there was a consistent decrease in putrescine and spermidine compared to controls. Changes in spermine were minimal. Although there was some variability in the change of putrescine in MG-63 cells, there was a clear overall trend toward decreased putrescine levels (Figure 1C–E). This is the same pattern seen in neuroblastoma [16] and other cancer models in which putrescine and spermidine are the two PAs most affected by DFMO, whereas spermine typically does not significantly change.

3.3. Alpha-difluoromethylornithine Induces Cell Cycle Arrest

Cells treated with DFMO did not appear apoptotic, suggesting that their decreased confluency was secondary to decreased proliferation rather than cell death. This prompted cell cycle analysis with flow cytometry after a six-day treatment with DFMO. Saos-2 cells showed profound G_2/M cell cycle arrest with the percentage of cells in this phase increasing from 21% to 43% while the cells in G_1 nearly halved from 46% to 27% (Figure 2E,F). U-2 OS cells showed combined G_1 and G_2/M cell cycle arrest with cells in G_2/M increasing from 19% to 29% and cells in G_1 increasing from 44% to 54% (Figure 2C,D). Thus it appeared that cell cycle arrest was, at least in part, the reason for the decreased cell proliferation in Saos-2 and U-2 OS. No cell cycle arrest was evident in MG-63 (Figure 2A,B).

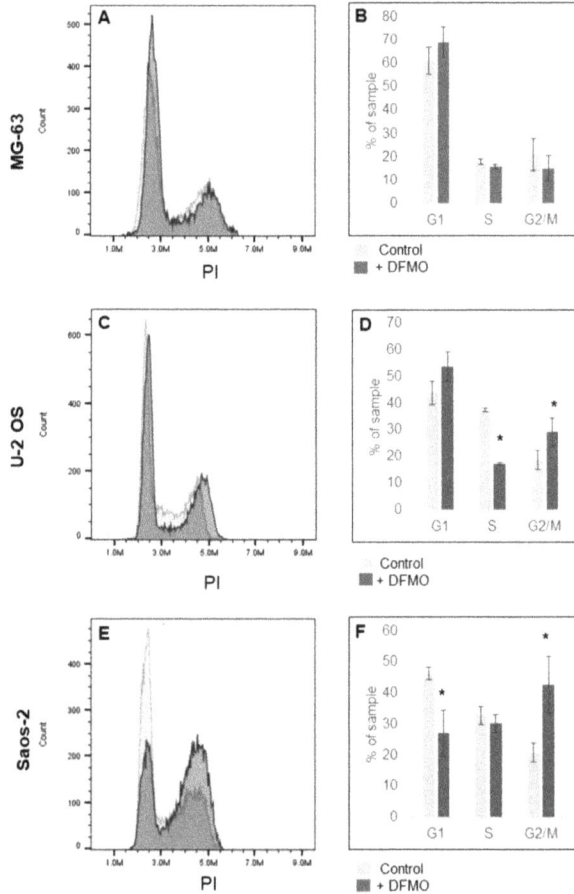

Figure 2. Effects of DFMO on cell cycle phase distribution in human OS cells. (**A,C,E**) Representative flow cytometry cell cycle analysis of MG-63, U-2 OS and Saos-2 cell lines stained with propidium iodide (PI) after 6 days of 5 mM DFMO exposure. Images are representative of three independent experiments ($N = 3$). (**B,D,F**) Percentage of cells in each phase of cell cycle averaged over at least three serial experiments ($N = 3$). U-2 OS and Saos-2 demonstrate G/2M cell cycle arrest after DFMO exposure. * denotes statistically significant changes in cell proliferation compared with control ($p < 0.05$).

3.4. Alpha-difluoromethylornithine Induces Differentiation

To evaluate an alternative mechanism by which osteosarcoma cells could have decreased proliferation without undergoing cell cycle arrest, we evaluated markers of osteogenic differentiation after exposure to DFMO for six days. Cells were grown in standard RPMI media throughout the experiment. After exposure to DFMO, Western blot was used to measure alkaline phosphatase (Alk Phos), collagen 1a1 (Col1a1) and osteopontin (OPN). Alk Phos is an early osteogenic differentiation marker whereas OPN and Col1a1 are late osteogenic differentiation markers. Alk Phos protein levels did not increase in any of our tested cell lines (Figure 3A) whereas the OPN precursor protein levels consistently increased in the presence of DFMO treatment in MG-63, U-2 OS and Saos-2 cells (Figure 3B). In addition, Col1a1 protein levels were increased in MG-63 and U-2 OS cells (Figure 3C). These results indicate that DFMO treatment results in terminal differentiation of some osteosarcoma cell lines. Remarkably, this occurred even in the absence of osteogenic differentiation media.

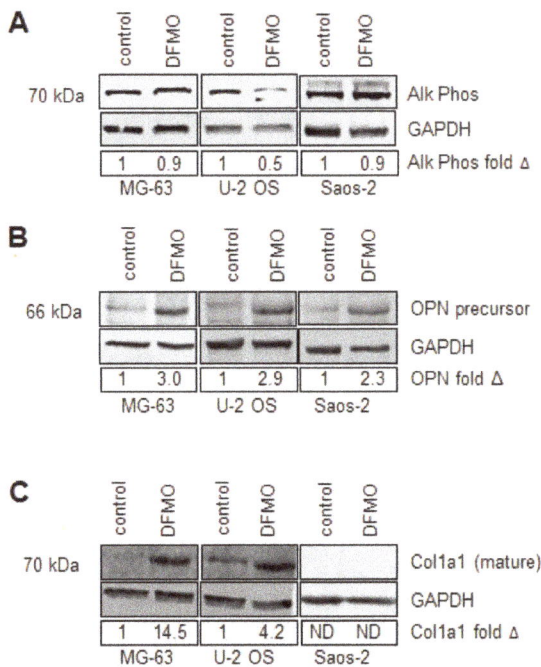

Figure 3. Differentiation of DFMO-treated human OS cells. Effect on the early differentiation marker (**A**) Alkaline phosphatase (Alk Phos) and late differentiation markers (**B**) osteopontin (OPN) and (**C**) collagen, type 1, alpha 1 (Col1a1), after six days of exposure to 5 mM DFMO. All cell lines showed increased levels of OPN, and MG-63 and U-2 OS displayed increased expression of Col1a1, demonstrating terminal osteogenic differentiation. Fold changes in protein levels (indicated below each blot) represent the average of quantified Western blot images from three independent experiments (*N* = 3).

3.5. Cell Recovery is Delayed by Alpha-difluoromethylornithine

The differentiation driven by DFMO led us to evaluate whether the inhibition of cell proliferation would be sustained after removal of the drug. After six days, DFMO-treated cells and untreated control cells were washed and reseeded into 96-well plates. After being allowed to settle overnight, cell viability was measured at 24, 48 and 72 h. In all tested cell lines there was marked delay in recovery of DFMO-treated cells (Figure 4A–C). This was especially pronounced in MG-63, in which treated cells showed minimal growth change over 72 h. Due to assay limitations beyond the 72 h time point with the RealTime-Glo™ reagent, we performed a similar experiment using a hemocytometer and trypan blue to determine the cell viability seven days after DFMO was removed from cells. Strikingly, even seven days after DFMO removal, the treated cells showed only minimal recovery (Figure 4D), clearly suggesting that DFMO prolongs cell growth inhibition, even in the absence of the drug.

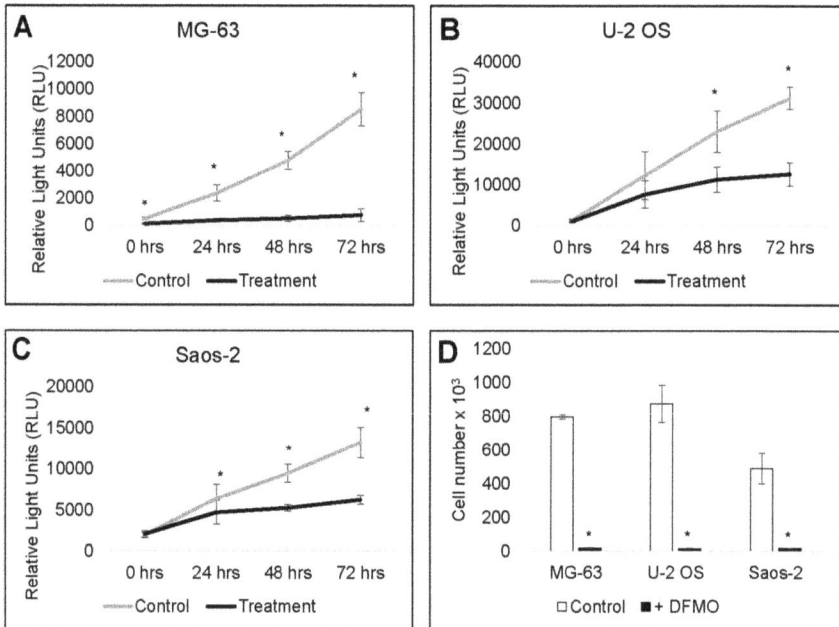

Figure 4. DFMO delays cell recovery in human OS cells. MG-63, U-2 OS and Saos-2 cells were exposed to 5 mM DFMO for six days and then reseeded in standard media. (**A–C**) Effect of treatment on cell recovery, 24, 48 and 72 h after removal of DFMO. Quantification of cell viability was determined by measuring relative light units (RLU) after the addition of RealTime-Glo™ MT Viability assay reagent. Data represent three independent experiments done in triplicate ($N = 9$). Error bars represent standard deviation. MG-63 (**A**), U-2 OS (**B**) and Saos-2 (**C**) cell viability was markedly decreased after DFMO exposure; (**D**) to evaluate cell recovery beyond the 72 h time point, treated cells (six days with DFMO) or untreated control cells were washed, reseeded in standard media, and counted with a hemocytometer and trypan blue, seven days after DFMO removal. At this time point, previously treated cells still showed only minimal recovery compared to controls. Data represent three independent experiments done in triplicate ($N = 9$). Error bars represent standard error. * denotes statistically significant changes in cell proliferation compared with control ($p < 0.05$).

4. Discussion

Osteosarcoma is the most common childhood bone tumor. In patients with metastatic or refractory disease, overall survival is only around 30% despite aggressive chemotherapy and surgical resection. In the last three decades there have been no significant developments that have improved survival, and new therapies are greatly needed. An important aspect of treatment both upfront and at relapse is to get a patient into a complete remission (CR) with chemotherapy and resection such that there is no visible evidence of disease. In the setting of relapse this may not provide a cure, however, it can extend life for years before another relapse occurs. Presumably then, there are residual, unmeasurably small amounts of viable osteosarcoma cells after completion of therapy which later reestablish themselves leading to recurrence. Hence, a prophylactic or maintenance regimen that could cause these cells to differentiate and lose their stem cell potential would prevent relapses. An oral medication with few adverse effects that could potentially be taken for long periods of time would be ideal. DFMO would be just such a treatment. Further, though DFMO does not seem to have much effect as an adjunct to upfront aggressive therapy in several cancer models, it has been shown to be effective at preventing cancer development in patients at high risk of developing primary cancer [19–21,24] or recurrence [24,25,28].

c-MYC is a direct transcriptional activator of the *ODC1* gene, which leads to ODC overexpression and PA-dependent cell proliferation. Our data demonstrate that the ODC inhibitor DFMO suppresses cell proliferation in three OS cell lines which express c-MYC and ODC (Supplementary Figure S1), through a combination of processes including cell cycle arrest and differentiation. The DFMO-mediated induction of differentiation has previously been observed in Friend's murine erythroleukemia (MEL) cells [34]. Although early differentiation markers were unchanged, those that are seen later in osteogenic maturation were increased with DFMO treatment, demonstrating that DFMO leads to terminal differentiation of these OS cell lines. When differentiation has been evaluated previously as a potential treatment target of OS therapy, there have been concerning results. Potent osteogenic bone morphogenetic proteins (BMPs) used to induce differentiation in some of these same cell lines actually resulted in increased proliferation in vitro and increased tumor growth in orthotopic in vivo models. The hypothesized mechanism was that differentiation defects in human OS cells only allow them to reach an early progenitor stage that is stuck in an early proliferative phase. This results in increased proliferation when exposed to osteogenic stimuli. However, when later steps of the differentiation pathway were induced, these cells were able to differentiate and they showed decreased proliferative activity [35]. Our results suggest that PA inhibition activates later steps of the differentiation pathway and pushes OS cells through the early progenitor stage to terminal differentiation, which results in decreased proliferation.

Furthermore, the effects of DFMO on OS cell growth remain for at least seven days after removal of DFMO. When OS cells exposed to DFMO for six days were reseeded in standard media, they continued to have astonishingly depressed growth for up to seven days. ODC is a rapidly metabolized enzyme; thus, once DFMO is removed, functional ODC and PAs quickly re-accumulate. The prolonged effects of DFMO on cell proliferation suggest that it might also induce effects independent of PA depletion.

From current clinical trials we have learned that extended treatment with DFMO in children is feasible and well tolerated [28]. The prolonged effects of DFMO exposure as observed in this study are particularly encouraging for progression into OS animal models and future clinical trials, as it suggests that even when there are times during which DFMO is not at a therapeutic level, there will be continued effects. This may also justify an easier, less frequent dosing schedule.

5. Conclusions

In conclusion, our results demonstrate that inhibition of PA synthesis with DFMO induces cell cycle arrest and terminal differentiation in human OS cells. These effects are persistent even after removal of DFMO. These findings suggest a role for DFMO in OS therapy, particularly in the setting of preventing relapse. Even when patients have achieved a radiographic complete remission with aggressive chemotherapy and surgical resection, at times, low levels of viable OS cells remain, resulting in later relapse. If these residual cells could be pushed into cell cycle arrest or toward terminal differentiation, relapse could potentially be prevented or delayed. Thus, further studies are needed to determine whether our results persist in vivo in OS mouse models.

Supplementary Materials: The following are available online at http://www.mdpi.com/2076-3271/6/3/65/s1. Figure S1: c-MYC and ODC1 expression in OS cell lines, Figure S2: DFMO IC-50 in OS cell lines.

Author Contributions: R.R.W. and A.S.B. conceived the experimental design of the paper. R.R.W. performed all experiments with technical support from C.R.S. and K.L.U. D.G. performed bio-informatic RNA expression analyses. R.R.W. and A.S.B. wrote the manuscript. All authors approved the final version of the manuscript.

Funding: This work was supported by a Helen DeVos Children's Hospital (HDVCH) Research Grant (RC107855) to Rebecca R. Weicht and André S. Bachmann and by Michigan State University Discretional Funds to André S. Bachmann.

Acknowledgments: We thank Patrick Woster (Medical University of South Carolina, Charleston, SC, USA) for providing DFMO and Otto Phanstiel (University of Central Florida, FL, USA) for providing the PA internal HPLC standards. The authors gratefully acknowledge Rachael Sheridan, manager of the Flow Cytometry Core Facility (Van Andel Research Institute, MI, USA), for performing flow cytometry analysis.

Conflicts of Interest: André S. Bachmann is the sole inventor of a U.S. patent (US 9,072,778) issued on 7 July 2015 entitled "Treatment Regimen for N-MYC, C-MYC, and L-MYC amplified and overexpressed tumors". No potential conflicts of interest were disclosed by the other authors.

References

1. Moore, D.D.; Luu, H.H. Osteosarcoma. *Cancer Treat. Res.* **2014**, *162*, 65–92. [PubMed]
2. Ottaviani, G.; Jaffe, N. The epidemiology of osteosarcoma. *Cancer Treat. Res.* **2009**, *152*, 3–13. [PubMed]
3. Roberts, R.D.; Wedekind, M.F.; Setty, B.A. Chemotherapy regimens for patients with newly diagnosed malignant bone tumors. In *Malignant Pediatric Bone—Tumorstreatment & Management*; Cripe, T.P., Yeager, N.D., Eds.; Springer International Publishing: Berlin, Germany, 2015; pp. 83–107.
4. Pegg, A.E. Recent advances in the biochemistry of polyamines in eukaryotes. *Biochem. J.* **1986**, *234*, 249–262. [CrossRef] [PubMed]
5. Pegg, A.E. Mammalian polyamine metabolism and function. *IUBMB Life* **2009**, *61*, 880–894. [CrossRef] [PubMed]
6. Pegg, A.E. Functions of polyamines in mammals. *J. Biol. Chem.* **2016**, *291*, 14904–14912. [CrossRef] [PubMed]
7. Pegg, A.E.; Feith, D.J. Polyamines and neoplastic growth. *Biochem. Soc. Trans.* **2007**, *35*, 295–299. [CrossRef] [PubMed]
8. Soda, K. The mechanisms by which polyamines accelerate tumor spread. *J. Exp. Clin. Cancer Res.* **2011**, *30*, 95. [CrossRef] [PubMed]
9. Wallace, H.M.; Fraser, A.V.; Hughes, A. A perspective of polyamine metabolism. *Biochem. J.* **2003**, *376*, 1–14. [CrossRef] [PubMed]
10. Bachmann, A.S. The role of polyamines in human cancer: Prospects for drug combination therapies. *Hawaii Med. J.* **2004**, *63*, 371–374. [PubMed]
11. Casero, R.A., Jr.; Marton, L.J. Targeting polyamine metabolism and function in cancer and other hyperproliferative diseases. *Nat. Rev. Drug Discov.* **2007**, *6*, 373–390. [CrossRef] [PubMed]
12. Geerts, D.; Koster, J.; Albert, D.; Koomoa, D.L.; Feith, D.J.; Pegg, A.E.; Volckmann, R.; Caron, H.; Versteeg, R.; Bachmann, A.S. The polyamine metabolism genes ornithine decarboxylase and antizyme 2 predict aggressive behavior in neuroblastomas with and without MYCN amplification. *Int. J. Cancer* **2010**, *126*, 2012–2024. [PubMed]
13. Gerner, E.W.; Meyskens, F.L., Jr. Polyamines and cancer: Old molecules, new understanding. *Nat. Rev. Cancer* **2004**, *4*, 781–792. [CrossRef] [PubMed]
14. Koomoa, D.L.; Geerts, D.; Lange, I.; Koster, J.; Pegg, A.E.; Feith, D.J.; Bachmann, A.S. DFMO/eflornithine inhibits migration and invasion downstream of MYCN and involves p27kip1 activity in neuroblastoma. *Int. J. Oncol.* **2013**, *42*, 1219–1228. [CrossRef] [PubMed]
15. Murray-Stewart, T.R.; Woster, P.M.; Casero, R.A., Jr. Targeting polyamine metabolism for cancer therapy and prevention. *Biochem. J.* **2016**, *473*, 2937–2953. [CrossRef] [PubMed]
16. Wallick, C.J.; Gamper, I.; Thorne, M.; Feith, D.J.; Takasaki, K.Y.; Wilson, S.M.; Seki, J.A.; Pegg, A.E.; Byus, C.V.; Bachmann, A.S. Key role for p27kip1, retinoblastoma protein RB, and MYCN in polyamine inhibitor-induced g1 cell cycle arrest in MYCN-amplified human neuroblastoma cells. *Oncogene* **2005**, *24*, 5606–5618. [CrossRef] [PubMed]
17. Pegg, A.E. Regulation of ornithine decarboxylase. *J. Biol. Chem.* **2006**, *281*, 14529–14532. [CrossRef] [PubMed]
18. Seiler, N. Thirty years of polyamine-related approaches to cancer therapy. Retrospect and prospect. Parts 1&2. Selective enzyme inhibitor & structural analogues and derivatives. *Curr. Drug Targets* **2003**, *4*, 537–585. [PubMed]
19. Meyskens, F.L., Jr.; Gerner, E.W. Development of difluoromethylornithine (DFMO) as a chemoprevention agent. *Clin. Cancer Res.* **1999**, *5*, 945–951. [PubMed]

20. Meyskens, F.L., Jr.; McLaren, C.E. Chemoprevention, risk reduction, therapeutic prevention, or preventive therapy? *J. Natl. Cancer Inst.* **2010**, *102*, 1815–1817. [CrossRef] [PubMed]

21. Meyskens, F.L., Jr.; McLaren, C.E.; Pelot, D.; Fujikawa-Brooks, S.; Carpenter, P.M.; Hawk, E.; Kelloff, G.; Lawson, M.J.; Kidao, J.; McCracken, J.; et al. Difluoromethylornithine plus sulindac for the prevention of sporadic colorectal adenomas: A randomized placebo-controlled, double-blind trial. *Cancer Prev. Res.* **2008**, *1*, 32–38. [CrossRef] [PubMed]

22. Meyskens, F.L., Jr.; Simoneau, A.R.; Gerner, E.W. Chemoprevention of prostate cancer with the polyamine synthesis inhibitor difluoromethylornithine. *Recent Res. Cancer Res.* **2014**, *202*, 115–120.

23. Pegg, A.E.; Shantz, L.M.; Coleman, C.S. Ornithine decarboxylase as a target for chemoprevention. *J. Cell Biochem. Suppl.* **1995**, *22*, 132–138. [CrossRef] [PubMed]

24. Bachmann, A.S.; Levin, V.A. Clinical applications of polyamine-based therapeutics. In *Polyamine Drug Discovery*; Woster, P.M., Casero, R.A., Jr., Eds.; Royal Society of Chemistry: London, UK, 2012; pp. 257–276.

25. Bachmann, A.S.; Geerts, D.; Sholler, G. Neuroblastoma: Ornithine decarboxylase and polyamines are novel targets for therapeutic intervention. In *Pediatric Cancer, Neuroblastoma: Diagnosis, Therapy, and Prognosis*; Hayat, M.A., Ed.; Springer: Berlin, Germany, 2012; Volume 1, pp. 91–103.

26. Koomoa, D.L.; Yco, L.P.; Borsics, T.; Wallick, C.J.; Bachmann, A.S. Ornithine decarboxylase inhibition by {alpha}-difluoromethylornithine activates opposing signaling pathways via phosphorylation of both Akt/protein kinase B and P27kip1 in neuroblastoma. *Cancer Res.* **2008**, *68*, 9825–9831. [CrossRef] [PubMed]

27. Rounbehler, R.J.; Li, W.; Hall, M.A.; Yang, C.; Fallahi, M.; Cleveland, J.L. Targeting ornithine decarboxylase impairs development of MYCN-amplified neuroblastoma. *Cancer Res.* **2009**, *69*, 547–553. [CrossRef] [PubMed]

28. Saulnier Sholler, G.L.; Gerner, E.W.; Bergendahl, G.; MacArthur, R.B.; VanderWerff, A.; Ashikaga, T.; Bond, J.P.; Ferguson, W.; Roberts, W.; Wada, R.K.; et al. A phase I trial of DFMO targeting polyamine addiction in patients with relapsed/refractory neuroblastoma. *PLoS ONE* **2015**, *10*, e0127246. [CrossRef] [PubMed]

29. Lee, M.J.; Chen, Y.; Huang, Y.P.; Hsu, Y.C.; Chiang, L.H.; Chen, T.Y.; Wang, G.J. Exogenous polyamines promote osteogenic differentiation by reciprocally regulating osteogenic and adipogenic gene expression. *J. Cell Biochem.* **2013**, *114*, 2718–2728. [CrossRef] [PubMed]

30. Tjabringa, G.S.; Zandieh-Doulabi, B.; Helder, M.N.; Knippenberg, M.; Wuisman, P.I.; Klein-Nulend, J. The polyamine spermine regulates osteogenic differentiation in adipose stem cells. *J. Cell Mol. Med.* **2008**, *12*, 1710–1717. [CrossRef] [PubMed]

31. Tsai, Y.H.; Lin, K.L.; Huang, Y.P.; Hsu, Y.C.; Chen, C.H.; Chen, Y.; Sie, M.H.; Wang, G.J.; Lee, M.J. Suppression of ornithine decarboxylase promotes osteogenic differentiation of human bone marrow-derived mesenchymal stem cells. *FEBS Lett.* **2015**, *589*, 2058–2065. [CrossRef] [PubMed]

32. Kim, H.I.; Schultz, C.R.; Buras, A.L.; Friedman, E.; Fedorko, A.; Seamon, L.; Chandramouli, G.V.R.; Maxwell, G.L.; Bachmann, A.S.; Risinger, J.I. Ornithine decarboxylase as a therapeutic target for endometrial cancer. *PLoS ONE* **2017**, *12*, e0189044. [CrossRef] [PubMed]

33. Minocha, S.C.; Minocha, R.; Robie, C.A. High-performance liquid chromatographic method for the determination of dansyl-polyamines. *J. Chromatogr.* **1990**, *511*, 177–183. [CrossRef]

34. Choudhary, S.K.; Sharma, D.; Dixit, A. D,l-alpha-difluoromethylornithine, an irreversible inhibitor of ornithine decarboxylase, induces differentiation in mel cells. *Cell Biol. Int.* **1999**, *23*, 489–495. [CrossRef] [PubMed]

35. Luo, X.; Chen, J.; Song, W.X.; Tang, N.; Luo, J.; Deng, Z.L.; Sharff, K.A.; He, G.; Bi, Y.; He, B.C.; et al. Osteogenic BMPS promote tumor growth of human osteosarcomas that harbor differentiation defects. *Lab. Invest.* **2008**, *88*, 1264–1277. [CrossRef] [PubMed]

Article

Knocking out Ornithine Decarboxylase Antizyme 1 (*OAZ1*) Improves Recombinant Protein Expression in the HEK293 Cell Line

Laura Abaandou [1,2] and Joseph Shiloach [1,*]

1   Biotechnology Core Laboratory, National Institute of Diabetes and Digestive and Kidney Diseases, National Institutes of Health, Bethesda, MD 20892, USA; laura.abaandou@nih.gov
2   Department of Chemistry and Biochemistry, George Mason University, Fairfax, VA 22030, USA
*   Correspondence: Josephs@niddk.nih.gov; Tel.: +1-301-496-9719

Received: 2 April 2018; Accepted: 1 June 2018; Published: 8 June 2018

Abstract: Creating efficient cell lines is a priority for the biopharmaceutical industry, which produces biologicals for various uses. A recent approach to achieving this goal is the use of non-coding RNAs, microRNA (miRNA) and small interfering RNA (siRNA), to identify key genes that can potentially improve production or growth. The ornithine decarboxylase antizyme 1 (*OAZ1*) gene, a negative regulator of polyamine biosynthesis, was identified in a genome-wide siRNA screen as a potential engineering target, because its knock down by siRNA increased recombinant protein expression from human embryonic kidney 293 (HEK293) cells by two-fold. To investigate this further, the *OAZ1* gene in HEK293 cells was knocked out using CRISPR genome editing. The *OAZ1* knockout cell lines displayed up to four-fold higher expression of both stably and transiently expressed proteins, with comparable growth and metabolic activity to the parental cell line; and an approximately three-fold increase in intracellular polyamine content. The results indicate that genetic inactivation of *OAZ1* in HEK293 cells is an effective strategy to improve recombinant protein expression in HEK293 cells.

Keywords: protein expression; antizyme 1; ornithine decarboxylase; CRISPR; human embryonic kidney 293 (HEK293)

---

1. Introduction

The production of recombinant proteins from mammalian cells for therapeutic and research purposes has always been associated with the challenge of improving production. As a result, volumetric productivity has been increased significantly in the last two decades by media optimization, careful process control [1–3], genetic modification [4–6], and cell line selection [1]. Chinese hamster ovary (CHO) cells are currently the main industrial producers of recombinant therapeutic proteins [7,8] because of their efficient growth and high expression characteristics. These and other non-human producers, such as baby hamster kidney (BHK21) cells and murine myeloma cells (NS0 and Sp2/0) [9], are limited by their ability to correctly perform post-translational modifications, which are important for proper activity of the produced protein [10]. Therefore, efforts are ongoing to improve growth and expression of recombinant protein from human cell lines, such as human embryonic kidney 293 (HEK293) and fibrosarcoma HT-1080 [11]. Recent research conducted by Su et al. looked at the possibility of improving protein expression from HEK293 cells by utilizing non-coding RNA, such as microRNA (miRNA) and small interfering RNA (siRNA) [12–14]. The study was conducted in two directions; first, the focus was on finding microRNA that have a direct effect on protein expression; and the second was focused on finding specific genes whose inactivation by siRNA improved expression. This work was performed by implementing two independent high throughput screenings of the non-coding RNAs. By screening the effect of 23,000 different siRNAs, ten genes whose knock-down triggered higher expression of recombinant luciferase were selected for further studies. Among the ten identified genes, ornithine decarboxylase antizyme (*OAZ1*) attracted special attention, because its inhibition improved the expression of luciferase with minimal effects on cell growth and viability. Additionally, the role of *OAZ1* in polyamine metabolism has been well studied [15,16], making it a promising target to explore the connection between *OAZ1* inhibition and the level of different polyamines. The *OAZ1* protein negatively regulates polyamine biosynthesis by degradation of ornithine decarboxylase. This enzyme catalyzes the decarboxylation of ornithine to form putrescine, a committed step in polyamine biosynthesis. This connection was investigated [17], and preliminary information showed that transient inhibition of *OAZ1* in HEK293 cells expressing luciferase by siRNA was associated with increased luciferase expression and a higher intracellular concentration of putrescine. Creating a permanent cell line lacking *OAZ1* should, therefore, be the next step in evaluating the potential of the *OAZ1* knockout cell line as an efficient producer of recombinant proteins from HEK293, and perhaps other mammalian cell lines.

Several gene editing tools are currently available, such as transcription activator-like effector nuclease (TALENS), zinc finger nucleases (ZFNs), and the clustered regularly interspaced short palindromic repeats (CRISPR)/Cas system, which consists of a Cas endonuclease directed to cleave a target sequence by a guide RNA. The CRISPR system, with its unprecedented level of simplicity, efficiency, and ability to carry out multiplexed mutations [18], was chosen to create an *OAZ1* deleted cell line. In this report, we evaluate the growth and production capabilities of the created cell line for production of recombinant proteins in both stable and transient transfection systems.

2. Materials and Methods

Cell line: HEK293 stably transfected with the luciferase gene of *Photinus pyralis* under a cytomegalovirus (CMV) promoter (CMV-Luc2-Hygro HEK293, Promega ID# CAS140901, Madison, WI, USA). This cell line will be referred to as the parental cell line in the text.

Cell culture: Cells were grown in adherent cultures in Dulbecco's Modified Eagle Medium (DMEM Gibco cat# 11995–040, Grand Island, NY, USA) supplemented with 10% fetal bovine serum (FBS, Atlanta Biologicals cat# S11150H, Flowery Branch, GA, USA), 100 units/mL of penicillin, and 100 µg/mL of streptomycin (Gibco cat# 15140–122, Grand Island, NY, USA). The cultures were incubated in a humidity-controlled incubator at 37 °C and 5% CO_2.

2.1. Construction of CRISPR/Cas9 Lentiviral Particles with Single Guide RNAs Targeting the OAZ1 Gene

Three all-in-one CRISPR/Cas9 lentiviral particles with single guide RNAs (sgRNAs) targeting the second exonic region of the *OAZ1* gene were designed and purified by Sigma Aldrich. All-in-one particles encode the sgRNA, Cas9 endonuclease, puromycin resistance, and green fluorescence protein (GFP), driven by a U6 promoter. The sequence of the target regions in the three guide RNAs are as follows:

taacccgggtccggggcctcgg-Sigma Aldrich cat# HS0000288774 (774)
gatcggctgaatgtaacagagg-Sigma Aldrich cat# HS0000288775 (775)
agacgccaaacgcattaactgg-Sigma Aldrich cat# HS0000288778 (778)

2.2. Transduction of Human Embryonic Kidney 293 Luciferase Expressing Cells with Lentiviral Particles and Isolation of Single Colonies

Ninety-six well plates (Corning cat #356717, Kennebunk, ME, USA) were coated with 1% Matrigel (Corning cat# 354248) and seeded with 75,000 CMV-Luc2-Hygro HEK293 cells per well in 30 µL of culture media. Plates were incubated overnight in a humidified incubator at 37 °C, 5% CO_2. Lentiviral particles, three *OAZ1* targeting lentiviral constructs, and CRISPR-Lenti non-targeting control transduction particles (Sigma-Aldrich # CRISPR12V, St Louis, MO, USA) were added the following day at a previously optimized multiplicity of infection (MOI) of 5, in 50 µL media and incubated overnight as above. The media were replaced for an additional overnight incubation. The media were replaced with selection media, containing 1 µg/mL puromycin, on day three post-transduction. The selection media were replaced every other day until confluence was achieved. Limiting dilutions were carried out from each well, as previously described [19,20], in 96-well plates. The wells were scored for the presence of GFP expressing single colonies over a period of two weeks. The wells containing single colonies were propagated and sub-cultured into larger vessels, until enough cells were available for assays (confluent T-25 flask).

2.3. Luciferase Assay for Selection of Highly Expressing Clones

Cell viability and luciferase activity were determined using the CellTiter-Glo luminescent cell viability assay and the One-Glo luciferase assay system (Promega cat# G7570 and cat# E6110, Madison, WI, USA, respectively), following manufacturer's protocol. Briefly, cells at a confluence of 80–90% in a 96-well plate were re-suspended in 100 µL media and transferred to a white opaque 96-well plate (Greiner Bio-one, cat# 655088, Frickenhausen, Germany) in quadruplets. Then, 100 µL of CellTiter-Glo reagent was added to two out of the four quads, mixed for 2 min on a shaker, and allowed to sit at room temperature for 10 min. At approximately 6 min into the 10-min incubation, 100 µL of One-Glo reagent was added to the remaining two wells, mixed briefly, and kept at room temperature until the end of the 10-min incubation period. The luminescence was read using the SpectraMax® microplate reader (Molecular devices, Wals, Austria) at an integration time of 250 ms.

2.4. Sequencing of the Ornithine Decarboxylase Antizyme 1 Gene in Parental and Mutant Strains

Total genomic DNA was isolated using the DNeasy kit (Qiagen, cat#69506, Hilden, Germany), following the manufacturers protocol. The genomic region flanking the CRISPR target site for the *OAZ1* gene was polymerase chain reaction (PCR) amplified, using the Phusion® High-Fidelity PCR Master Mix with HF Buffer (NEB, cat# MO531S, Ipswitch, MA, USA), and the following primers: forward primer—cagcagcagtgagagttcca; reverse primer—gcttttggagagcaatgggag. Amplicons were gel-extracted using the QIAquick® Gel Extraction Kit (QIAGEN, Ref# 28704, Hilden, Germany), and sequenced using capillary DNA sequencing. Sequences were aligned with the parental sequence using the Clustal Omega (European Bioinformatics Institute, Cambridge, UK) alignment program.

2.5. Growth Characterization Determination

Seven T-25 flasks per cell line were each inoculated with 7×10^5 cells in 5 mL culture media and incubated, as described above. The cells were enumerated each day for seven days, using the CEDEX HiREs cell analyzer/counter (Roche # 7766, Basel, Switzerland), and the levels of glucose and lactate were quantified using the YSI 2900 bio-analyzer (Yellow Springs Instrument Co., Yellow Springs, OH, USA).

2.6. Antizyme 1 and Luciferase Western Blot

Parental and CRISPR-treated cells in confluent six-well plates were lysed using radioimmunoprecipitation assay (RIPA) cell lysis buffer (Thermo Scientific, # 89900, Rockford, IL, USA) supplemented with 1× protease and phosphatase inhibitor (Thermo Scientific #1861281, Rockford, IL, USA), and total protein was quantified at A280 using the Nanodrop One^c (Thermo Scientific, Rockford, IL, USA). Cell lysate samples containing equal amounts of protein (340 µg) and recombinant firefly luciferase protein (abcam # ab100961, Cambridge, MA, USA) were electrophoresed on a 4–12% bis-tris gel (ThermoFisher, # NP0322BOX, Rockford, IL, USA). Separated proteins were transferred onto a nitrocellulose membrane and antizyme 1 (AZ1) or luciferase was detected using rabbit anti-AZ1 polyclonal antibodies targeting the *N*-terminal region of the AZ1 protein (Sigma-Aldrich, cat# SAB1307119, St Louis, MO, USA)/HRP conjugated anti-rabbit secondary antibodies (ThermoFisher Scientific # 65–6120, Rockford, IL, USA); or mouse anti-luciferase polyclonal antibodies (ThermoFisher, # PA1–179, Rockford, IL, USA)/HRP conjugated goat anti-mouse antibodies (KPL # 474–1806, Gaithersburg, MD, USA). Mouse anti-β-actin monoclonal antibodies (Sigma # A2228) and HRP conjugated goat anti-mouse antibodies (KPL # 474–1806, Gaithersburg, MD, USA) were used to detect β-actin in identical samples. The signal was developed using the SuperSignal® west Pico Chemiluminescent substrate (ThermoFisher, # OD187429, Rockford, IL, USA), and visualized in a LAS-4000 Mini Luminescent Image Analyzer (GE healthcare, # 28955810, Marlborough, MA, USA). The intensity of the Western blot bands were determined using ImageJ software and the luciferase/actin ratio, and fold increase was reported.

2.7. Quantitative PCR Analysis

Real time quantitative PCR was done using the SYBR GREEN protocol. Total RNA was extracted from *OAZ1* knockout and parental cells using the RNeasy kit (QIAGEN, cat# 74101, Hilden, Germany). First strand cDNA was synthesized from total RNA using the Maxima First strand cDNA synthesis kit for real time quantitative PCR (RT-qPCR) (Thermo Scientific, # K1642, Vilnius Lithuania). Then, 20 ng of cDNA was amplified in a qPCR using the following primers, *OAZ1*: forward primer—GGAACCGTAGACTCGCTCAT, reverse primer—TCGGAGTGAGCGTTTATTTG; and Luc: forward primer—GTGGTGTGCA GCGAGAATAG, reverse primer—CGCTCGTTGTAGATGTCGTTAG.

Threshold and threshold cycle (Ct) values were determined automatically by the RQ Manager™ Software (Applied Biosystems, Foster City, CA, USA) using default parameters. The comparative cycle threshold ($2^{-\Delta\Delta Ct}$ method) was used to analyze the expression levels of genes examined in this study. The abundance of each gene transcript was normalized by glyceraldehyde phosphate dehydrogenase (*GAPDH*) or β Actin (*ACTB*) gene expression levels and expressed in arbitrary units (AU). The relative quantization of gene expression was performed in triplicates for each sample.

2.8. Transfection of Cells with Secreted Alkaline Phosphatase Plasmid and Quantification of SEAP Activity

Twenty-four well plates were seeded with 200,000 cells per well in culture media one day prior to transfection. The following day, cells were transfected with 500 ng/well of the pSELECT-zeo-SEAP plasmid, encoding the embryonic secreted alkaline phosphatase (SEAP) driven by a EF-1α/HTLV composite promoter (InvivoGen # psetz-seap, San Diego, CA, USA), using the Lipofectamine® 2000 transfection reagent (Invitrogen # 11668-019, Carlsbad, CA, USA), and following the manufacturer's protocol. After two days, cell culture media were replaced with selection media (culture media supplemented with zeocin 200 µg/mL), and samples were collected for analysis on day three and day five. Alkaline phosphatase activity in the culture supernatant was quantified using the Quanti-Blue colorimetric assay (InvivoGen #rep-qb1, San Diego, CA, USA), cells were enumerated as described above, and total and specific SEAP activity was determined.

2.9. Polyamine Quantification

Five million cells per strain were washed twice in phosphate buffered saline (PBS) and pelleted at 1200 rpm for 5 min. The cells were disrupted with 2.3 mm zirconium beads in a mixture of methanol/water (50:50; *v/v*) acidified with 0.1% formic acid. Polyamines in cell extract were quantified using Ultra-High Performance Liquid Chromatography (UHPLC) analysis (Agilent 1290, Agilent Technologies, Santa Clara CA, USA), coupled to hybrid triple quadrupole/ion trap mass spectrometer (QTRAP 5500 from AB Sciex, Vaughan, ON, Canada), on Zorbax Eclipse Plus C18, rapid resolution high density (RRHD) column (2.1 × 50 mm 1.8 micron, Agilent Technologies). The chromatography was performed using ultrapure water containing 0.1% formic acid with 1 mM perfluoro heptanoic acid as solvent B and 0.1% formic acid plus 1 mM perfluoro heptanoic acid in 100% methanol as solvent A. The Liquid chromatograph tandem mass-spectrometry (LC-MS/MS) was run for 8.0 min with a flow rate of 300 µL/min. The gradient elution was performed at 50% solvent A for 0.00–4.50 min, for 4.50–5.25 min 100%, 5.25–6.75 min 50%, and 6.75–8.00 min 50%. The mass spectra were acquired using a Turbo Spray Ionization of 2500 V in positive ion mode, and multiple reaction monitoring (MRM). The curtain gas (nitrogen), CAD (collision activated dissociation), nebulizing, and heating gas were set to 40 psi, medium, 50 psi, and 60 psi, respectively. The temperature of the source was fixed at 500 °C. The mass spectrometer was set to have a dwell time of 50 ms. LC-MS/MS data were processed using Analyst 1.6.1 software (AB Sciex, Vaughan, ON, Canada). The results were compared to internal standard of 1,6-hexanediamine and standard curves for putrescine, spermidine, and spermine protein quantification were carried out in parallel using the modified Lowry assay.

One batch of three samples (5×10^5 cells) of each cell line, prepared from the same passage, was analyzed. The polyamine content and protein concentration was determined for each sample.

2.10. Statistical Analysis

Mean values and standard deviation or standard error were calculated using standard methods. *p*-values were determined using the Chi-square test.

3. Results

3.1. Creating HEK293 OAZ1-Deficient Cell Line

HEK293 luciferase expressing cells were treated with three different lentiviral CRISPR constructs, targeting the coding region of the *OAZ1*, in order to knock out the gene. From the three transduced pools, eight single-cell derived clones were isolated using limiting dilutions. The activity of the constitutively expressed luciferase was measured in the isolated clones and in the parental cell line (Table 1). Among the isolated clones, 775–1 and 775–3 showed 7-fold and 2.8-fold higher specific luminescence, respectively, when compared with the parental cell line, and were selected for further investigation.

Table 1. Luciferase enzymatic activity and fold increase in the luciferase activity of isolated single colonies, parental cell line, and negative control cell line. Luciferase activity and cell density were quantified in a luminescence-based assay. Specific luminescence represents luminescence per cell. Parental: human embryonic kidney 293 (HEK293) cell line constitutively expressing luciferase; 774–1, 774–2, 774–3, 775–1, 775–2, 775–3, 778–1: isolated single colonies. Specific luminescence values represent the average of duplicate samples and the standard deviation. RLU = Relative luminescence units.

Single Colony-Derived Clone	Specific Luminescence (RLU/Cell)	Fold Increase in Specific Luciferase Activity
774–1	3.0 ± 1.0	0.6 ± 0.3
774–2	2.2 ± 0.1	0.4 ± 0.0
774–3	6.2 ± 1.0	1.1 ± 0.2
775–1	40.1 ± 2.4	7.0 ± 0.4
775–2	3.2 ± 0.6	0.6 ± 0.2
775–3	16.1 ± 0.7	2.8 ± 0.1
775–4	4.5 ± 3.2	0.8 ± 0.6
778–1	6.4 ± 0.1	1.1 ± 0.0
Parental	5.8 ± 1.9	1.0 ± 0.0
Negative control	4.4 ± 1.2	0.8 ± 0.2

The CRISPR driven disruption of *OAZ1* in the selected cell lines was confirmed by alignment of the CRIPSR treated sequences with the parental sequence. The results shown in Figure 1a revealed the presence of a consecutive nine-nucleotide deletion in the 775–1 *OAZ1* and a single base insertion in the 775–3 *OAZ1*, both of which were confirmed, using PCR, to be bi-allelic (results not shown). Transcriptional and translational analyses determined them to be nonsense mutations, with predicted truncated AZ1 protein of 103 amino acids (775–1) and 128 amino acids (775–3), compared with the 228-full-length protein (Supplemental Figure 1). Consistent with the above findings, RT-qPCR quantification of *OAZ1* mRNA revealed lower levels of *OAZ1* mRNA in both knockout cell lines. The 775–1 cell line contained 70% *OAZ1* mRNA, while 775–3 contained about 30% of the *OAZ1* mRNA (Figure 1b). These findings were confirmed by Western blot, in which AZ1 antibodies did not detect the protein in cell lysates of 775–1 and 773–1 (Figure 1c). The results show that the *OAZ1* gene was functionally knocked out in both the 775–1 and 775–3 cell lines.

(a)

```
Parental    5' GGTTCTGAATTGGGCTGCACCCCCTGGACAGGATGATCGGCTGAATGTAACAGAGGAACT 3'
775-1          GGTTCTGAATTGGGCTGCACCCCCTGGACAGGATGATCGGCTGAATGTAA---------T

Parental    5' GGGTTCTGAATTGGGCTGCACCCCCTGGACAGGATGATCGGCTGAATGTAA-CAGAGGAA 3'
775-3          GGGTTCTGAATTGGGCTGCACCCCCTGGACAGGATGATCGGCTGAATGTAAACAGAGGAA
```

(b) (c)

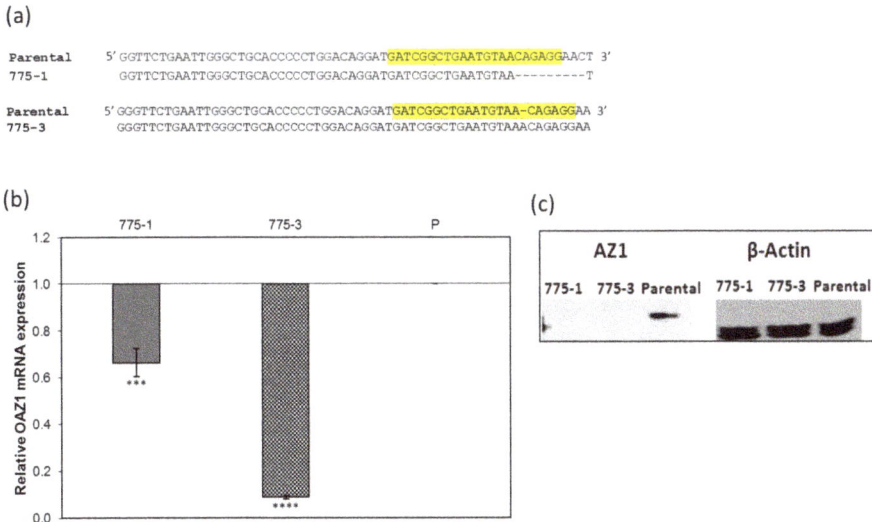

Figure 1. Clustered regularly interspaced short palindromic repeats (CRISPR)-mediated disruption of the ornithine decarboxylase antizyme (*OAZ1*) gene in human embryonic kidney 293 (HEK293)-luc cells and the generation of *OAZ1* knockout cell lines. (**a**) Clustal Omega DNA sequence alignment of the CRISPR *OAZ1* guide RNA (gRNA) target and surrounding regions, showing mutations in the *OAZ1* gene in the CRISPR-treated cell lines, 775–1 and 775–3. Highlighted region is the gRNA target sequence; (**b**) Real time quantitative polymerase chain reaction (RT-qPCR) of reverse-transcribed total cellular mRNA, using primers targeting *OAZ1* cDNA, confirms lower transcriptional levels of *OAZ1* mRNA in the CRISPR-treated cell lines, 775–1 and 775–3, of 0.7-fold (****, p-value < 0.00001) and 0.3-fold (***, p-value < 0.001) respectively, compared with the parental cell line. β-actin mRNA levels were used as endogenous controls. The assay was carried out in triplicates, and the results represent the average of three independent assays, with error bars representing the standard deviation; (**c**) Western blot analysis confirms *OAZ1* knockout in the 775–1 and 775–3 cell lines. Polyclonal antibodies against the N-terminal of antizyme 1 (AZ1), were used to detect the protein in 340 μg of electrophoresed whole cell lysates after transfer onto a nitrocellulose membrane, while monoclonal antibodies against the loading control β-actin were used to detect the protein in identical samples.

3.2. Growth Properties and Metabolic Activity of the OAZ1 Knockout Cell Lines

Seven-day growth kinetics and metabolic activity of the 775–1, 775–3 clones, and parental cell lines are shown in Figure 2. Growth rate and viability are shown in Figure 2a, with all cell lines achieving confluency on day six, and exhibiting similar doubling times of approximately 24 h. The *OAZ1* knockout cell lines showed similar viabilities to the parental cell line. The results show that the *OAZ1* inactivation and the increased burden of improved protein expression had minimal effects on the growth and viability of the host cell. The glucose consumption and the lactate production of the knockout and parental cell line were also similar, in accordance with the growth data (Figure 2b).

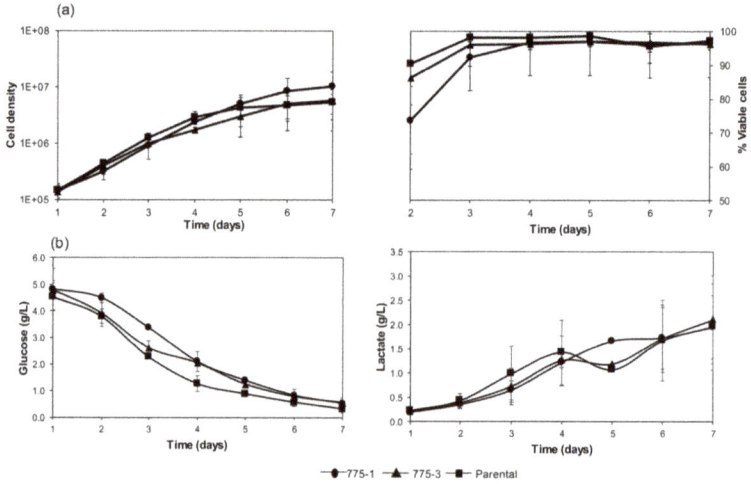

Figure 2. Seven-day growth characterization of the *OAZ1-* cell lines, 775–1 and 775–3, and the parental cell line. (**a**) Proliferation and viability of the *OAZ1-* cell lines 775–1 (●), 775–3 (▲), and the parental (■) cell line. Cells were plated at a density of 7×10^5 cells in T-25 flasks and enumerated using trypan blue exclusion over a seven-day period; and (**b**) glucose and lactate concentration in the culture media quantified using a bioanalyzer over the same seven-day period. Data from two independent growth studies are shown, with error bars representing the standard deviation.

3.3. Transcription Expression and Luciferase Activity in the OAZ1 Knockout Cell Line

Comparison of total and specific luciferase activity from the *OAZ1* knockout and the parental cells are shown in Figure 3a,b. Specific luciferase activity was not affected by the cell growth and, as expected, there was correlation between total activity and cell density. Quantification of total luciferase activity, messenger RNA level by RT-qPCR, and luciferase protein by immunoblot in samples at equivalent growth stages are shown in Figure 4. Higher specific and total luciferase activity observed in the *OAZ1* knockout cell lines (Figure 4a) were the result of an increase in both mRNA (Figure 4b) and protein expression (Figure 4c).

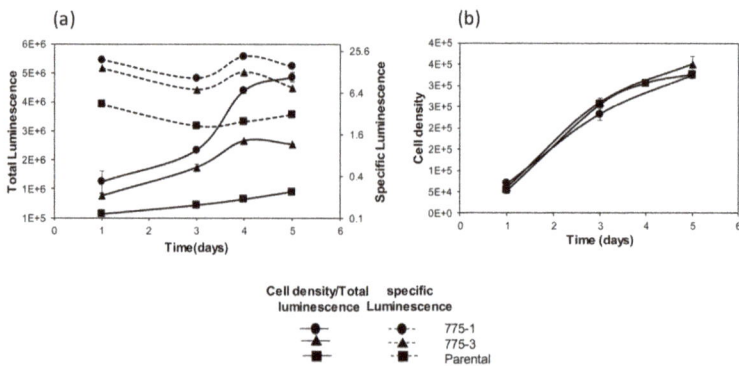

Figure 3. Time course growth and luciferase activity from parental and *OAZ1* deleted cells. Cells were seeded in duplicates, in 12-well plates at 1×10^5 cells per well. (**a**) Total and specific luciferase activity quantified for a five-day period; and (**b**) cell density determined by luminescence based assay for the same five-day period. Graphs represent one of three independent studies and error bars represent standard deviation of duplicate measurements.

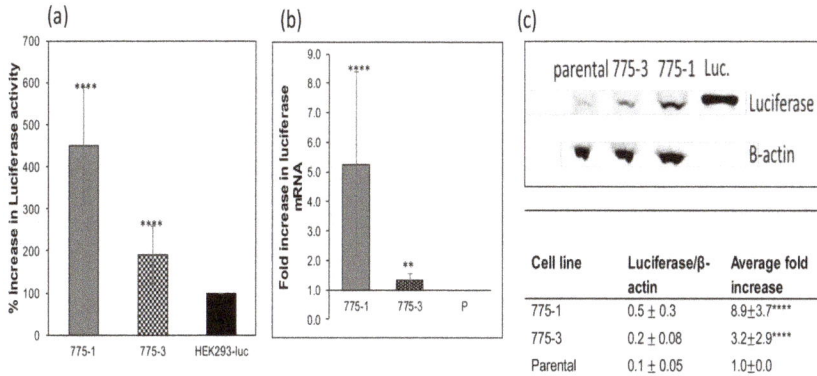

Figure 4. Luciferase activity, transcription, and expression. (a) Luciferase was quantified and reported as described above. The results represent the average of at least five independent assays, each carried out in duplicates, and error bars represent the standard deviation (****, *p*-value < 0.0001). (b) RT-qPCR of reverse-transcribed total cellular mRNA using primers targeting luciferase cDNA confirms higher transcriptional levels of luciferase mRNA in the *OAZ1*- cell lines of approximately 500% for 775–1 (****, *p*-value < 0.0001) and approximately 150% for 775–3 (**, *p*-value < 0.05), compared to the parental cell line. The assay was carried out in triplicates, and the results represent the average of three independent assays, with error bars representing the standard deviation. (c) Western blot analysis of 340 μg of whole cell lysates and 100 ng of recombinant luciferase protein for luciferase and β-actin expression, using anti-luciferase and anti-β-actin primary antibodies showing higher levels of intracellular luciferase protein in the knockout cell lines, 775–1 and 775–3, than in the parental cell line. ImageJ program was used to quantify the band intensities. β-actin-normalized luciferase band intensities were used to determine the fold increase in luciferase expression in the *OAZ1*- cell lines, 775–1 (8.9-fold; ****, *p*-value < 0.00001) and 775–3 (3.2-fold; ****, *p*-value < 0.00001), over the parental cell line. The results represent the average and standard deviation of three independent experiments, with a representative blot shown.

3.4. Expression of Secreted Alkaline Phosphatase in the OAZ1-Deficient HEK 293 Cell Line

To evaluate the protein expression capability of the HEK *OAZ1*-deficient cell line, the cells were transiently transfected with plasmid encoding SEAP. Alkaline phosphatase activity was quantified in cell culture supernatant on the third and fifth day post-transfection, together with viable cell density. Consistent with the activity of the constitutive luciferase, the alkaline phosphatase activity was higher in the knockout cell lines, with a 3.5-fold and 1.5-fold increase in the 775–1 and 775–3 cell lines, respectively (Figure 5).

3.5. Polyamine Concentration in the Parental and the OAZ1-Defficint Cell Line

The known function of antizyme 1 is negative regulation of intracellular polyamine concentration. The effect of the knockout of *OAZ1* on cellular polyamine levels was thus investigated. Cells grown to confluence were enumerated, and the total amount of polyamine in the equivalent numbers of cells was quantified using UHPLC. Three batches of cells from each cell line were analyzed. The picomole amount of putrescine, spermine, and spermidine per microgram protein are summarized in Table 2. Compared with the parental cells, higher concentration of intracellular polyamines was detected in the *OAZ1* knockout cell lines; it was 2-fold higher in the 775–1 and 3.5-fold higher in the 775–3, with the most significant difference observed in the level of putrescine, where it was 10-fold higher in 775–1, and 88-fold higher in the 775–3 mutants.

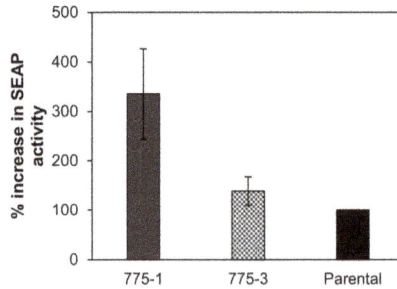

Figure 5. Activity of transiently transfected secreted alkaline phosphatase in *OAZ1-* and parental cell lines. Alkaline phosphatase (AP) activity assayed on day three and day five in the cell culture supernatant of 775–1, 775–3, and parental cells transfected with a plasmid encoding human fetal secreted AP (SEAP). SEAP activity is reported as the percent increase in specific activity (activity per cell), in the *OAZ1* cell lines of (335 ± 92)% for 775–1 and (138 ± 29)% for 775–3, relative to the that of the parental cell line. The results represent the average of two independent transfections, with a total of four biological replicates. Error bars represent the standard deviation.

Table 2. Intracellular polyamine levels in ornithine decarboxylase antizyme 1 deficient (*OAZ1-*) and parental cell lines. The intracellular levels of the polyamines putrescine, spermine, and spermidine in 5×10^5 pelleted cells were quantified chromatographically, with simultaneous quantification of total protein, using the modified Lowry assay. Polyamine content is reported relative to the total protein content of the sample. The results represent the average of three samples from a single batch of each cell line prepared from the same passage, and the standard deviation.

Cell Line	Polyamines (pmol/µg Protein)			
	Total Polyamine	Putrescine	Spermidine	Spermine
775–1	20.3 ± 2.1 ****	2.5 ± 0.7 ***	12.8 ± 2.4 **	5.0 ± 1.3
775–3	38.1 ± 2.0 ****	21.9 ± 1.3 ***	13.4 ± 0.6 **	2.8 ± 0.4
Parental	11.7 ± 1.3	0.2 ± 0.0	8.2 ± 0.4	3.3 ± 0.9

** p-value < 0.05; *** p-value < 0.001; **** p-value < 0.0001.

4. Discussion

Mammalian cell lines are the producers of choice for many recombinant therapeutic proteins. Among the currently utilized producing cell lines, human cell lines are becoming more relevant because of their innate ability to perform the correct post-translation modifications needed for stable and functional proteins [21,22]. However, compared with their non-human counterparts, their productivity is relatively low, necessitating the development of efficient cell lines using genetic engineering, and the improvement of growth and expression strategies.

In an effort to identify genes potentially affecting recombinant protein expression, a high throughput screen evaluating the effect of siRNA gene knock down on luciferase expression from HEK293 cells was conducted [17]. From a total of approximately 20,000 evaluated genes, *OAZ1* was identified as a promising candidate, whose deletion could improve protein expression, and was confirmed by transient transfection of siRNA against *OAZ1*.

The work presented here describes the creation of the HEK293 cell line with *OAZ1* deletion using CRISPR genome editing. Two single clone-derived HEK293 luciferase expressing cell lines deficient in the *OAZ1* gene product, antizyme 1, were created by targeting the second exon of *OAZ1* for disruption. The premise of targeting exon 2 (*OAZ1* has six exons) was to cause disruptions early in the sequence, which would likely result in a truncated or functionally inactive protein, because the ornithine decarboxylase (ODC) binding site is in the internal (122–144) and C-terminal (211–218) portions of the protein [23]. As a result, two mutants with the predicated truncated antizyme 1 sequences of 103 and 128 amino acids, structurally incapable of binding ODC, were selected.

Compared with the parental cell line, both engineered cell lines demonstrated higher expression of luciferase, a stably transfected cytoplasmic protein, and alkaline phosphatase, a transiently transfected secreted protein. The growth kinetics and the metabolic activity of the mutant cell lines were comparable to the parental cell line and, as expected, their intracellular polyamine content was higher.

The improved expression of recombinant proteins exhibited by these *OAZ1* knockout cell lines can be attributed to the observed increase in the level of polyamines, which is likely the result of a missing functional *OAZ1* gene. Antizyme 1, the protein product of *OAZ1*, is known to regulate polyamine production by inhibiting ODC [15,16]. Inhibiting AZ1 has been shown to cause an increase in intracellular polyamines levels, particularly putrescine [24]. The disproportionate effect of *OAZ1* knockout on the different polyamines can be attributed to the *OAZ1*-independent catabolism of spermine and spermidine, but not of putrescine [25]. Adding exogenous polyamines to cell culture of HEK293 has also been shown to enhance recombinant protein expression in a concentration dependent manner [17].

Improved expression of recombinant proteins from the *OAZ1* deficient cells was observed both in cells stably expressing luciferase, and from the same cells transiently transfected with secreted alkaline phosphatase, although to a lesser extent. This suggests that the presented approach for improved expression can be applied to different recombinant proteins, and perhaps different cell lines, although likely with different levels of enhancement depending on the properties of the expressed proteins. The increase in luciferase activity of the knockout cell lines is accompanied by an increase in both mRNA transcription and protein expression. By comparing the increase of mRNA and protein, it was concluded that the increase in expression was engendered at the transcriptional level, with little to no post-transcriptional effects. This observation is different from the initial results obtained from the siRNA silencing of the *OAZ1* gene, in which there was no increase in luciferase mRNA [17].

Improved protein expression is often accompanied by undesirable side effects, such as growth and metabolic disadvantages [26,27], caused by increased metabolic load on the cells. This was not observed in the *OAZ1* deficient or knockout cell lines, where no significant differences in growth and nutrient utilization were observed. These observations can be explained by the known role of *OAZ1* as a negative regulator of cell growth [28], meaning its absence might offset the possible negative effects on cell growth engendered by improved protein expression. It is also possible that the effect of metabolic load is cell line dependent [29]. A peculiar phenomenon is that although both knockout cell lines have higher intracellular polyamine concentrations than the parental cell line, the observed relationship between polyamine concentration and recombinant protein expression is not proportional. The lower producing cell line, 775–3, has at least a two-fold higher polyamine concentration than the higher producing cell line, 775–1. This discrepancy can be explained by possible toxic effects of elevated intracellular polyamine levels, which were not high enough to cause significant cytotoxicity, on the rate of protein synthesis [30]. The previous report by Xiao et al. also showed that improved expression by addition of external polyamines is concentration-dependent up to an optimal concentration, above which higher concentrations have negative effects on the expression [17].

The work presented here was done in anchorage dependent cells growing in serum supplemented media, which are, therefore, limited in their capacity to scale up. The next step would be to investigate the effect of the *OAZ1* deletion in cells growing in suspension, which are more relevant to large-scale recombinant protein production.

Supplementary Materials: The following are available online at http://www.mdpi.com/2076-3271/6/2/48/s1.

Author Contributions: J.S. conceived and supervised the experiments; L.A. designed and performed the experiments. L.A. and J.S. wrote the manuscript.

Acknowledgments: The authors thank Sarah Inwood for critical reading of the manuscript. The research was supported by the Intramural Research Program of the National Institute of Diabetes and Digestive and Kidney Diseases (NIDDK/NIH).

Conflicts of Interest: The authors declare no conflict of interest.

References

1. Wurm, F.M. Production of recombinant protein therapeutics in cultivated mammalian cells. *Nat. Biotechnol.* **2004**, *22*, 1393–1398. [CrossRef] [PubMed]
2. Hacker, D.L.; De Jesus, M.; Wurm, F.M. 25 years of recombinant proteins from reactor-grown cells—Where do we go from here? *Biotechnol. Adv.* **2009**, *27*, 1023–1027. [CrossRef] [PubMed]
3. Li, F.; Vijayasankaran, N.; Shen, A.; Kiss, R.; Amanullah, A. Cell culture processes for monoclonal antibody production. *mAbs* **2010**, *2*, 466–479. [CrossRef] [PubMed]
4. Fischer, S.; Marquart, K.F.; Pieper, L.A.; Fieder, J.; Gamer, M.; Gorr, I.; Schulz, P.; Bradl, H. miRNA engineering of CHO cells facilitates production of difficult-to-express proteins and increases success in cell line development. *Biotechnol. Bioeng.* **2017**, *114*, 1495–1510. [CrossRef] [PubMed]
5. Kim, J.Y.; Kim, Y.-G.; Lee, G.M. CHO cells in biotechnology for production of recombinant proteins: Current state and further potential. *Appl. Microbiol. Biotechnol.* **2012**, *93*, 917–930. [CrossRef] [PubMed]
6. Krämer, O.; Klausing, S.; Noll, T. Methods in mammalian cell line engineering: From random mutagenesis to sequence-specific approaches. *Appl. Microbiol. Biotechnol.* **2010**, *88*, 425–436. [CrossRef] [PubMed]
7. Butler, M.; Spearman, M. The choice of mammalian cell host and possibilities for glycosylation engineering. *Curr. Opin. Biotechnol.* **2014**, *30*, 107–112. [CrossRef] [PubMed]
8. Durocher, Y.; Butler, M. Expression systems for therapeutic glycoprotein production. *Curr. Opin. Biotechnol.* **2009**, *20*, 700–707. [CrossRef] [PubMed]
9. Estes, S.; Melville, M. Mammalian cell line developments in speed and efficiency. *Adv. Biochem. Eng. Biotechnol.* **2014**, *139*, 11–33. [PubMed]
10. Patnaik, S.K.; Stanley, P. Lectin-resistant CHO glycosylation mutants. *Methods Enzymol.* **2006**, *416*, 159–182. [PubMed]
11. Ghaderi, D.; Taylor, R.E.; Padler-Karavani, V.; Diaz, S.; Varki, A. Implications of the presence of N-glycolylneuraminic acid in recombinant therapeutic glycoproteins. *Nat. Biotechnol.* **2010**, *28*, 863–867. [CrossRef] [PubMed]
12. Xiao, S.; Chen, Y.C.; Betenbaugh, M.J.; Martin, S.E.; Shiloach, J. MiRNA mimic screen for improved expression of functional neurotensin receptor from HEK293 cells. *Biotechnol. Bioeng.* **2015**, *112*, 1632–1643. [CrossRef] [PubMed]
13. Inwood, S.; Buehler, E.; Betenbaugh, M.; Lal, M.; Shiloach, J. Identifying HIPK1 as Target of miR-22-3p Enhancing Recombinant Protein Production from HEK 293 Cell by Using Microarray and HTP siRNA Screen. *Biotechnol. J.* **2017**, *13*, 1700342. [CrossRef] [PubMed]
14. Druz, A.; Chen, Y.C.; Guha, R.; Betenbaugh, M.; Martin, S.E.; Shiloach, J. Large-scale screening identifies a novel microRNA, miR-15a-3p, which induces apoptosis in human cancer cell lines. *RNA Biol.* **2013**, *10*, 287–300. [CrossRef] [PubMed]
15. Coffino, P. Polyamines in spermiogenesis: Not now, darling. *Proc. Natl. Acad. Sci. USA* **2000**, *97*, 4421–4423. [CrossRef] [PubMed]
16. Hayashi, S.; Murakami, Y.; Matsufuji, S. Ornithine decarboxylase antizyme: A novel type of regulatory protein. *Trends Biochem. Sci.* **1996**, *21*, 27–30. [CrossRef]
17. Xiao, S.; Chen, Y.C.; Buehler, E.; Mandal, S.; Mandal, A.; Betenbaugh, M.; Park, M.H.; Martin, S.; Shiloach, J. Genome-scale RNA interference screen identifies antizyme 1 (OAZ1) as a target for improvement of recombinant protein production in mammalian cells. *Biotechnol. Bioeng.* **2016**, *113*, 2403–2415. [CrossRef] [PubMed]
18. Gupta, R.M.; Musunuru, K. Expanding the genetic editing tool kit: ZFNs, TALENs, and CRISPR-Cas9. *J. Clin. Investig.* **2014**, *124*, 4154–4161. [CrossRef] [PubMed]
19. Goding, J.W. Antibody production by hybridomas. *J. Immunol. Methods* **1980**, *39*, 285–308. [CrossRef]
20. Fuller, S.A.; Takahashi, M.; Hurrell, J.G. Cloning of hybridoma cell lines by limiting dilution. *Curr. Protocols Mol. Biol.* **2001**, *1*. [CrossRef]
21. Leavitt, R.W.; Braithwaite, C.; Jensen, M.M. Colicin V38 and microcin C38 produced by *Escherichia coli* strain 38. *Avian Dis.* **1997**, *41*, 568–577. [CrossRef] [PubMed]
22. Wallick, S.C.; Kabat, E.A.; Morrison, S.L. Glycosylation of a VH residue of a monoclonal antibody against alpha (1—6) dextran increases its affinity for antigen. *J. Exp. Med.* **1988**, *168*, 1099–1109. [CrossRef] [PubMed]

23. Ichiba, T.; Matsufuji, S.; Miyazaki, Y.; Murakami, Y.; Tanaka, K.; Ichihara, A.; Hayashi, S. Functional regions of ornithine decarboxylase antizyme. *Biochem. Biophys. Res. Commun.* **1994**, *200*, 1721–1727. [CrossRef] [PubMed]

24. Kim, S.W.; Mangold, U.; Waghorne, C.; Mobascher, A.; Shantz, L.; Banyard, J.; Zetter, B.R. Regulation of cell proliferation by the antizyme inhibitor: Evidence for an antizyme-independent mechanism. *J. Cell Sci.* **2006**, *119*, 2583–2591. [CrossRef] [PubMed]

25. Pegg, A.E. Spermidine/spermine-N^1-acetyltransferase: A key metabolic regulator. *Am. J. Physiol. Endocrinol. Metab.* **2008**, *294*, E995–E1010. [CrossRef] [PubMed]

26. Kidane, A.H.; Guan, Y.; Evans, P.M.; Kaderbhai, M.A.; Kemp, R.B. Comparison of heat flux in wild type and genetically-engineered Chinese hamster ovary cells. *J. Therm. Anal.* **1997**, *49*, 771–783. [CrossRef]

27. Schröder, M.; Friedl, P. Overexpression of recombinant human antithrombin III in Chinese hamster ovary cells results in malformation and decreased secretion of recombinant protein. *Biotechnol. Bioeng.* **1997**, *53*, 547–559. [CrossRef]

28. Bercovich, Z.; Snapir, Z.; Keren-Paz, A.; Kahana, C. Antizyme affects cell proliferation and viability solely through regulating cellular polyamines. *J. Biol. Chem.* **2011**, *286*, 33778–33783. [CrossRef] [PubMed]

29. Yallop, G.A.; Svendsen, I. *Recombinant Protein Production with Prokaryotic and Eukaryotic Cells*; Kluwer Academic Publishers: Dordrecht, The Netherlands, 2000.

30. Mitchell, J.L.; Diveley, R.R., Jr.; Bareyal-Leyser, A.; Mitchell, J.L. Abnormal accumulation and toxicity of polyamines in a difluoromethylornithine-resistant HTC cell variant. *Biochim. Biophys. Acta* **1992**, *1136*, 136–142. [CrossRef]

medical sciences

MDPI

Article

The Role of Cadaverine Synthesis on Pneumococcal Capsule and Protein Expression

Mary F. Nakamya [1,†], Moses B. Ayoola [1,†], Seongbin Park [1], Leslie A. Shack [1], Edwin Swiatlo [2] and Bindu Nanduri [1,3,*]

[1] Department of Basic Sciences, College of Veterinary Medicine, P.O. Box 6100, Mississippi State, MS 39762, USA; mfn35@msstate.edu (M.F.N.); mba185@msstate.edu (M.B.A.); sp1679@msstate.edu (S.P.); shack@cvm.msstate.edu (L.A.S.)
[2] Section of Infectious Diseases, Southeast Louisiana Veterans Health Care System, New Orleans, LA 70112, USA; edwin.swiatlo@va.gov
[3] Institute for Genomics, Biocomputing & Biotechnology, Mississippi State University, Mississippi State, MS 39762, USA
* Correspondence: bnanduri@cvm.msstate.edu; Tel.: +1-662-325-5859; Fax: +1-662-325-1031
† These authors contributed equally to this work.

Received: 10 December 2017; Accepted: 9 January 2018; Published: 19 January 2018

Abstract: Invasive infections caused by *Streptococcus pneumoniae*, a commensal in the nasopharynx, pose significant risk to human health. Limited serotype coverage by the available polysaccharide-based conjugate vaccines coupled with increasing incidence of antibiotic resistance complicates therapeutic strategies. Bacterial physiology and metabolism that allows pathogens to adapt to the host are a promising avenue for the discovery of novel therapeutics. Intracellular polyamine concentrations are tightly regulated by biosynthesis, transport and degradation. We previously reported that deletion of *cadA*, a gene that encodes for lysine decarboxylase, an enzyme that catalyzes cadaverine synthesis results in an attenuated phenotype. Here, we report the impact of *cadA* deletion on pneumococcal capsule and protein expression. Our data show that genes for polyamine biosynthesis and transport are downregulated in Δ*cadA*. Immunoblot assays show reduced capsule in Δ*cadA*. Reduced capsule synthesis could be due to reduced transcription and availability of precursors for synthesis. The capsule is the predominant virulence factor in pneumococci and is critical for evading opsonophagocytosis and its loss in Δ*cadA* could explain the reported attenuation in vivo. Results from this study show that capsule synthesis in pneumococci is regulated by polyamine metabolism, which can be targeted for developing novel therapies.

Keywords: *Streptococcus pneumoniae*; polyamines; pneumococcal pneumonia; proteomics; capsule; complementation; metabolism; cadaverine

1. Introduction

Streptococcus pneumoniae (pneumococcus) is a Gram-positive encapsulated pathogen that resides asymptomatically in the nasopharynx of healthy humans. In children, elderly, and immunocompromised individuals pneumococci can become pathogenic, causing mild to severe infections such as sinusitis, meningitis, community acquired pneumonia and septicemia [1]. Pneumococcal infections cause approximately 2 million deaths globally with about 900,000 reported cases in the United States alone, resulting in approximately 400,000 hospitalizations annually [2]. There are over 90 pneumococcal serotypes with unique capsular polysaccharide structures [3] and only 24 serotypes are included in the current existing polysaccharide-based vaccines, PCV13 and PPSV23 [4], combined. Increased resistance to antibiotics such as penicillin, cephalosporins, and fluoroquinolones also complicates treatment [5,6]. The diversity of pneumococcal serotypes, coupled with genomic plasticity and the increasing selection for non-vaccine serotypes, mandates the development of novel protein-based vaccines that are conserved across serotypes and drives the search for new antimicrobial targets.

Polyamines are small, ubiquitous polycationic molecules with hydrocarbon backbones that are positively charged at physiological pH. Polyamines participate in many important biological functions in both eukaryotes and prokaryotes [7]. They interact with negatively charged molecules such as nucleic acids, proteins, and modulate DNA replication, transcription and translation [8]. The most common cellular polyamines in prokaryotes are diamines, putrescine (1,4-diaminobutane) and cadaverine (1,5-diaminopentane), and a triamine, spermidine (*N*-(3-Aminopropyl)- 1,4-diaminobutane) [9–11]. Bacteria regulate intracellular polyamine levels by biosynthesis, import of extracellular polyamines through adenosine triphosphate (ATP)-binding cassette transporters (ABC transporters) or antiporters and catabolism.

Polyamines are known to modulate virulence of pathogenic bacteria [12]. Spermine influences biofilm formation in *Neisseria gonorrhoeae* [13], *Bacillus subtilis* [14], *Escherichia coli* [15], *Yersinia pestis* [16] and *Vibrio cholerae* [17]. Polyamines have been linked to escape from phagolysosomes, bacteriocin production, toxin activity and stress responses in many pathogenic bacteria [18–21]. Cadaverine inhibits multiplication of *Shigella flexneri* by preventing lysis of phagolysosome [20]. Cadaverine has also been shown to inhibit adherence of toxin producing *E. coli* [22]. Different types of polyamines have different effects on growth and virulence. Therefore, studies that focus on the impact of polyamines in bacterial pathogens and their role in infection are warranted. Polyamine transport and synthesis genes are highly conserved across pneumococcal serotypes [23], while polyamines catabolism is poorly annotated in the genome. Our earlier studies show that intact polyamine transport and synthesis genes are necessary for virulence of *S. pneumoniae* in murine models of colonization, pneumococcal pneumonia and sepsis [24]. Deletion of genes that encode lysine decarboxylase (*cadA*), and spermidine synthase (*speE*), enzymes that catalyze cadaverine and spermidine synthesis, and polyamine transporter (*potABCD*) in pneumococci results in an attenuated phenotype in vivo [24]. We also demonstrated the therapeutic potential of targeting polyamine metabolism genes in pneumococci. Immunization with PotD, the extracellular substrate binding subunit of the polyamine transport operon *potABCD*, affords protection in mice against colonization, pneumonia and sepsis [25,26]. However, the impact of the deletion of polyamine biosynthesis on pneumococcal protein expression, including virulence factor expression that ultimately regulates survival in the host in a polyamine dependent manner is not known [18].

Here, we investigated the role of *cadA*, on pneumococcal capsule and protein expression. We hypothesized that the attenuated phenotype of lysine decarboxylase deficient pneumococci is due to reduced capsule production, rendering the bacterium more susceptible to host defense. Our results show loss of thecapsule and downregulation of synthesis of putrescine and spermidine as well as transport in Δ*cadA*. Our data strongly suggests that polyamine metabolism plays a significant role in the regulation of capsular polysaccharide biosynthesis in pneumococci, and is an attractive target for developing novel therapeutics.

2. Materials and Methods

2.1. Bacterial Strains, and Growth Conditions

Streptococcus pneumoniae serotype 4 strain TIGR4 was used in this study [27]. All strains were grown in Todd-Hewitt broth supplemented with 0.5% yeast extract (THY) or on 5% sheep blood agar plates (BAP) in 5% CO_2. An isogenic mutant of TIGR4 deficient in *cadA* was generated by polymerase chain reaction (PCR)-ligation mutagenesis as described previously [24]. Briefly, PCR primers were designed to amplify upstream (600 nt 5′ to the start codon) and downstream (600 nt 3′ from the transcription termination site) of *cadA* from TIGR4 chromosomal DNA (Table 1). Genomic pieces were joined by gene splicing, and insertion of spectinomycin resistance gene (*spec*) amplified from pORI280 [28] by overlap extension (SOEing) PCR [24]. The recombinant product was transformed into TIGR4 as described previously [29]. Transformants were selected on BAP with spectinomycin (100 μg/mL) and *cadA* gene deletion was confirmed by sequencing. To establish that cadaverine is responsible for loss of the capsule, we complemented the Δ*cadA* mutant in trans by cloning *cadA* gene amplified from TIGR4 (Table 1) into pABG5 vector for analysis of transcription [30], and complemented transformants were selected on BAP with kanamycin (50 μg/mL) and confirmed by PCR.

Table 1. Sequences of primers used in this study.

Primer	Sequence * (5′→3′)	Experiment
cadAF1	AGCAAATATAAACCCGAGTAAAAA	Mutagenesis
cadAR1	CAGGTACCGCTTGTGACCTGGAACATC	Mutagenesis
cadAF2	CAGAGCTCGTTTCGGTTTGCGATTTT	Mutagenesis
cadAR2	GATCTTCCGTCCCTTGGAG	Mutagenesis
cadAF-XbaI	TTCCCCGGGCCGTGCGAAAATCATCGCC	Complementation
cadAR-SacI	ATTCGAGGAAGACAGAGGTGTACTATTC	Complementation
gyrBF	CCGTCCTGCTGTTGAGACC	qRT-PCR
gyrBR	GTGAAGACCACCTGAAACCTTG	qRT-PCR
potDF	AAACCTGAAAATGCTCTCCAAAATG	qRT-PCR
potDR	CCTTATCTTCCTTTGTTTCCTCTGG	qRT-PCR
cps4AF	TCAAGTCAAGTCAGAATACCGATTTG	qRT-PCR
cps4AR	TCAAAGACACTATTTAGGACAATGGC	qRT-PCR
speEF	TGCGGATGATTTCGTCTACAATG	qRT-PCR
speER	CCAGTTCAGGATAGAGGGTTAATAC	qRT-PCR
aguAF	GCTTAGTCCTGGTCGCAATC	qRT-PCR
aguAR	CTGGGGATCATTTTCGTCAT	qRT-PCR
lys9F	GGCTTGACTGCTCTTCTTGG	qRT-PCR
lys9R	AGTAAGAACCTGGCGCAGAA	qRT-PCR
nspCF	ATGTATTTGCGCCTGCTTTC	qRT-PCR
nspCR	TGGTGCACAAGGGTCATAGA	qRT-PCR

* underlined sequence complementary to *Streptococcus pneumoniae* TIGR4 chromosomal DNA. qRT-PCR: quantitative reverse transcription-PCR.

2.2. In Vitro Growth of TIGR4 and ΔcadA

TIGR4 and Δ*cadA* were inoculated into THY (10^5 colony forming units (CFU)/mL) and growth was monitored by measuring optical density at 600 nm ($OD_{600\,nm}$) using a Cytation 5 multifunction plate reader (BioTek, Winooski, VT, USA) at 37 °C with 5% CO_2. We used GrowthRates [31], a software tool that uses the output of plate reader files to automatically monitor growth rate in exponential phase, lag phase and maximal OD to compare TIGR4 and Δ*cadA* growth curves. We also measured the viability of TIGR4 and Δ*cadA* in THY by plating cells on BAP every 2 h for CFU enumeration. Morphology was compared by Gram staining of mid-log phase ($OD_{600\,nm}$ 0.4) cultures of bacteria.

2.3. Measurement of Capsular Polysaccharides

Capsular polysaccharide (CPS) was quantified by immunoblotting, as described [32]. An isogenic capsular variant of TIGR4 (T4R) in which the *cps* locus is replaced with the Janus cassette resulting in an unencapsulated phenotype [33] was used as a control. Briefly, bacterial strains were cultured in THY supplemented with 10% fetal bovine serum to an $OD_{600\,nm}$ of 0.2, plated on BAP for CFU enumeration and 1 mL bacteria was pelleted and stored at −20 °C until further use. The CFUs for all strains were ~9.0 × 10^7/mL. CPS was extracted in a lysis buffer (4% deoxycholate, 50 μg/mL DNAseI and 50 μg/mL RNAse) at 37 °C for 10 min and centrifuged at 18,000× *g* for 10 min. Four threefold serial dilutions of the supernatant in phosphate-buffered saline (PBS) were spotted in duplicate (2 μL) on 0.2-μm-pore-size nitrocellulose membrane (Thermo Fisher Scientific, Waltham, MA, USA) with suction and air dried at 60 °C for 15 min. The membranes were blocked and incubated with rabbit anti-serotype 4 serum (Cedarlane, Burlington, NC, USA) at 1:1000 and a horseradish peroxidase (HRP) conjugated goat anti-rabbit secondary antibody (Thermo Fisher Scientific, Waltham, MA, USA) at 1:10,000. Membranes were developed with enhanced chemiluminiscence (ECL) detection (Thermo Fisher Scientific, Waltham, MA, USA) and scanned using a ChemiDoc XRS+ with Image Lab software (Bio-Rad, Hercules, CA, USA).

2.4. Proteomics

Total proteins were isolated from mid-log phase ($OD_{600 \, nm}$ 0.4) TIGR4, and $\Delta cadA$ (*n* = 4) cultured in THY and subjected to liquid chromatography–tandem mass spectrometry (LC-MS/MS) analysis as described earlier [24,34]. Briefly, proteins were isolated from bacterial pellets sonicated in NP-40 lysis buffer (0.5% NP-40, 150 mM NaCl, 20 mM $CaCl_2 \cdot 2H_2O$, 50 mM Tris, pH 7.4) supplemented with 1X protease inhibitor cocktail/ethylenediaminetetraacetic acid (EDTA) using a Covaris S220 focused-ultrasonicator (Covaris, Woburn, MA, USA). Protein concentration from the supernatant was determined using a Pierce bicinchoninic acid assay (BCA) Protein Assay Kit (Thermo Fisher Scientific, Waltham, MA, USA) and 30 μg was precipitated with methanol and chloroform (4:1), solubilized in 8 M urea, reduced (0.005 M dithiothreitol (DTT) at 65 °C for 10 min) and alkylated (0.01 M iodoacetamide at 37 °C for 30 min) and digested with porcine trypsin (2 μg at 37 °C, overnight, 50:1 ratio of protein: trypsin, Promega Corporation, Madison, WI, USA). Tryptic peptides were desalted using a C18 spin column (Thermo Fisher Scientific, Waltham, MA, USA) and analyzed by linear trap quadropole (LTQ) Orbitrap Velos mass spectrometer (Thermo Fisher Scientific, Waltham, MA, USA) equipped with an Advion nanomate electrospray ionization (ESI) source (Advion, Ithaca, NY, USA). Peptides (500 ng) were eluted from a C18 column (100-μm id × 2 cm, Thermo Fisher Scientific) onto an analytical column (75-μm ID × 10 cm, C18, Thermo Fisher Scientific) using a 180 min gradient with 99.9% acetonitrile, 0.1% formic acid at a flow rate of 400 nL/min and introduced into an LTQ-Orbit rap. Data dependent scanning was performed by the Xcalibur v 2.1.0 software [35] using a survey mass scan at 60,000 resolution in the Orbitrap analyzer scanning mass/charge (*m/z*) 400–1600, followed by collision-induced dissociation (CID) tandem mass spectrometry (MS/MS) of the 14 most intense ions in the linear ion trap analyzer. Precursor ions were selected by the monoisotopic precursor selection (MIPS) setting with selection or rejection of ions held to a ±10 ppm window. Dynamic exclusion was set to place any selected *m/z* on an exclusion list for 45 s after a single MS/MS. Tandem mass spectra were searched against a *Streptococcus pneumoniae* serotype4 strain ATCC BAA/TIGR4 fasta protein database downloaded from UniProtKB to which common contaminant proteins (e.g., human keratins obtained at ftp://ftp.thegpm.org/fasta/cRAP) were appended. All MS/MS spectra were searched using Thermo Proteome Discoverer 1.3 (Thermo Fisher Scientific) considering fully tryptic peptides with up to two missed cleavage sites. Variable modifications considered during the search included methionine oxidation (15.995 Da), and cysteine carbamidomethylation (57.021 Da). Peptides were identified at 99% confidence with XCorr score cutoffs [36] based on a reversed database search. The protein and peptide identification results were visualized with Scaffold v 3.6.1 (Proteome Software Inc., Portland, OR, USA). Protein identifications with a minimum of two peptides identified at 0.1% peptide false discovery rate (FDR) were deemed correct. Significant changes in protein expression between $\Delta cadA$ and TIGR4 were identified by Fisher's exact test at a *p*-value of ≤ 0.054 and fold change of ±1.3. Fold changes in protein expression were calculated using weighted normalized spectra with 0.5 imputation value. Various bioinformatics resources such as DAVID [37], KEGG [38] and STRING [39] were utilized to determine the functions of the identified proteins. The PRoteomics IDEntifications (PRIDE) database is a centralized, standards compliant, public data repository for proteomics data. The mass spectrometry proteomics data from this study is deposited to the ProteomeXchange Consortium via the PRIDE partner repository [40] with the dataset identifier PXD008621.

2.5. Quantitative Real Time PCR

Gene expression in $\Delta cadA$ was measured by quantitative reverse transcription-PCR (qRT-PCR). The primers used for qRT-PCR are listed in Table 1. All primers were validated by performing a melt curve analysis with SYBR Green (Thermo Fisher Scientific Waltham, MA, USA) to ensure the amplification of a single specific product. In brief, total RNA was purified from mid-log phase TIGR4 and Δ*cadA* grown in THY (*n* = 3) using the RNeasy Midi kit and QIAcube (Qiagen, Valencia, CA, USA). Purified total RNA (7.5 ng/reaction) was transcribed into cDNA and qRT-PCR was performed using the SuperScript III Platinum SYBR Green One-Step qRT-PCR Kit (Thermo Fisher Scientific, Waltham, MA, USA) as previously described [34]. Relative quantification of gene expression was determined by using the Stratagene Mx3005P qPCR system (Agilent, Santa Clara, CA, USA). Expression of target genes, *speE*, *potD*, *cps4A*, *aguA*, *lys9*, and *nspC* was normalized to the expression of *gyrB* and fold change determined by the comparative C_T method.

3. Results

3.1. Impact of Δ*cadA* on Pneumococcal Growth

S. *pneumoniae* TIGR4 and Δ*cadA* had comparable growth kinetics in THY, as reported previously [24]. The deletion mutant had a shorter lag phase than TIGR4 and had a lower cell density at stationary phase (Figure 1A(i)). There is evidence that pneumococcal serotypes with high colonization prevalence have a short lag phase in vitro when cultured in complete medium compared to invasive serotypes [41]. The shorter lag phase in Δ*cadA* could have implications for its invasiveness that needs confirmation in future studies. Since polyamines impact a number of cellular processes that can modulate growth, it is possible that deletion of Δ*cadA* alters some of these processes that ultimately results in reduced cell density. However, there is no significant difference in the exponential growth rate between TIGR4 (0.011 min^{-1}) and Δ*cadA* (0.015 min^{-1}). Both TIGR4 and Δ*cadA* had comparable CFUs during growth in THY (Figure 1A(ii)). Mid-log phase TIGR4 (Figure 1B(i)) and Δ*cadA* cells (Figure 1B(ii)) showed no difference morphology as both exhibited the characteristic diplococci morphology. These results are similar to our earlier observation that deletion of lysine decarboxylase has no qualitative difference in pneumococcal growth in vitro. However, lysine decarboxylase is indispensable for survival in murine models of colonization, pneumonia and sepsis [24].

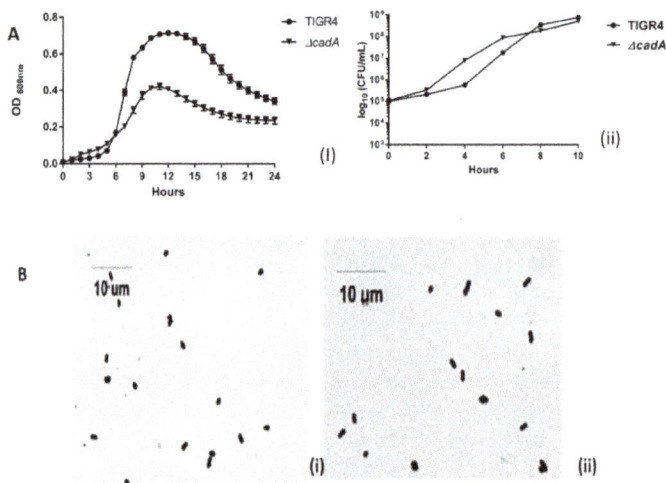

Figure 1. Growth of TIGR4 and Δ*cadA* and Gram stain morphology in vitro. (**A**) Growth of TIGR4 and Δ*cadA* in THY (*n* = 3) was monitored by measuring absorbance 600 nm (**i**) and viability (**ii**) was estimated by plating on blood agar plates (BAP) for colony forming units (CFU) enumeration. (**B**) Morphology of TIGR4 (**i**) and Δ*cadA* (**ii**) was observed by Gram staining.

3.2. Lysine Decarboxylase is Required for Capsule Production in S. pneumoniae

Isogenic deletion of lysine decarboxylase in *S. pneumoniae* TIGR4 led to attenuation in murine models of colonization, pneumococcal pneumonia and sepsis [24]. The capsule renders pneumococci resistant to opsonophagocytosis and is essential for pneumococcal virulence [42]. Loss of the capsule associated with impaired cadaverine synthesis could explain the observed attenuation of virulence in *S. pneumoniae* Δ*cadA*. We compared total CPS from TIGR4, T4R, Δ*cadA* and Δ*cadA* complemented strains (Δ*cadA* (comp)) using serotype 4-specific CPS antibodies. Our results (Figure 2) clearly show that deletion of *cadA* results in loss of the capsule in two independently-derived mutants that lack lysine decarboxylase (Δ*cadA1* [24] and Δ*cadA2* (this study)) compared to TIGR4. Both Δ*cadA* deletion strains exhibit loss of the capsule (Figure 2) ruling out the possibility that the observed phenotype is due to a random change in the genome elsewhere and not specific to *cadA* deletion. This was further confirmed by complementation. CPS in Δ*cadA* is fully restored to the levels comparable to that of TIGR4 by complementation with pABG5-*cadA* construct. These results clearly demonstrate that deletion of lysine decarboxylase in *S. pneumoniae* results in the loss of CPS. Loss of the capsule could render Δ*cadA* susceptible to host defenses resulting in an attenuated phenotype in murine models of colonization and invasive disease. Impact of lysine decarboxylase on capsule synthesis in vivo needs to be evaluated in future studies.

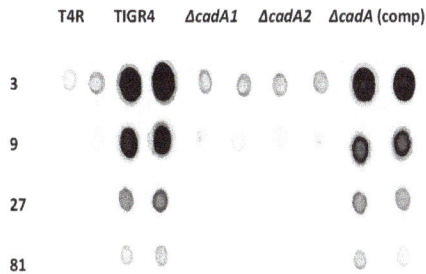

Figure 2. Immunoblot analysis of capsular polysaccharides in TIGR4 and mutant strains. All strains were cultured in THY supplemented with fetal bovine serum (FBS) to mid-log phase. Total capsular polysaccharide (CPS) isolated from equal number of cells for each strain, and 3× dilutions were spotted onto a nitrocellulose membrane. Membranes were probed with rabbit anti-serotype 4 sera and horseradish peroxidase (HRP)-conjugated goat anti-rabbit secondary antibody. Membranes were developed with enhanced chemiluminiscence (ECL) detection and scanned using a ChemiDoc XRS+ with Image Lab software (Bio-Rad, Hercules, CA, USA). Data from representative immunoblot from two independent colonies for each strain are shown.

3.3. Lysine Decarboxylase Effects on Pneumococcal Protein Expression

To identify pneumococcal molecular mechanisms that are responsive to lysine decarboxylase, we carried out mass spectrometry based proteomics with TIGR4 and Δ*cadA*. A total of 772 proteins were identified (Table 1) which represents 34.5% of the annotated protein coding genes in the TIGR4 genome [27]. We identified significant changes in the expression of 132 proteins of which 52 are upregulated and 80 are downregulated in Δ*cadA* compared to TIGR4 (Table 2). Molecular functions and pathways represented by the differentially expressed proteins are discussed in the following sections and shown in Table 2.

Table 2. Significant changes in Δ*cadA* proteome compared to TIGR4.

Description	Protein	Δ*cadA*/TIGR4 (Fold Change)	Function
N-carbamoylputrescine amidase	*SP_0922	−10.0	Putrescine biosynthesis
Carboxynorspermidine decarboxylase	NspC	−5.0	Spermidine biosynthesis
Homoserine dehydrogenase	Hom	−1.4	Lysine biosynthesis
4-hydroxy-tetrahydrodipicolinate synthase	DapA	−10.0	Lysine biosynthesis
4-hydroxy-tetrahydrodipicolinate reductase	DapB	−2.5	Lysine biosynthesis
N-acetyldiaminopimelate deacetylase	SP_2096	−2.5	Lysine biosynthesis
Saccharopine dehydrogenase	Lys9	−25.0	Lysine biosynthesis
Aspartate-semialdehyde dehydrogenase	Asd	−25.0	Lysine biosynthesis
2,3,4,5-tetrahydropyridine-2-carboxylate *N*-Succinyl transferase	DapH	−1.7	Lysine biosynthesis
50S ribosomal protein L21	RplU	−5.0	Regulation of protein elongation
Ribosome maturation factor	RimP	-5.0	Regulation of protein maturation
Lysine-tRNA ligase	LysS	−3.3	Amino acid metabolism
Iron-compound ABC Transporter	FhuD	−50.0	Iron complex ABC transporter
Phosphate-binding protein PstS 2	PstS 2	41.0	Phosphate ion transport
Phosphate import ATP-binding protein PstB 3	PstB 3	36.0	Phosphate ion transport
Phosphate transport system permease protein	PstC	7.0	Phosphate ion transport
Phosphate-specific transport system accessory protein PhoU homolog	PhoU	43.0	Phosphate ion transport
ABC transporter, ATP-binding/permease protein	SP_2073	−3.3	Oligopeptide ABC transporter
Oligopeptide binding protein	OppA	−25.0	Oligopeptide ABC transporter
Oligopeptide transport ATP-binding protein	OppD	−1.4	Oligopeptide ABC transporter
Oligopeptide transport ATP-binding protein	OppF	−1.7	Oligopeptide ABC transporter
Oligopeptide transport system permease protein	OppB	−1.7	Oligopeptide ABC transporter
Manganese ABC transporter-substrate-binding lipoprotein	PsaA	2.4	Oxidative stress
Manganese ABC transporter, ATP -binding protein	PsaB	6.8	Oxidative stress
Penicillin-binding protein 2x	Pbp2X	−2.5	Peptidoglycan biosynthesis
Choline kinase	Pck	−2.0	Cell wall biosynthesis
UDP-glucose 4-epimerase	GalE-1	−1.3	Carbohydrate metabolism
Tagatose 1,6-diphosphate aldolase	LacD	1.4	Carbohydrate metabolism
Galactose-6-phosphate isomerase subunit	LacB	2.1	Carbohydrate metabolism
Catabolite control protein A	CcpA	−2.5	Carbohydrate metabolism
Bifunctional protein	GlmU	−1.7	UDP- GlcNAc synthesis
N-acetylglucosamine-6-phosphate deacetylase	NagA	1.4	*N*-acetylglucosamine degradation
N-acetylglucosamine-6-phosphate deaminase	NagB	2.1	*N*-acetylglucosamine degradation
Transketolase, C-terminal subunit	TktC	67.0	Pentose phosphate pathway
Transketolase, N-terminal subunit	TktN	46.0	Pentose phosphate pathway
Ascorbate-specific PTS, EIIC component	SgaT2	31.0	Ascorbate utilization
Ascorbate-specific PTS system, EIIB component	SgaB2	32.0	Ascorbate utilization
Phosphocarrier protein HPr	PtsH	21.0	Phosphotransferase system (PTS)

*: locus tag ID; ABC: ATP binding cassette; ATP: Adenosine triphosphate; UDP: uridine diphosphate; GlcNac: N-acetylglucosamine; PTS: phosphotransferase system.

3.3.1. Capsule Biosynthesis

CPS synthesis in TIGR4 is by the Wzy polymerase dependent mechanism. CPS synthesis is a multistep process that begins with the transfer of sugar-1-phosphate on the cytosolic side onto a C55 lipid undecaprenyl-phosphate (Und-P), followed by the addition of remaining sugars by glycosyl transferases to form a repeat unit. The repeat unit structure of serotype 4 CPS consists of galactose, *N*-acetylmannosamine, *N*-acetylfucosamine and *N*-acetylgalactosamine [43]. Und-P oligosaccharide repeat units are translocated to the outer face of the cytoplasmic membrane by Wzx transporter and polymerized into high molecular weight polysaccharide by Wzy polymerase. All genes involved in capsule biosynthesis are present as a single operon between *dexB* and *aliA* in pneumococcal genomes. The first four genes in the operon *cpsABCD* are important for modulation of synthesis and are common to all serotypes [44]. The rest of the genes in the operon are serotype specific. Changes in the expression of *cpsA* is a good measure of the transcription *cps* locus. We expected to see a significant reduction in Δ*cadA* in the expression of some of the enzymes that catalyze the multi-step CPS biosynthesis. Our proteomics data showed reduced expression of UDP-glucose-4-epimerase (Table 2), that catalyzes the conversion of UDP-glucose to UDP-galactose in CPS biosynthesis [45]. While the reduced expression of this protein is consistent with the observed loss of capsule, it is unlikely that this marginal change can explain the magnitude of the loss of CPS comparable to unencapsulated T4R (Figure 2) without additional changes in the pneumococcal proteome that directly or indirectly impact capsule synthesis. Expression of bifunctional protein GlmU (Table 2) was significantly downregulated. This enzyme catalyzes the last two sequential reactions in the de novo biosynthetic pathway for UDP-*N*-acetylglucosamine (UDP- GlcNAc). GlmU catalyzes the reaction that transfers an acetyl group from acetyl coenzyme A to glucosamine 6-phosphate to synthesize acetylated glucosamine 6-phosphate. NagB is an enzyme that catalyzes the conversion of glucosamine 6-phosphate to fructose 6-phosphate and is known to regulate GlmU [46]. In Δ*cadA* expression of NagA and NagB, two enzymes involved in UDP-GlcNAc degradation was significantly higher compared to TIGR4. Taken together, the net effect of changes in the expression of GlmU, NagA and NagB would result in lower concentrations of UDP-GlcNAc, a precursor for UDP-ManNAc, which is a constituent of serotype 4 CPS repeat unit, and could contribute to reduced CPS synthesis in Δ*cadA*.

3.3.2. Polyamine Biosynthesis

Deletion of lysine decarboxylase in *S. pneumoniae* resulted in a significant decrease in the expression of Lys9, Asd, DapA, DapB, Hom, DapH and SP_2096 involved in the biosynthesis of lysine, the substrate for *cadA* (Table 2). Expression of *N*-carbamoylputrescine amidase which catalyzes the synthesis of putrescine from *N*-carbamoylputrescine in the arginine and proline metabolism was significantly downregulated. Expression of NspC which catalyzes the synthesis of spermidine from carbamoyl spermidine was also significantly downregulated (Table 2). The net effect of these protein expression changes in polyamine biosynthesis pathways in Δ*cadA*, is expected to result in reduced intracellular concentrations of cadaverine, putrescine and spermidine (Figure 3). We reported reduced intracellular concentrations of cadaverine, putrescine and spermidine in Δ*cadA* [24] previously. Results from this study explain this observed impact on intracellular polyamine concentrations in Δ*cadA*.

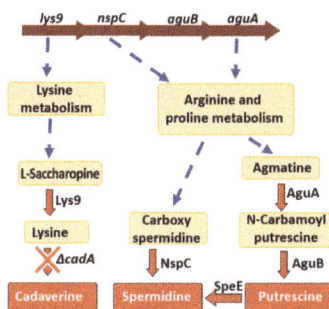

Figure 3. Impact of lysine decarboxylase on polyamine synthesis. Genes encoding the enzymes Lys9, NspC, AguB and AguA that catalyze reactions in the polyamine biosynthesis pathways are arranged as a single operon in the genome, and Transcription of this operon is downregulated in Δ*cadA*. Reactions that involve multiple steps are represented by a broken arrow. We identified reduced expression of *lys9*, *nspC*, *aguA* and *speE* in Δ*cadA* compared to TIGR4 by qRT-PCR. Expression of Lys9 and NspC proteins were reduced in lysine decarboxylase impaired pneumococci.

3.3.3. Peptidoglycan

The rigid, stable shape of bacteria is provided by the peptidoglycan layer which is made of *N*-acetylmuramic acid-(β-1, 4)-*N*-acetylglucosamine (MurNAc-GlcNAc) disaccharides cross-linked by peptides. The peptidoglycan layer, teichoic acid (TA) and lipoteichoic acid (LTA) constitute the cell wall of Gram-positive bacteria such as *S. pneumoniae* [47]. Both LTA and TA contain phosphorylcholine (PC), which plays a major role in *S. pneumoniae* adhesion and also forms attachment of choline binding surface proteins (CBPs) [48]. The peptidoglycan layer provides attachment for many structural components including the polysaccharide capsule [49] and it is important in the adhesion of *S. pneumoniae* to the host tissues. Our results show reduced expression of penicillin-binding protein 2X (Table 2), which is involved in peptidoglycan biosynthesis that contributes to bacterial cell division and growth [50,51]. Choline kinase, an enzyme which catalyzes the synthesis of PC from choline was downregulated in $\Delta cadA$ relative to TIGR4. In some Gram-positive pathogens, choline kinase is known to be important for the production of cell wall elements and LTA [52]. PC is necessary for the adherence of *S. pneumoniae* during the transition from colonization to invasive disease [48]. Reduced choline kinase expression in $\Delta cadA$ could also modulate virulence.

3.3.4. ABC Transporters

Our results show that expression of ABC transporters that bind metal ions is significantly altered in $\Delta cadA$. Manganese, a transition metal ion is a prosthetic group in superoxide dismutase, has direct antioxidant properties and is known to be important for pneumococcal physiology. An ABC-type permease PsaBCA [53] transports manganese and *psa* mutants are avirulent [54] due to their hypersensitivity to oxidative stress [51]. Expression of Manganese ABC transporter-substrate-binding lipoprotein (PsaA) and Manganese ABC transporter, ATP -binding protein (PsaB) is significantly higher in $\Delta cadA$ (Table 2). An ABC transporter potentially responsible for iron-siderophore transport (FhuD), specifically ferric hydroxamate is significantly lower in $\Delta cadA$ compared to TIGR4. Iron is a critical cofactor for many enzymes, and uptake and efflux of this critical micronutrient from different host niches that differ in the quantity and form of iron is an important aspect of pneumococcal pathogenesis. Reduced expression of FhuD could impair iron homeostasis and have an impact on virulence. An alternate explanation involves $\Delta cadA$ response to oxidative stress by reducing uptake of iron due to reduced expression of FhuD. High intracellular iron concentrations can result in an increase in oxidative stress [55] and $\Delta cadA$ with reduced intracellular polyamine concentrations would be more susceptible to oxidative stress [21]. Increased expression of PsaA and PsaB could also be in response to increased susceptibility to oxidative stress in $\Delta cadA$.

Expression of five ABC transporters, four of which are involved in oligopeptide trafficking (SP_2073, OppA, OppD, OppB and OppF) were downregulated in $\Delta cadA$ (Table 2). Reduced expression of oligopeptide transporter proteins could significantly alter transport of substrates that would in turn affect biosynthesis of amino acids, polyamines and other cellular components. There was reduced expression of lysine-transfer RNA (tRNA) ligase, a protein involved in amino acid metabolism. Expression of RplU, which binds to 23S ribosomal RNA (rRNA) in the initial stages of protein translation and ribosome maturation factor (RimP), which modulates final stages of translation was significantly reduced in $\Delta cadA$. Our results show that lysine decarboxylase deficiency results in reduced expression of a number of proteins involved in amino acid transport and metabolism which could impair pneumococcal growth and fitness. For survival in the host, pathogenic bacteria have to cope with phosphate (Pi) limiting or enriched host microenvironments. Phosphate is the component of nucleic acids, phospholipids and energy storage (ATP). Bacteria acquire phosphate by Pi-specific transport systems. Under Pi limitation Pho regulon is activated and results in Pi import through an ABC transporter complex PstSACB [56]. Our data showed increased expression of PstS2, PstB3, PstC and PhoU (Table 2) which can impact Pi homeostasis. Low Pi is known to alter virulence factor expression in a number of pathogenic bacteria including *E. coli* [57]. It is known that activation of Pho regulon in pathogenic bacteria under Pi starvation conditions, activates oxidative stress response through mechanisms that are yet to be described [56]. Increased expression of Pst system could be part of oxidative stress response and possibly contribute to altered virulence in Δ*cadA* through mechanism that are not known at present.

3.3.5. Pentose Phosphate Pathway

The ability of bacterial pathogens to survive in the host largely depends on acquiring nutrients and adapting their metabolism to different host microenvironments. Pneumococci have the ability to utilize a variety of carbohydrates as carbon sources via Embden-Meyerhof-Parnas (EMP) pathway (glycolysis) and pentose phosphate pathway (PPP) [27]. Pyruvate and ATP are the end products of glycolysis. The oxidative branch of PPP generates ribulose 5-phosphate and reduced nicotinamide adenine dinucleotide phosphate (NADPH) while the non-oxidative branch generates a number of sugar phosphates that provide precursors for nucleotide, amino acid and vitamin B6 synthesis. Our data supports increased carbon flux through the non-oxidative branch of PPP due to the observed increase in the expression of transketolase (Tkt), an enzyme that catalyzes the interconversion of sugar-phosphates in the pathway. Expression of all proteins SgaR2, SgaB2, SgaT2, including TktN and TktC, encoded in a single regulon belonging to BglG family transcriptional regulator is higher in Δ*cadA* (Table 2). We also identified a significantly higher expression for one of the two general proteins of PTS (phosphotransferase system), phosphocarrier protein HPr (PtsH) which could support the proper functioning of PTS transport systems in the BglG regulon in Δ*cadA*. A shift in metabolism towards PPP is often in response to oxidative stress to maintain NADH/NADPH redox homeostasis and to synthesize ribose-5-phosphate for DNA repair [58].

3.3.6. Carbohydrate Metabolism

Our results show increased expression of proteins that are involved in galactose and tagatose catabolism. Expression of LacB is higher in Δ*cadA* (Table 2) which generates the substrate tagatose 6-phosphate for the enzyme LacD. Enzymatic action of LacB and LacD would result in higher levels of glyceraldehyde 3-phosphate, which can be channeled into PPP by transaldolase and transketolase. We identified reduced expression of catabolite control protein A (CcpA), a protein that regulates carbon catabolite repression in our lysine decarboxylase mutant. CcpA is known to control the expression of a number of virulence factors in Gram-positive bacteria [59]. For instance, in *S. pneumoniae*, CcpA contributes to sugar metabolism and virulence [60].

3.4. Measurement of Gene Expression in Δ*cadA*

3.4.1. Capsule Biosynthesis

We did not identify major differences in the expression of proteins that catalyze different steps in CPS synthesis. To determine whether CPS synthesis is regulated at the transcriptional level, we compared *cps4A* mRNA expression between TIGR4 and Δ*cadA* by qRT-PCR. Our results show a significant reduction in the expression of *cps4A* (Table 3), which could explain reduced CPS synthesis in Δ*cadA*.

Table 3. Changes in gene expression in Δ*cadA* compared to TIGR4.

Gene	Description	Δ*cadA*/TIGR4 (Fold change)	*p*-Value
potD	Spermidine/putrescine ABC transporter, spermidine/putrescine-binding protein	−2.0	1.93E−04
speE	Spermidine synthase	−27.0	1.29E−06
cps4A	Capsular polysaccharide biosynthesis protein 4A	−2.0	2.03E−07
lys9	Saccharopine dehydrogenase	−26.0	3.83E−12
nspC	Carboxynorspermidine decarboxylase	−34.0	4.70E−10
aguA	Agmatine deiminase	−30.0	2.87E−12

3.4.2. Polyamine Synthesis and Transport

Our proteomics data identified significantly reduced expression of proteins involved in the biosynthesis of putrescine, spermidine and cadaverine (Figure 2). The genes encoding proteins NspC, Lys9, AguA and AguB are predicted to constitute a single operon in the genome [61]. Based on our proteomics results, we expected to see reduced expression of polyamine biosynthesis (*lys9*, *aguA*, and *nspC*) genes and our qRT-PCR data shows reduced expression of these genes in Δ*cadA* (Table 3). Impaired lysine decarboxylase also resulted in a significant reduction in the expression of spermidine synthase (Table 3). Pneumococci can compensate for reduced polyamine synthesis by increasing the import of extracellular polyamines. Extracellular polyamine uptake in pneumococci is predicted to be via a single ABC transporter, organized as a four gene operon, *potABCD* [62]. The proposed structure of the putrescine/spermidine transporter has PotD, an extracellular substrate binding domain that binds polyamines, PotB, and PotC which form transmembrane channels that transport polyamines and PotA that is a membrane associated cytosolic ATPase [23]. We did not detect polyamine transport proteins in our proteomics data. We measured the expression of *potD* mRNA in Δ*cadA* and observed a significant decrease (Table 3), which would further contribute to reduced intracellular concentrations of putrescine and spermidine, putative substrates for the PotABCD transporter.

4. Discussion

Polyamines are important for host-pathogen interactions during bacterial infections. Polyamines in pathogenic bacteria play an important role in physiological stress responses and adaptation to growth in vivo. Putrescine is a constituent of the cell wall in a number of Gram-negative bacteria, such as *Salmonella enterica*, *E. coli* and *Proteus mirabilis* [63,64], while cadaverine is covalently linked to the peptidoglycan layer of *Veillonella alcalescens* [65]. Spermidine modulates autolysis and ion trafficking across the cell membrane in *S. pneumoniae*, protecting pneumococci from cationic antimicrobial compounds [7]. Cadaverine regulates porins that control the permeability of membranes [30], and enables *E. coli* to survive acidic stress [66]. Current literature clearly demonstrates that deletion of polyamine synthesis and/or transport in pathogenic bacteria (including pneumococci) leads to reduced virulence in animal models [24,67,68]. To date, studies that describe specific host innate immune mechanisms induced by pathogenic bacteria with altered polyamine metabolism, or specific effects of impaired polyamine metabolism on pathogen molecular mechanisms are largely unknown. A few examples of specific roles of polyamines in bacterial pathogens include the following: in intracellular pathogen *Shigella*, cadaverine is shown to be important for reducing enterotoxic activity [69], inhibiting trans-epithelial migration of polymorphonuclear neutrophils, increasing survival in macrophages, and enhancing antioxidant defenses in vitro [20]. *Salmonella typhimurium* mutant deficient in polyamine biosynthesis has reduced invasive potential and survival in epithelial cells in vitro and is attenuated in a mouse model of typhoid fever [70]. In *Yersinia pestis*, loss of intracellular spermidine and putrescine affects biofilm formation and biosynthesis defective *Y. pestis* is less virulent in a murine model of bubonic plague [71]. When *Francisella tularensis* is cultured in the presence of spermine or spermidine prior to macrophage infection assays, there is reduced pro-inflammatory response in vitro [55]. Thus altered polyamine metabolism has varying impacts on bacterial virulence.

Invasive infections caused by *S. pneumoniae*, a commensal in the nasopharynx, pose significant risk to human health. The available polysaccharide-based conjugate vaccines are effective in reducing vaccine serotypes in population, but ultimately lead to serotype replacement [72] from a reservoir of more than 90 capsular serotypes [3]. The critical role of the upper respiratory microbiota and the consequence of its perturbation in various pathophysiological conditions mandates a cautious approach to drug and vaccine discovery. Anti-virulence strategies that target genes/proteins that are necessary for fitness during invasive infection can offer serotype-independent coverage without impacting nasopharyngeal colonization, that is, disarm, but not eradicate, pneumococci [68]. Specific aspects of bacterial physiology and metabolism that allow pathogens to adapt to the host are a promising avenue for the discovery of novel therapeutics. Intracellular polyamine concentrations are tightly regulated by biosynthesis, transport and degradation. Polyamine transport and synthesis genes are conserved in pneumococci. Deletion of the polyamine transport *(potABCD)* operon or spermidine synthase and lysine decarboxylase biosynthesis genes had no significant impact on pneumococcal growth in vitro [24]. However, enhanced bacterial clearance in murine models of colonization, pneumococcal pneumonia, and sepsis were seen with mutant strains [24]. In a murine model of pneumococcal pneumonia, Δ*potABCD* failed to elicit host defenses that are intact in TIGR4, and was cleared more efficiently by opsonophagocytosis by neutrophils [34]. TIGR4 cultured in vitro (in THY) had spermidine as the most abundant intracellular polyamine, followed by cadaverine and putrescine [24]. Polyamine transport and metabolism impaired Δ*potABCD*, Δ*cadA* and Δ*speE* showed reduced levels of spermidine, cadaverine and putrescine, relative to TIGR4. [24].

Here we report in vitro characterization of lysine decarboxylase deficient pneumococci. Our results clearly demonstrate loss of CPS in Δ*cadA*. This loss of the capsule is specific to the deletion of the *cadA* gene, as complementation of the wild type gene restored capsule, comparable to that of TIGR4. Invasive pneumococcal serotypes can resist complement mediated opsonophagocytosis by neutrophils due to the presence of the capsule [71,73]. All three complement pathways are activated for opsonophagocytosis [73,74]; classical, lectin and alternative. Activation of the classical pathway is by antibody (including non-specific immunoglobulin M (IgM) produced during infection), or by C-reactive protein (CRP), an acute phase protein. Steric inhibition of the interaction between complement components and the Fc portion of immunoglobulins by the capsule enables pneumococci to evade this host defense. The alternative pathway is constitutively activated at low levels by complement protein C3b, and capsule-mediated resistance to opsonophagocytosis also includes decreased cleavage of C3b to iC3b [75]. Pneumococci regulate capsule expression as they invade host tissues. In the initial stages of colonization of the nasopharynx, they express the capsule to evade mucus, and in the subsequent stage of adhesion, the capsule is downregulated to expose surface adhesins for adherence and colonization. The invasive phase requires upregulation of the capsule to resist opsonophagocytosis [76]. Some of the known regulatory mechanisms for CPS synthesis involve phase variation [77], deletion of *pgdA* and *adr*, acetylases of peptidoglycan and increased transcription of *cpsA* [78] and deletion of pyruvate oxidase (*spxB*) that all lead to increased CPS [79], mutations in *arcD* that result in reduced CPS [80], and tyrosine phosphorylation of CpsBCD proteins that modulate CPS levels [81]. Here, for the first time, we show that polyamines, specifically cadaverine modulate CPS synthesis. Inactivation of lysine decarboxylase in pneumococci results in reduced capsule synthesis. The mechanisms that result in reduced CPS synthesis are due to the combinatorial effects on the regulation of UDP-GlcNAc synthesis and degradation (Figure 4) that could reduce the availability of UDP-ManNAc of the CPS repeat unit in serotype 4 pneumococci with concomitant transcriptional downregulation of *cps4A* gene. A moderate reduction in *cpsA* expression in TIGR4 capsule promoter mutant resulted in twofold reduction in CPS [78]. In Δ*cadA*, there is a twofold reduction in the expression of *cps4A mRNA* (Table 3), which would be expected to have a greater impact on CPS synthesis. The intersection between central metabolism and virulence is becoming evident for many bacterial pathogens including pneumococci. As pneumococci sense and adapt to different host niches, they modulate capsule synthesis which requires a shift in the metabolism towards increased synthesis of precursors for CPS synthesis. A recent study showed that reduced acetyl-coA levels result in loss of capsule [56], supporting the link between central metabolism and capsule formation.

Figure 4. Mechanisms for reduced capsule synthesis in lysine decarboxylase deficient pneumococci. Deletion of lysine decarboxylase in pneumococci results in reduced capsule compared to wild type TIGR4 strain. Reduced capsule synthesis could be due to the reduced expression (shown in red) of capsular polysaccharide biosynthesis gene *cps4A*, the first gene in the *cps* locus in *S. pneumoniae* TIGR4 (**A**). The first four genes from the *cps* locus adjacent to the gene *dexB* that are conserved in all pneumococcal serotypes are shown. Our proteomics data indicates a shift in central metabolism (**B**) from glycolysis (yellow box) to the non-oxidative branch of pentose phosphate pathway (green box), due to increased expression (shown in blue) of transketolase (Tkt,). Reactions that involve multiple steps are represented by a broken line. Reduced expression of GlmU involved in the synthesis and increased expression of NagB involved in the degradation of UDP-GlcNAc could result in reduced concentration of UDP-GlcNAc. UDP-GlcNAc is a precursor for UDP-ManNAc, an acetylated sugar that is the constituent of the 4-sugar repeat unit of capsular polysaccharide in capsular serotype 4 (open oval). The net effect of these changes in pneumococcal gene and protein expression could result in the observed reduction in capsule biosynthesis in Δ*cadA*.

Given the well-established role of polyamines in bacterial adaptation to oxidative stress [24], the reduced intracellular polyamines due to reduced expression of enzymes catalyzing synthesis (proteomics data, Table 2) and transport (qRT-PCR data, Table 3) could explain the observed shift in metabolism that increases carbon flux through the pentose phosphate pathway. Reduced uptake of iron and upregulation of manganese transporter proteins, known pneumococcal virulence factors linked to oxidative stress response support this idea. It is possible that increased oxidative stress in Δ*cadA* results in metabolic adaptation that diverts the energy from capsule production towards stress responses. With data presented here, it is difficult to distinguish between correlation and causation due to impaired cadaverine synthesis. Future studies that utilize additional omics approaches and data analysis in an integrated manner are required to deconvolute the complexity of pneumococcal response to altered polyamine metabolism. Although we have begun to identify some of the molecular mechanisms that are responsive to polyamines in pneumococci that govern virulence, annotation of polyamine metabolism genes is limited in pneumococci. Our knowledge of polyamine metabolism in pathogenic bacteria and specific effects of polyamines on bacterial translation relies heavily on studies from *E. coli*. Basic biochemistry pertaining to substrate specificity of the transporters and biosynthesis genes is yet to be determined. Our data shows that reduced expression of putrescine and spermidine biosynthesis in Δ*cadA* could ultimately result in the altered intracellular polyamines that we reported earlier [24]. Future studies focused on determining individual contribution of these polyamine biosynthesis genes to pneumococcal virulence are necessary for a comprehensive description of the role of polyamine synthesis in pneumococcal pathogenesis. Specific effects of polyamines on protein synthesis in bacteria are known. Studies in *E. coli* have identified a "polyamine modulon", a set of 17 genes whose translation is enhanced by polyamines [82]. Mechanisms by which polyamines increase protein synthesis include formation of the initiation complex due to structural changes in the Shine-Dalgarno sequence and the initiation codon AUG, initiation from inefficient initiation codon, and +1 frame shifting and suppression of nonsense codons. Here we report altered expression of a number of pneumococcal proteins in response to impaired lysine decarboxylase. It is possible that some of these proteins are regulated by polyamines utilizing the mechanisms described in *E. coli*. Future studies focused on understanding the specific mechanisms utilized by polyamines to impact protein expression can result in mechanistic insights into the impact of altered metabolism on pneumococcal virulence. Despite these knowledge gaps, our previous work has clearly established the importance of polyamine transport and synthesis genes in pneumococcal pathogenesis. In this study we show that impaired polyamine synthesis results in reduced capsule synthesis. This foundational knowledge at the intersection of polyamine metabolism and pneumococcal virulence reinforces the need for future mechanistic studies focused on developing novel therapeutics targeting polyamine metabolism in pneumococci.

Supplementary Materials: The following are available online at www.mdpi.com/2076-3271/6/1/8/s1. Table S1: Proteins identified by mass spectrometry, Table S2: Significant changes in Δ*cadA* protein expression compared to TIGR4.

Acknowledgments: We thank Maria D. Basco for technical assistance and Moyim Kim for assistance with Figures 3 and 4. This work was supported by grant #P20GM103646 (Center for Biomedical Research Excellence in Pathogen Host Interactions) from the National Institute for General Medical Sciences. Mass spectrometry data was acquired by the University of Arizona Analytical and Biological Mass Spectrometry Facility supported by grants CA023074 and 1S10 RR028868-01.

Author Contributions: B.N. and E.S. conceived, supervised and designed the experiments. S.B.P., L.A.S., M.F.N., M.B.A. performed the experiments and drafted the manuscript. B.N., M.F.N. and M.B.A. analyzed the data. B.N. and E.S. contributed to the final draft. All authors approved the final version of the manuscript.

Conflicts of Interest: The authors declare no conflict of interest.

References

1. Bridy-Pappas, A.E.; Margolis, M.B.; Center, K.J.; Isaacman, D.J. *Streptococcus pneumoniae*: Description of the pathogen, disease epidemiology, treatment, and prevention. *Pharmacotherapy* **2005**, *25*, 1193–1212. [CrossRef] [PubMed]
2. Huang, S.S.; Johnson, K.M.; Ray, G.T.; Wroe, P.; Lieu, T.A.; Moore, M.R.; Zell, E.R.; Linder, J.A.; Grijalva, C.G.; Metlay, J.P.; et al. Healthcare utilization and cost of pneumococcal disease in the United States. *Vaccine* **2011**, *29*, 3398–3412. [CrossRef] [PubMed]
3. Weinberger, D.M.; Trzcinski, K.; Lu, Y.J.; Bogaert, D.; Brandes, A.; Galagan, J.; Anderson, P.W.; Malley, R.; Lipsitch, M. Pneumococcal capsular polysaccharide structure predicts serotype prevalence. *PLoS Pathog.* **2009**, *5*, e1000476. [CrossRef] [PubMed]
4. Hayward, S.; Thompson, L.A.; McEachern, A. Is 13-Valent Pneumococcal Conjugate Vaccine (PCV13) Combined With 23-Valent Pneumococcal Polysaccharide Vaccine (PPSV23) Superior to PPSV23 alone for reducing incidence or severity of pneumonia in older adults? A Clin-IQ. *J. Patient Cent. Res. Rev.* **2016**, *3*, 111–115. [CrossRef] [PubMed]
5. Felmingham, D. Comparative antimicrobial susceptibility of respiratory tract pathogens. *Chemotherapy* **2004**, *50* (Suppl. 1), 3–10. [CrossRef] [PubMed]
6. Doern, G.V.; Richter, S.S.; Miller, A.; Miller, N.; Rice, C.; Heilmann, K.; Beekmann, S. Antimicrobial resistance among *Streptococcus pneumoniae* in the United States: Have we begun to turn the corner on resistance to certain antimicrobial Classes? *Clin. Infect. Dis.* **2005**, *41*, 139–148. [CrossRef] [PubMed]
7. Igarashi, K.; Kashiwagi, K. Polyamines: Mysterious modulators of cellular functions. *Biochem. Biophys. Res. Commun.* **2000**, *271*, 559–564. [CrossRef] [PubMed]
8. Igarashi, K.; Kashiwagi, K. Characteristics of cellular polyamine transport in prokaryotes and eukaryotes. *Plant Physiol. Biochem.* **2010**, *48*, 506–512. [CrossRef] [PubMed]
9. Michael, A.J. Exploring polyamine biosynthetic diversity through comparative and functional genomics. *Methods Mol. Biol.* **2011**, *720*, 39–50. [PubMed]

10. Michael, A.J. Biosynthesis of polyamines and polyamine-containing molecules. *Biochem. J.* **2016**, *473*, 2315–2329. [CrossRef] [PubMed]

11. Tabor, C.W.; Tabor, H. Polyamines in microorganisms. *Microbiol. Rev.* **1985**, *49*, 81–99. [PubMed]

12. Di Martino, M.L.; Campilongo, R.; Casalino, M.; Micheli, G.; Colonna, B.; Prosseda, G. Polyamines: Emerging players in bacteria-host interactions. *Int. J. Med. Microbiol.* **2013**, *303*, 484–491. [CrossRef] [PubMed]

13. Goytia, M.; Dhulipala, V.L.; Shafer, W.M. Spermine impairs biofilm formation by *Neisseria gonorrhoeae*. *FEMS Microbiol. Lett.* **2013**, *343*, 64–69. [CrossRef] [PubMed]

14. Burrell, M.; Hanfrey, C.C.; Murray, E.J.; Stanley-Wall, N.R.; Michael, A.J. Evolution and multiplicity of arginine decarboxylases in polyamine biosynthesis and essential role in *Bacillus subtilis* biofilm formation. *J. Biol. Chem.* **2010**, *285*, 39224–39238. [CrossRef] [PubMed]

15. Sakamoto, A.; Terui, Y.; Yamamoto, T.; Kasahara, T.; Nakamura, M.; Tomitori, H.; Yamamoto, K.; Ishihama, A.; Michael, A.J.; Igarashi, K.; et al. Enhanced biofilm formation and/or cell viability by polyamines through stimulation of response regulators UvrY and CpxR in the two-component signal transducing systems, and ribosome recycling factor. *Int. J. Biochem. Cell Biol.* **2012**, *44*, 1877–1886. [CrossRef] [PubMed]

16. Patel, C.N.; Wortham, B.W.; Lines, J.L.; Fetherston, J.D.; Perry, R.D.; Oliveira, M.A. Polyamines are essential for the formation of plague biofilm. *J. Bacteriol.* **2006**, *188*, 2355–2363. [CrossRef] [PubMed]

17. Karatan, E.; Duncan, T.R.; Watnick, P.I. NspS, a predicted polyamine sensor, mediates activation of *Vibrio cholerae* biofilm formation by norspermidine. *J. Bacteriol.* **2005**, *187*, 7434–7443. [CrossRef] [PubMed]

18. Wortham, B.W.; Patel, C.N.; Oliveira, M.A. Review article Polyamines in bacteria: Pleiotropic effects yet specific mechanisms. *Adv. Exp. Med. Biol.* **2007**, *603*, 106–115. [PubMed]

19. Pan, Y.H.; Liao, C.C.; Kuo, C.C.; Duan, K.J.; Liang, P.H.; Yuan, H.S.; Hu, S.T.; Chak, K.F. The critical roles of polyamines in regulating ColE7 production and restricting ColE7 uptake of the colicin-producing *Escherichia coli*. *J. Biol. Chem.* **2006**, *281*, 13083–13091. [CrossRef] [PubMed]

20. Fernandez, I.M.; Silva, M.; Schuch, R.; Walker, W.A.; Siber, A.M.; Maurelli, A.T.; McCormick, B.A. Cadaverine prevents the escape of *Shigella flexneri* from the phagolysosome: A connection between bacterial dissemination and neutrophil transepithelial signaling. *J. Infect. Dis.* **2001**, *184*, 743–753. [CrossRef] [PubMed]

21. Shah, P.; Swiatlo, E. A multifaceted role for polyamines in bacterial pathogens. *Mol. Microbiol.* **2008**, *68*, 4–16. [CrossRef] [PubMed]

22. Torres, A.G.; Vazquez-Juarez, R.C.; Tutt, C.B.; Garcia-Gallegos, J.G. Pathoadaptive mutation that mediates adherence of shiga toxin-producing *Escherichia coli* O111. *Infect. Immun.* **2005**, *73*, 4766–4776. [CrossRef] [PubMed]

23. Ware, D.; Jiang, Y.; Lin, W.; Swiatlo, E. Involvement of potD in *Streptococcus pneumoniae* polyamine transport and pathogenesis. *Infect. Immun.* **2006**, *74*, 352–361. [CrossRef] [PubMed]

24. Shah, P.; Nanduri, B.; Swiatlo, E.; Ma, Y.; Pendarvis, K. Polyamine biosynthesis and transport mechanisms are crucial for fitness and pathogenesis of *Streptococcus pneumoniae*. *Microbiology* **2011**, *157 Pt 2*, 504–515. [CrossRef] [PubMed]

25. Shah, P.; Swiatlo, E. Immunization with polyamine transport protein PotD protects mice against systemic infection with *Streptococcus pneumoniae*. *Infect. Immun.* **2006**, *74*, 5888–5892. [CrossRef] [PubMed]

26. Shah, P.; Briles, D.E.; King, J.; Hale, Y.; Swiatlo, E. Mucosal immunization with polyamine transport protein D (PotD) protects mice against nasopharyngeal colonization with *Streptococcus pneumoniae*. *Exp. Biol. Med.* **2009**, *234*, 403–409. [CrossRef] [PubMed]

27. Tettelin, H.; Nelson, K.E.; Paulsen, I.T.; Eisen, J.A.; Read, T.D.; Peterson, S.; Heidelberg, J.; DeBoy, R.T.; Haft, D.H.; Dodson, R.J.; et al. Complete Genome Sequence of a Virulent Isolate of *Streptococcus pneumoniae*. *Science* **2001**, *293*, 498–506. [CrossRef] [PubMed]

28. Kees Leenhouts, G.V.J.K. A lactococcal pWV01-based integration toolbox for bacteria. *Methods Cell Sci.* **1998**, *20*, 35–50. [CrossRef]

29. Bricker, A.L.; Camilli, A. Transformation of a type 4 encapsulated strain of *Streptococcus pneumoniae*. *FEMS Microbiol. Lett.* **1999**, *172*, 131–135. [CrossRef] [PubMed]

30. Granok, A.B.; Parsonage, D.; Ross, R.P.; Caparon, M.G. The RofA Binding Site in *Streptococcus pyogenes* is utilized in multiple transcriptional pathways. *J. Bacteriol.* **2000**, *182*, 1529–1540. [CrossRef] [PubMed]

31. Hall, B.G.; Acar, H.; Nandipati, A.; Barlow, M. Growth rates made easy. *Mol. Biol. Evol.* **2014**, *31*, 232–238. [CrossRef] [PubMed]

32. Eberhardt, A.; Hoyland, C.N.; Vollmer, D.; Bisle, S.; Cleverley, R.M.; Johnsborg, O.; Havarstein, L.S.; Lewis, R.J.; Vollmer, W. Attachment of capsular polysaccharide to the cell wall in *Streptococcus pneumoniae*. *Microb. Drug Resist.* **2012**, *18*, 240–255. [CrossRef] [PubMed]

33. Rychli, K.; Guinane, C.M.; Daly, K.; Hill, C.; Cotter, P.D. Generation of nonpolar deletion mutants in *Listeria monocytogenes* using the "SOEing" method. In *Listeria Monocytogenes: Methods and Protocols*; Jordan, K., Fox, E.M., Wagner, M., Eds.; Springer: New York, NY, USA, 2014; pp. 187–200.

34. Rai, A.N.; Thornton, J.A.; Stokes, J.; Sunesara, I.; Swiatlo, E.; Nanduri, B. Polyamine transporter in *Streptococcus pneumoniae* is essential for evading early innate immune responses in pneumococcal pneumonia. *Sci. Rep.* **2016**, *6*, 26964. [CrossRef] [PubMed]

35. Andon, N.L.; Hollingworth, S.; Koller, A.; Greenland, A.J.; Yates, J.R.; Haynes, P.A. Proteomic characterization of wheat amyloplasts using identification of proteins by tandem mass spectrometry. *Proteomics* **2002**, *2*, 1156–1168. [CrossRef]

36. Qian, W.J.; Jacobs, J.M.; Camp, D.G., 2nd; Monroe, M.E.; Moore, R.J.; Gritsenko, M.A.; Calvano, S.E.; Lowry, S.F.; Xiao, W.; Moldawer, L.L.; et al. Comparative proteome analyses of human plasma following in vivo lipopolysaccharide administration using multidimensional separations coupled with tandem mass spectrometry. *Proteomics* **2005**, *5*, 572–584. [CrossRef] [PubMed]

37. Huang, D.W.; Sherman, B.T.; Lempicki, R.A. Systematic and integrative analysis of large gene lists using DAVID bioinformatics resources. *Nat. Protoc.* **2008**, *4*, 44–57. [CrossRef] [PubMed]

38. Yoshida, M.; Kashiwagi, K.; Kawai, G.; Ishihama, A.; Igarashi, K. Polyamine enhancement of the synthesis of adenylate cyclase at the translational level and the consequential stimulation of the synthesis of the RNA polymerase sigma 28 subunit. *J. Biol. Chem.* **2001**, *276*, 16289–16295. [CrossRef] [PubMed]

39. Szklarczyk, D.; Morris, J.H.; Cook, H.; Kuhn, M.; Wyder, S.; Simonovic, M.; Santos, A.; Doncheva, N.T.; Roth, A.; Bork, P.; et al. The STRING database in 2017: Quality-controlled protein-protein association networks, made broadly accessible. *Nucleic Acids Res.* **2017**, *45*, D362–D368. [CrossRef] [PubMed]

40. Vizcaino, J.A.; Csordas, A.; del-Toro, N.; Dianes, J.A.; Griss, J.; Lavidas, I.; Mayer, G.; Perez-Riverol, Y.; Reisinger, F.; Ternent, T.; et al. 2016 update of the PRIDE database and its related tools. *Nucleic Acids Res.* **2016**, *44*, D447–D456. [CrossRef] [PubMed]

41. Battig, P.; Hathaway, L.J.; Hofer, S.; Muhlemann, K. Serotype-specific invasiveness and colonization prevalence in *Streptococcus pneumoniae* correlate with the lag phase during in vitro growth. *Microbes Infect.* **2006**, *8*, 2612–2617. [CrossRef] [PubMed]

42. Melin, M.; Trzcinski, K.; Meri, S.; Kayhty, H.; Vakevainen, M. The capsular serotype of *Streptococcus pneumoniae* is more important than the genetic background for resistance to complement. *Infect. Immun.* **2010**, *78*, 5262–5270. [CrossRef] [PubMed]

43. Bentley, S.D.; Aanensen, D.M.; Mavroidi, A.; Saunders, D.; Rabbinowitsch, E.; Collins, M.; Donohoe, K.; Harris, D.; Murphy, L.; Quail, M.A.; et al. Genetic analysis of the capsular biosynthetic locus from all 90 pneumococcal serotypes. *PLoS Genet.* **2006**, *2*, e31. [CrossRef] [PubMed]

44. Yother, J. Capsules of *Streptococcus pneumoniae* and other bacteria: Paradigms for polysaccharide biosynthesis and regulation. *Annu. Rev. Microbiol.* **2011**, *65*, 563–581. [CrossRef] [PubMed]

45. Zeng, Y.; He, Y.; Wang, K.Y.; Wang, J.; Zeng, Y.K.; Chen, Y.X.; Chen, D.; Geng, Y.; OuYang, P. cpsJ gene of *Streptococcus iniae* is involved in capsular polysaccharide synthesis and virulence. *Antonie Van Leeuwenhoek* **2016**, *109*, 1483–1492. [CrossRef] [PubMed]

46. Rodriguez-Diaz, J.; Rubio-Del-Campo, A.; Yebra, M.J. Regulatory insights into the production of UDP-N-acetylglucosamine by *Lactobacillus casei*. *Bioengineered* **2012**, *3*, 339–342. [CrossRef] [PubMed]

47. Reichmann, N.T.; Grundling, A. Location, synthesis and function of glycolipids and polyglycerolphosphate lipoteichoic acid in Gram-positive bacteria of the phylum *Firmicutes*. *FEMS Microbiol. Lett.* **2011**, *319*, 97–105. [CrossRef] [PubMed]

48. Molina, R.; Gonzalez, A.; Stelter, M.; Perez-Dorado, I.; Kahn, R.; Morales, M.; Moscoso, M.; Campuzano, S.; Campillo, N.E.; Mobashery, S.; et al. Crystal structure of CbpF, a bifunctional choline-binding protein and autolysis regulator from *Streptococcus pneumoniae*. *EMBO Rep.* **2009**, *10*, 246–251. [CrossRef] [PubMed]

49. Navarre, W.W.; Ton-That, H.; Faull, K.F.; Schneewind, O. Multiple enzymatic activities of the murein hydrolase from staphylococcal phage. Identification of a D-alanyl-glycine endopeptidase activity. *J. Biol. Chem.* **1999**, *274*, 15847–15856. [CrossRef] [PubMed]

50. Schweizer, I.; Peters, K.; Stahlmann, C.; Hakenbeck, R.; Denapaite, D. Penicillin-binding protein 2X of *Streptococcus pneumoniae:* The mutation Ala707Asp within the C-terminal PASTA2 domain leads to destabilization. *Microb. Drug Resist.* **2014**, *20*, 250–257. [CrossRef] [PubMed]

51. Zhang, W.; Jones, V.C.; Scherman, M.S.; Mahapatra, S.; Crick, D.; Bhamidi, S.; Xin, Y.; McNeil, M.R.; Ma, Y. Expression, essentiality, and a microtiter plate assay for mycobacterial GlmU, the bifunctional glucosamine-1-phosphate acetyltransferase and *N*-acetylglucosamine-1-phosphate uridyltransferase. *Int. J. Biochem. Cell Biol.* **2008**, *40*, 2560–2571. [CrossRef] [PubMed]

52. Zimmerman, T.; Ibrahim, S. Choline kinase, a novel drug target for the inhibition of *Streptococcus pneumoniae. Antibiotics* **2017**, *6*, E20. [CrossRef] [PubMed]

53. McAllister, L.J.; Tseng, H.J.; Ogunniyi, A.D.; Jennings, M.P.; McEwan, A.G.; Paton, J.C. Molecular analysis of the psa permease complex of *Streptococcus pneumoniae. Mol. Microbiol.* **2004**, *53*, 889–901. [CrossRef] [PubMed]

54. Yesilkaya, H.; Kadioglu, A.; Gingles, N.; Alexander, J.E.; Mitchell, T.J.; Andrew, P.W. Role of manganese-containing superoxide dismutase in oxidative stress and virulence of *Streptococcus pneumoniae. Infect. Immun.* **2000**, *68*, 2819–2826. [CrossRef] [PubMed]

55. Imlay, J.A. Iron-sulphur clusters and the problem with oxygen. *Mol. Microbiol.* **2006**, *59*, 1073–1082. [CrossRef] [PubMed]

56. Chekabab, S.M.; Harel, J.; Dozois, C.M. Interplay between genetic regulation of phosphate homeostasis and bacterial virulence. *Virulence* **2014**, *5*, 786–793. [CrossRef] [PubMed]

57. Chekabab, S.M.; Jubelin, G.; Dozois, C.M.; Harel, J. PhoB activates *Escherichia coli* O157:H7 virulence factors in response to inorganic phosphate limitation. *PLoS ONE* **2014**, *9*, e94285. [CrossRef] [PubMed]

58. Stincone, A.; Prigione, A.; Cramer, T.; Wamelink, M.M.; Campbell, K.; Cheung, E.; Olin-Sandoval, V.; Gruning, N.M.; Kruger, A.; Tauqeer Alam, M.; et al. The return of metabolism: Biochemistry and physiology of the pentose phosphate pathway. *Biol. Rev. Camb. Philos. Soc.* **2015**, *90*, 927–963. [CrossRef] [PubMed]

59. Warner, J.B.; Lolkema, J.S. CcpA-dependent carbon catabolite repression in bacteria. *Microbiol. Mol. Biol. Rev.* **2003**, *67*, 475–490. [CrossRef] [PubMed]

60. Iyer, R.; Baliga, N.S.; Camilli, A. Catabolite control protein A (CcpA) contributes to virulence and regulation of sugar metabolism in *Streptococcus pneumoniae. J. Bacteriol.* **2005**, *187*, 8340–8349. [CrossRef] [PubMed]

61. Mao, F.; Dam, P.; Chou, J.; Olman, V.; Xu, Y. DOOR: A database for prokaryotic operons. *Nucleic Acids Res.* **2009**, *37*, D459–D463. [CrossRef] [PubMed]

62. Ware, D.; Watt, J.; Swiatlo, E. Utilization of putrescine by *Streptococcus pneumoniae* during growth in choline-limited medium. *J. Microbiol.* **2005**, *43*, 398–405. [PubMed]

63. Koski, P.; Vaara, M. Polyamines as Constituents of the Outer Membranes of *Escherichia coli* and *Salmonella typhimurium. J. Bacteriol.* **1991**, *173*, 3695–3699. [CrossRef] [PubMed]

64. Yethon, J.A.; Vinogradov, E.; Perry, M.B.; Whitfield, C. Mutation of the lipopolysaccharide core glycosyltransferase encoded by waaG destabilizes the outer membrane of *Escherichia coli* by interfering with core phosphorylation. *J. Biotechnol.* **2000**, *182*, 5620–5623. [CrossRef]

65. Kamio, Y. Structural specificity of diamines covalently linked to peptidoglycan for cell growth of *Veillonella alcalescens* and *Selenomonas ruminantium. J. Bacteriol.* **1987**, *169*, 4837–4840. [CrossRef] [PubMed]

66. Samartzidou, H.; Mehrazin, M.; Xu, Z.; Benedik, M.J.; Delcour, A.H. Cadaverine inhibition of porin plays a role in cell survival at acidic pH. *J. Bacteriol.* **2003**, *185*, 13–19. [CrossRef] [PubMed]

67. Paterson, G.K.; Blue, C.E.; Mitchell, T.J. Role of two-component systems in the virulence of *Streptococcus pneumoniae. J. Med. Microbiol.* **2006**, *55 Pt 4*, 355–363. [CrossRef] [PubMed]

68. McDaniel, L.S.; Swiatlo, E. Should pneumococcal vaccines eliminate nasopharyngeal colonization? *MBio* **2016**, *7*, e00545-16. [CrossRef] [PubMed]

69. Maurelli, A.T.; Fernandez, R.E.; Bloch, C.A.; Rode, C.K.; Fasano, A. Black holes and bacterial pathogenicity: A large genomic deletion that enhances the virulence of *Shigella* spp. and enteroinvasive *Escherichia coli. Proc. Natl. Acad. Sci. USA* **1998**, *95*, 3943–3948. [CrossRef] [PubMed]

70. Duggan, J.M.; You, D.; Cleaver, J.O.; Larson, D.T.; Garza, R.J.; Guzman Pruneda, F.A.; Tuvim, M.J.; Zhang, J.; Dickey, B.F.; Evans, S.E. Synergistic interactions of TLR2/6 and TLR9 induce a high level of resistance to lung infection in mice. *J. Immunol.* **2011**, *186*, 5916–5926. [CrossRef] [PubMed]

71. Wortham, B.W.; Oliveira, M.A.; Fetherston, J.D.; Perry, R.D. Polyamines are required for the expression of key Hms proteins important for *Yersinia pestis* biofilm formation. *Environ. Microbiol.* **2010**, *12*, 2034–2047. [CrossRef] [PubMed]

72. Feldman, C.; Anderson, R. Recent advances in our understanding of *Streptococcus pneumoniae* infection. *F1000Prime Rep.* **2014**, *6*, 82. [CrossRef] [PubMed]

73. Hardy, G.G.; Magee, A.D.; Ventura, C.L.; Caimano, M.J.; Yother, J. Essential role for cellular phosphoglucomutase in virulence of type 3 *Streptococcus pneumoniae. Infect. Immun.* **2001**, *69*, 2309–2317. [CrossRef] [PubMed]

74. Brown, E.J. Interaction of gram-positive microorganisms with complement. *Curr. Top. Microbiol. Immunol.* **1985**, *121*, 159–187. [PubMed]

75. Paterson, G.K.; Orihuela, C.J. Pneumococci: Immunology of the innate host response. *Respirology* **2010**, *15*, 1057–1063. [CrossRef] [PubMed]

76. Shenoy, A.T.; Orihuela, C.J. Anatomical site-specific contributions of pneumococcal virulence determinants. *Pneumonia* **2016**, *8*, 7. [CrossRef] [PubMed]

77. Weiser, J.N.; Austrian, R.; Sreenivasan, P.K.; Masure, H.R. Phase variation in pneumococcal opacity: Relationship between colonial morphology and nasopharyngeal colonization. *Infect. Immun.* **1994**, *62*, 2582–2589. [PubMed]

78. Shainheit, M.G.; Mule, M.; Camilli, A. The core promoter of the capsule operon of *Streptococcus pneumoniae* is necessary for colonization and invasive disease. *Infect. Immun.* **2014**, *82*, 694–705. [CrossRef] [PubMed]

79. Carvalho, S.M.; Farshchi Andisi, V.; Gradstedt, H.; Neef, J.; Kuipers, O.P.; Neves, A.R.; Bijlsma, J.J. Pyruvate oxidase influences the sugar utilization pattern and capsule production in *Streptococcus pneumoniae. PLoS ONE* **2013**, *8*, e68277. [CrossRef] [PubMed]

80. Gupta, R.; Yang, J.; Dong, Y.; Swiatlo, E.; Zhang, J.R.; Metzger, D.W.; Bai, G. Deletion of arcD in *Streptococcus pneumoniae* D39 impairs its capsule and attenuates virulence. *Infect. Immun.* **2013**, *81*, 3903–3911. [CrossRef] [PubMed]

81. Echlin, H.; Frank, M.W.; Iverson, A.; Chang, T.C.; Johnson, M.D.; Rock, C.O.; Rosch, J.W. Pyruvate Oxidase as a Critical Link between Metabolism and Capsule Biosynthesis in *Streptococcus pneumoniae. PLoS Pathog.* **2016**, *12*, e1005951. [CrossRef] [PubMed]

82. Igarashi, K.; Kashiwagi, K. Modulation of protein synthesis by polyamines. *IUBMB Life* **2015**, *67*, 160–169. [CrossRef] [PubMed]

medical sciences

MDPI

Article

A Novel Polyamine-Targeted Therapy for BRAF Mutant Melanoma Tumors

Molly C. Peters [1], Allyson Minton [1], Otto Phanstiel IV [2] and Susan K. Gilmour [1,*]

[1] Lankenau Institute for Medical Research, 100 Lancaster Avenue, Wynnewood, PA 19096, USA; PetersM@mlhs.org (M.C.P.); Mintonar@msn.com (A.M.)
[2] Biomolecular Research Annex, University of Central Florida, 12722 Research Parkway, Orlando, FL 32826-3227, USA; Otto.Phanstiel@ucf.edu
*   Correspondence: GilmourS@mlhs.org; Tel.: +1-484-476-8429

Received: 29 November 2017; Accepted: 28 December 2017; Published: 5 January 2018

Abstract: Mutant serine/threonine protein kinase B-Raf (BRAF) protein is expressed in over half of all melanoma tumors. Although BRAF inhibitors (BRAFi) elicit rapid anti-tumor responses in the majority of patients with mutant BRAF melanoma, the tumors inevitably relapse after a short time. We hypothesized that polyamines are essential for tumor survival in mutant BRAF melanomas. These tumors rely on both polyamine biosynthesis and an upregulated polyamine transport system (PTS) to maintain their high intracellular polyamine levels. We evaluated the effect of a novel arylpolyamine (AP) compound that is cytotoxic upon cellular entry via the increased PTS activity of melanoma cells with different *BRAF* mutational status. Mutant BRAF melanoma cells demonstrated greater PTS activity and increased sensitivity to AP compared to wild type BRAF (BRAF^{WT}) melanoma cells. Treatment with an inhibitor of polyamine biosynthesis, α-difluoromethylornithine (DFMO), further upregulated PTS activity in mutant BRAF cells and increased their sensitivity to AP. Furthermore, viability assays of 3D spheroid cultures of mutant BRAF melanoma cells demonstrated greater resistance to the BRAFi, PLX4720, compared to 2D monolayer cultures. However, co-treatment with AP restored the sensitivity of melanoma spheroids to PLX4720. These data indicate that mutant BRAF melanoma cells are more dependent on the PTS compared to BRAF^{WT} melanoma cells, resulting in greater sensitivity to the PTS-targeted cytotoxic AP compound.

Keywords: polyamines; α-difluoromethylornithine; polyamine transport system; melanoma; mutant BRAF

1. Introduction

Melanoma is a highly aggressive tumor with poor prognosis in the metastatic stage. Multiple oncogenic mutations (including genes encoding serine/threonine protein kinase B-Raf (BRAF), the neuroblastoma RAS homolog (*NRAS*), and the proto-oncogene receptor tyrosine protein kinase KIT) drive this highly heterogeneous disease, with mutations in the *BRAF* gene detected in half of all melanoma tumors [1]. The treatment of metastatic melanoma has been revolutionized over the last decade with the discovery of highly prevalent *BRAF* mutations, which drive constitutive activation of the RAS-RAF-MEK-ERK pathway and promote uncontrolled proliferation [1]. Ninety percent of reported *BRAF* mutations result in substitution of glutamic acid for valine at amino acid 600 (the V600E mutation) [2,3]. The subsequent rapid development of selective inhibitors of mutant BRAF^{V600E} proteins (vemurafenib and dabrafenib) demonstrated a major advance in the treatment of melanoma patients harboring the BRAF^{V600E} mutation. However, nearly 100% of the patients exhibit disease progression within seven months after treatment with BRAF inhibitors [4–6]. Thus, new ways to overcome the acquired resistance to these inhibitors are urgently needed to increase survival in melanoma patients.

An alternative approach is to target a downstream pathway that is essential for survival of oncogene-addicted tumor cells. While oncogenes indeed drive proliferation, they do so via downstream effector molecules. For example, downstream of extracellular regulated kinase (ERK) signaling is c-MYC (myelocytomatosis viral proto-oncogene homolog), a known regulator of ornithine decarboxylase (ODC) transcription and polyamine biosynthesis [7]. The native polyamines (putrescine, spermidine and spermine) are amino acid-derived polycations that have been implicated in a wide array of biological processes, including cellular proliferation, differentiation, chromatin remodeling, hypusination of the eukaryotic initiation factor-5A (eIF-5A) and apoptosis [8]. Multiple oncogene-encoded proteins, including c-MYC and RAS, are known to upregulate key polyamine biosynthetic enzymes [7,9,10] as well as the cellular uptake of polyamines by activating the polyamine transport system (PTS) [11–14]. Compared to normal cells, tumor cells have been shown to contain elevated levels of polyamines [15–18]. These intracellular polyamine levels are maintained via tightly-regulated biosynthetic, catabolic, and uptake and export pathways [19]. Polyamine uptake is upregulated in many tumor types, especially in melanoma tumor cells when compared to normal cells [11,20]. Thus, melanoma tumor cells notoriously replete with multiple oncogenic mutations have a greatly increased need for polyamines compared to normal cells to meet their increased metabolic needs [20].

Our objective was to exploit the oncogene-induced polyamine transport activity in melanoma cells by selectively targeting the PTS with a novel arylmethyl-polyamine (AP) compound (Figure 1, [21]). The two-armed design of AP predicated upon a naphthyl core provides PTS hyperselectivity and high potency [21]. Key to our drug design is that both exogenous polyamines and polyamine-based drugs are imported into tumors via a specific uptake system [8,21,22]. Here, we show that polyamine uptake is increased in mutant BRAF^{V600E} melanoma cells, and that AP treatment significantly increases cell death in BRAF^{V600E} melanoma cells compared to BRAF^{WT} melanoma cells. Furthermore, we show that BRAF inhibitor-resistance in melanoma tumor spheroid cultures can be overcome by treatment with AP. These studies provide valuable insights into developing more effective treatment strategies to restore sensitivity of melanoma tumor cells to BRAF inhibitors. In short, the mutant BRAF-driven polyamine addiction can be targeted by cytotoxic polyamine compounds, which selectively target melanoma cells with high polyamine import activity.

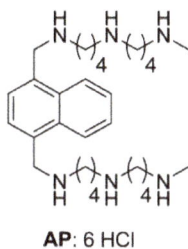

AP: 6 HCl

Figure 1. Structure of the arylpolyamine (AP).

2. Materials and Methods

2.1. Cell Lines and Reagents

All human melanoma cell lines including WM983B, WM3734, WM3743, WM989, WM88, WM3451, WM3211, and 1205Lu were obtained as kind gifts from Dr. Meenhard Herlyn (The Wistar Institute, Philadelphia, PA, USA). These cells were maintained in MCDB153 (Sigma-Aldrich, St. Louis, MO, USA) and Leibovitz's L-15 (Mediatech Inc, Manassas, VA, USA) medium (4:1 ratio) supplemented with 2% fetal calf serum and 2 mmol/L CaCl$_2$. B16F10 cells were obtained from the American

Type Culture Collection (Manassas, VA, USA) and maintained in Dulbecco's Modified Eagle Medium (DMEM) (Invitrogen, Waltham, MA, USA) supplemented with 10% fetal bovine serum and 100 U/mL Penicillin/Streptomycin. The YUMM1.7 cell line (kindly provided by Marcus Bosenburg, Yale University, New Haven, CT, USA) that harbors a $BRAF^{V600E}$ mutation and inactivation of the Phosphatase and tensin homolog (*PTEN*) gene was maintained in DMEM/F12 (Invitrogen, Waltham, MA, USA) medium supplemented with 10% fetal bovine serum and 100 U/mL Penicillin/Streptomycin.

PLX4720, (S1152; Selleckchem, Houston, TX, USA) a derivative related to PLX4032/Vemurafenib (Plexxikon, Berkeley, CA, USA) was prepared as a 50 mM stock solution in dimethyl sulfoxide and stored at $-20\ ^{\circ}$C. The synthesis of the AP compound has been described previously [21]. The compound was dissolved in phosphate buffered saline (PBS) to provide an initial stock (10 mM), which was filtered through a 0.2 um filter to ensure sterility. Subsequent dilutions were made in PBS to generate the desired stock solutions.

2.2. 3D Spheroid Culture

The nanoscale scaffolding NanoCulture plates (NCP) were purchased from (Organogenix Inc, Woburn, MA, USA). The base of each NCP is constructed with a transparent cyclo-olefin resinous sheet with a nanoscale indented pattern. To form spheroids, 1205Lu human melanoma cells were seeded in a 96-well NCP at 1×10^4 cells/well in MCDB153 (Sigma-Aldrich, St. Louis, MO, USA) and Leibovitz's L-15 (Mediatech Inc, Manassas, VA, USA) medium (4:1 ratio) supplemented with 2% heat-inactivated fetal calf serum and 2 mM $CaCl_2$ and incubated in a conventional cell incubator at $37\ ^{\circ}$C in an atmosphere of 5% CO_2 and normal O_2 levels. When visible spheroids began to form on day 3 after the cells were seeded on the NCPs, treatment with PLX4720 and/or AP was initiated. After drug treatment for 48 h, the spheroid cultures were assayed for cell viability.

2.3. Cell Viability Assay

Cell proliferation assays were conducted in 96-well plates at 25–30% starting confluence to determine the effect of exposure to increasing concentrations of PLX4720 or AP with or without 1 mM α-difluoromethylornithine (DFMO) for 72 h. Cell viability was assessed using the EZQuant Cell Quantifying Kit (Alstem, Richmond, CA, USA) in which the tetrazolium salt WST-8 is reduced by the metabolic activity of live cells to formazan dye. For spheroids treated with PLX4720 and/or AP, viability of the spheroid cells was estimated by quantification of the adenosine triphosphate present using a CellTiter-Glo Luminescent Cell Viability Assay (Promega Co., Madison, WI, USA). The 72 h half maximal inhibitory concentration (IC50) values for AP were calculated using nonlinear regression (sigmoidal dose response) of the plot of percentage inhibition versus the log of inhibitor concentration in GraphPad Prism (v5; GraphPad Software, Inc., La Jolla, CA, USA). The IC50 value is defined as the concentration of the compound required to inhibit 50% cell viability compared to an untreated control.

2.4. Radiolabeled Spermidine Transport Assays

Polyamine transport in tumor cells was evaluated essentially as described previously [23,24]. Radioactive spermidine (Net-522, Spermidine Trihydrochloride, [Terminal Methylenes-^{3}H(N)], specific activity 16.6 Ci/mmol; Perkin Elmer, Boston, MA, USA) was used. Cells were plated in 96 well plates and grown to approximately 80% confluence. Half of the cells were treated with 1 mM DFMO for 40 h. After repeated washing with PBS, ^{3}H-spermidine was added at 0.5 μM and incubated for 60 min at $37\ ^{\circ}$C. Cells were then washed with cold PBS containing 50 μM spermidine and lysed in 0.1% sodium dodecyl sulfate solution at $37\ ^{\circ}$C for 30 min with mixing. Cell lysates were then aliquoted for scintillation counting and for protein assay using a microplate Bio-Rad protein assay (Bio-Rad, Hercules, CA, USA). Results were expressed as counts per minute (CPM)/μg protein.

2.5. Statistical Analysis

All in vitro experiments were performed at least in triplicate, and data were compiled from two to three separate experiments. Analyses were done using a one-way analysis of variance with a Tukey test for statistical significance or a Students *t*-test. In all cases, values of $p \leq 0.05$ were regarded as being statistically significant.

3. Results

3.1. Human Mutant BRAF^{V600E} Melanoma Cells Are More Sensitive to Cytotoxic Effects of AP Than BRAF^{WT} Melanoma Cells

A panel of human melanoma cell lines with different BRAF mutational status was screened for their sensitivity to the BRAF inhibitor PLX4720. We confirmed previous findings that mutant BRAF^{V600E} melanoma cells, including WM983B, WM3734, 1205Lu, WM989, and WM88, demonstrated marked sensitivity to PLX4720 (IC50 values ≤ 3.0 μM), whereas BRAF^{WT} melanoma cells, including WM3451, WM3743, and WM3211, demonstrated relative resistance to treatment with PLX4720 (IC50 > 3.0 μM) [25]. This approach allowed us to rank the relative sensitivity of each cell line to the BRAF inhibitor, PLX4720. Thus, the sensitivity of these BRAF^{V600E} melanoma cells to BRAF inhibition with PLX4720 reflected their functional dependence on mutant BRAF signaling to sustain their proliferation and viability.

Likewise, we tested whether mutant BRAF^{V600E} cells were more sensitive to increasing concentrations of the cytotoxic polyamine transport ligand, AP (Figure 1), compared to BRAF^{WT} melanoma cells. Table 1 shows that mutant BRAF^{V600E} melanoma cells demonstrated greater sensitivity to AP (IC50 < 2.5 μM) than BRAF^{WT} melanoma cells (IC50 > 4.0 μM). This observation was reflected by the greater polyamine transport activity in BRAF^{V600E} melanoma cells compared to BRAF^{WT} melanoma cells (Figure 2A and Table 1). Since AP accumulated at a faster rate in BRAF^{V600E} human melanoma cells with higher polyamine transport rates compared to BRAF^{WT} human melanoma cells (Figure 2A), BRAF^{V600E} melanoma cells were significantly ($p < 0.01$) more sensitive to AP exposure than BRAF^{WT} melanoma cells (Figure 2B). In summary, cell lines with high polyamine import activity were more sensitive to the cytotoxic polyamine compound.

Table 1. Polyamine transport activity and sensitivity to AP in human melanoma cells with different BRAF mutational status cultured ± DFMO [a].

Cell Line	BRAF Mutational Status	− DFMO		+ DFMO	
		IC50 (μM AP)	PTS Activity [b] (cpm ^{3}H Spd/μg Protein)	IC50 (μM AP)	PTS Activity [b] (cpm ^{3}H Spd/μg Protein)
WM983B	V600E	2.6	487 ± 64	0.9	706 ± 78 *
WM3734	V600E	2.2	444 ± 46	1.2	654 ± 59 *
1205Lu	V600E	0.7	230 ± 22	0.6	299 ± 54
WM989	V600E	1.2	304 ± 28	1.0	348 ± 48
WM88	V600E	0.8	302 ± 18	0.2	343 ± 59
WM3451	WT	9.0	130 ± 21	5.1	157 ± 34
WM3743	WT	8.9	117 ± 23	10.9	110 ± 25
WM3211	WT	4.5	278 ± 58	5.2	251 ± 46

[a] The mean values for half maximal inhibitory concentration (IC50) for AP and polyamine transport system (PTS) activity for human melanoma cells expressing mutant BRAF^{V600E} are compared with that for melanoma cells expressing wild type (WT) BRAF under conditions where cells were cultured without added DFMO or with 1 mM α-difluoromethylornithine (DFMO). [b] PTS activity expressed as counts per minute (CPM) ^{3}H Spermidine (Spd)/μg protein ± standard deviation. PTS assays and cell viability assays were performed at least three times with each cell line. * $p \leq 0.001$ when compared to the PTS activity in the absence of DFMO.

It is well known that lowering intracellular levels of polyamines with inhibitors of polyamine biosynthesis can increase uptake of extracellular polyamines as well as exogenous polyamine analogues [20,26]. WM983B and WM3743 melanoma cells were pretreated for 40 h with 1 mM

DFMO, an inhibitor of ODC, the first and rate-limiting enzyme in polyamine biosynthesis, before measuring their polyamine transport activity. DFMO treatment dramatically increased polyamine transport activity in BRAF^{V600E} WM983B melanoma cells, but not in the BRAF^{WT} WM3743 melanoma cells (Supplementary Materials Figure S1A). In general, polyamine depletion with DFMO treatment enhanced polyamine uptake more in the screened human BRAF^{V600E} melanoma cells compared to that seen in BRAF^{WT} melanoma cells (Table 1, Figure 2A). Since DFMO treatment increases polyamine transport activity, we tested whether co-treatment with AP and DFMO will increase the sensitivity of melanoma cells to AP. For instance, BRAF^{V600E} WM983B melanoma cells are significantly ($p \leq 0.0001$) more sensitive to AP treatment when co-treated with DFMO (IC50 = 0.7 μM) compared to that with AP alone (IC50 = 2.3 μM) (Supplementary Materials Figure S1C). In contrast, sensitivity to AP was not increased in DFMO-co-treated BRAF^{WT} melanoma cells (Table 1, Figure 2B). These data indicate that human BRAF^{V600E} melanoma cells demonstrate greater polyamine transport activity and increased sensitivity to AP compared to BRAF^{WT} melanoma cells, and their sensitivity can be increased by inhibition of polyamine biosynthesis with DFMO.

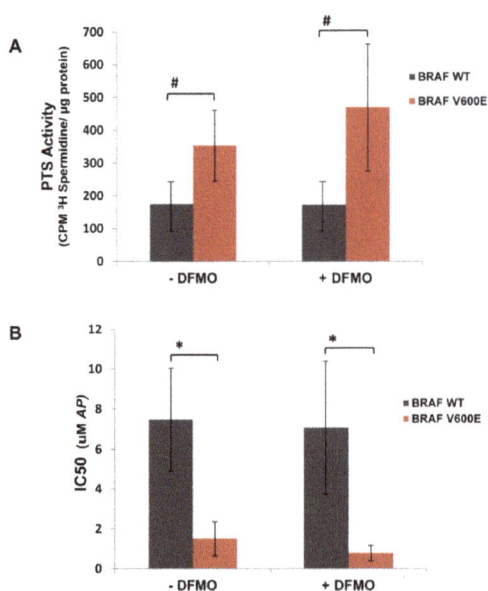

Figure 2. Greater PTS activity and increased sensitivity to AP in mutant BRAF^{V600E} human melanoma cells compared to wild type (WT) BRAF^{WT} cells. (**A**) BRAF^{V600E} human melanoma cells (WM983B, WM3734, 1205Lu, WM989, and WM88) and BRAF^{WT} human melanoma cells (WM3451, WM3743, and WM3211) were cultured with and without 1 mM DFMO for 40 h and then pulsed with 0.5 μM ^{3}H-spermidine for 60 min at 37 °C. Cell lysates were assayed for CPM ^{3}H-spermidine per mg protein by scintillation counting. The mean PTS activity ± SD for BRAF^{WT} melanoma cells is compared with that of BRAF^{V600E} melanoma cells under conditions where cells were cultured without added DFMO or with 1 mM DFMO. (**B**) BRAF^{V600E} human melanoma cells (WM983B, WM3734, 1205Lu, WM989, and WM88) and BRAF^{WT} human melanoma cells (WM3451, WM3743, and WM3211) were treated with increasing doses of AP with or without 1 mM DFMO, using 5–6 samples per dose of AP. After 72 h of culture, cell survival was determined via EZQuant Cell Quantifying assay (Alstem, Richmond, CA, USA). AP IC50 values were calculated by GraphPad Prism 6. The mean AP IC50 values ± SD for BRAF^{WT} melanoma cells is compared with that of BRAF^{V600E} melanoma cells under conditions where cells were cultured without added DFMO or with 1 mM DFMO; # $p \leq 0.05$; * $p < 0.01$.

3.2. AP Is More Cytotoxic to BRAF^{V600E} Murine Melanoma Cells Than BRAF^{WT} Melanoma Cells

Since human melanoma cells possess multiple oncogenic mutations in addition to $BRAF^{V600E}$, we compared AP cytotoxicity and PTS activity in the murine B16F10 melanoma cell line that is $BRAF^{WT}$ with $BRAF^{V600E}$ YUMM1.7 cell line that was derived from a melanoma tumor that spontaneously developed in a $BRAF^{V600E}$/PTEN^{null} transgenic mouse [27]. As expected, YUMM1.7 cells were very sensitive to PLX4720 with a lower IC50 compared to B16F10 cells (Figure 3A). In addition, $BRAF^{V600E}$ YUMM1.7 cells were significantly ($p < 0.0001$) more sensitive to AP and had a much lower IC50 value for AP compared to $BRAF^{WT}$ B16F10 cells (Figure 3B). In particular, DFMO co-treatment increased the sensitivity of YUMM1.7 cells to AP (IC50 = 0.8 μM AP without DFMO and IC50 = 0.2 μM AP with DFMO co-treatment), and this correlated with a marked DFMO-induction of PTS activity in $BRAF^{V600E}$ YUMM1.7 cells (Figure 3D). In contrast, $BRAF^{WT}$ B16F10 cells demonstrated no significant induction in polyamine uptake following DFMO treatment (Figure 3D). However, B16F10 cells retrovirally infected to express the mutant $BRAF^{V600E}$ protein exhibited a similar PTS activity profile as that seen with YUMM1.7 cells. DFMO treatment was shown to enhance polyamine uptake in the B16F10-$BRAF^{V600E}$ cells as was seen with YUMM1.7 cells (Figure 3D). AP was also more cytotoxic in B16F10-$BRAF^{V600E}$ cells (IC50 = 24.4 μM) compared to control-infected B16F10-pBABE cells that were infected with retrovirus expressing the empty plasmid (IC50 = 36.4 μM). These data suggest that melanoma cells with a mutant $BRAF^{V600E}$ protein are more dependent on the polyamine uptake system compared to cells with a $BRAF^{WT}$ protein, resulting in greater sensitivity to the PTS-targeted cytotoxic AP compound that enters and kills melanoma cells via the polyamine transport system.

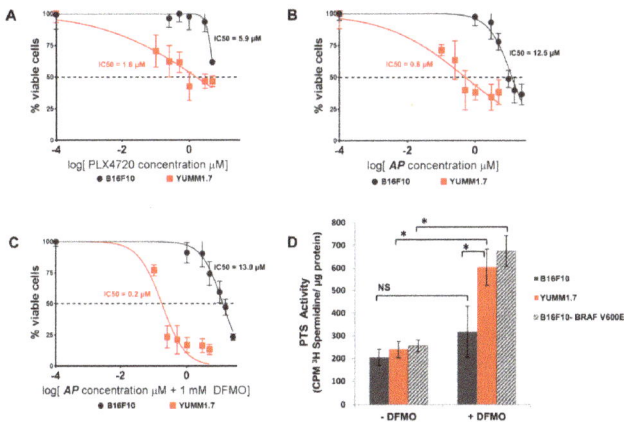

Figure 3. BRAF^{V600E} murine melanoma cells are more sensitive to AP than BRAF^{WT} melanoma cells. (**A**) Murine BRAF^{V600E} YUMM1.7 melanoma cells and BRAF^{WT} B16F10 melanoma cells were treated with increasing doses of PLX4720. After 72 h of culture, cell survival was determined via EZQuant Cell Quantifying assay. IC50 values were calculated by GraphPad Prism 6; $p = 0.0013$. (**B**) Murine BRAF^{V600E} YUMM1.7 melanoma cells and BRAF^{WT} B16F10 melanoma cells were treated with increasing doses of AP. After 72 h of culture, cell survival was determined via EZQuant Cell Quantifying assay. IC50 values were calculated by GraphPad Prism 6; $p < 0.0001$. (**C**) Murine BRAF^{V600E} YUMM1.7 melanoma cells and BRAF^{WT} B16F10 melanoma cells were treated with increasing doses of AP ± 1 mM DFMO. After 72 h of culture, cell survival was determined via EZQuant Cell Quantifying assay. IC50 values were calculated by GraphPad Prism 6; $p < 0.0001$. (**D**) YUMM1.7 and B16F10 melanoma cells and B16F10 cells retrovirally infected to express the mutant BRAF^{V600E} protein were cultured with and withoutS 1 mM DFMO for 40 h and then pulsed with 0.5 μM ^{3}H-spermidine for 60 min at 37 °C. Cells were washed with cold PBS containing 50 μM spermidine, and cell lysates were assayed for CPM ^{3}H-spermidine per mg protein by scintillation counting; * $p < 0.0001$; NS: not significant.

3.3. Increased Resistance of Spheroid Melanoma Cells to PLX4720 Is Overcome with AP Co-Treatment

Because growth of cells in a 3D culture system has been found to be more representative of the in vivo microenvironment, we cultured 1205Lu human melanoma cells using a nanoscale, scaffold-based NCP in which tumor cells easily form 3D spheroids [28]. Although these tumor spheroid cultures are grown in ambient air, the spheroid microenvironment closely resembles that in tumors with a hypoxic core and is more relevant for drug sensitivity compared to that seen with monolayer cultures [28–30]. Similar to previous reports [31], BRAF^{V600E} mutant 1205Lu melanoma cells grown as spheroids on NCPs were more resistant to 48 h treatment with PLX4720 (25 μM) compared to the same cells grown in 2D monolayer cultures in ambient air (Figure 4). Both spheroid and monolayer cultures were similarly sensitive to 48 h treatment with a high concentration of AP (25 μM) alone. We then tested the effect of AP treatment on the PLX4720-resistant phenotype of the 1205Lu spheroid and monolayer cultures. Co-treatment with both PLX4720 (25 μM) and AP (25 μM) led to a dramatic reduction in cell viability in the spheroid cultures unlike monolayer cultures that showed no further reduction in cell viability when compared to PLX4720 treatment alone (Figure 4). Thus, the increased resistance of the melanoma spheroid cultures to PLX4720 was eliminated with AP co-treatment.

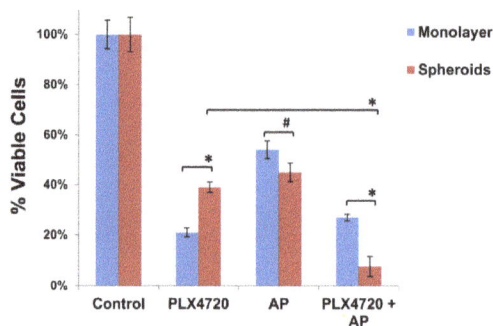

Figure 4. Increased resistance of spheroid melanoma cells to PLX4720 is overcome with AP co-treatment. BRAF^{V600E} mutant 1205Lu melanoma cells were seeded at 1×10^4 cells in each well of 24-well NanoCulture plates (NCPs). When spheroids were formed on day 3, the 3D cultures of spheroids were treated with PLX4720 (25 μM) and/or AP (25 μM). 2D monolayer cultures of 1205Lu melanoma cells were also treated with PLX4720 (25 μM) and/or AP (25 μM). After drug treatment for 48 h, the viability of spheroids and monolayer cultures was assayed using the CellTiter-Glo Luminescent Cell Viability Assay. The percent cell survival in each treatment group was calculated relative to cells treated with medium only under the same conditions. As controls, the growth of cells without drug treatment under each condition was normalized as 100% separately. The means are presented $\pm$ SD; * $p < 0.0001$; # $p = 0.0028$.

4. Discussion

Melanoma is challenging to treat due to its genetic heterogeneity, and successful therapy requires targeting multiple molecular vulnerabilities. Our data show that melanoma tumor cells expressing mutant BRAF^{V600E} exhibit a high demand for polyamine growth factors and a greatly upregulated PTS. Utilizing the PTS for drug delivery, the AP compound attacks the melanoma cells via one of its key modes of survival. Indeed, we propose that polyamines are essential for the survival of melanomas. Polyamine levels are dramatically elevated in tumor cells compared to normal cells, often the result of oncogenic induction [11,20]. Previous studies have shown that the c-MYC and RAS can upregulate polyamine biosynthesis [9,10] and increase cellular uptake of polyamines by inducing PTS activity [12–14]. Although melanoma cells are notoriously replete with multiple oncogenic mutations, more than half of all melanoma tumors express a mutant BRAF protein [20]. Our data suggest that

BRAF[V600E] melanoma tumors have a greatly increased metabolic need for polyamines compared to normal cells. We have exploited the BRAF[V600E]-induced PTS activity in metastatic melanoma cells by targeting the PTS with AP.

Putrescine, spermidine, and spermine play key roles in cellular proliferation, signal transduction, gene expression, and autophagic states that contribute to tumor survival [32–35]. These endogenous polyamines and the polyamine-based AP compete to be imported into tumors via the PTS [21]. However, studies suggest that arylmethyl-polyamines similar to AP have enhanced cytotoxic potency via their multiple electrostatic interactions with DNA [36] and topoisomerase II [37]. Since AP selectively targets tumor cells with high polyamine transport rates, normal cells are significantly less sensitive to AP since they have low PTS activity [21]. Studies have shown that polyamine biosynthesis and cellular uptake are induced in hypoxic regions of tumors and in tumor spheroids [38]. Moreover, depletion of polyamines during hypoxia resulted in increased apoptosis [38], indicating that polyamines play an essential role in the ability of tumor cells to adapt to hypoxic stress and reactive oxygen species. Indeed, polyamines are known to exert anti-oxidant functions [39]. Although the melanoma spheroid cultures in this study were cultured with ambient air, it is well documented that the cells at the center of spheroids are hypoxic, thus modeling the heterogeneous 3D structure of in vivo melanoma tumors that often contain hypoxic regions [28,31]. Solid tumors contain poorly vascularized, hypoxic regions that contribute to tumor progression by activating a hypoxia stress response via hypoxia inducible factor-1α that promotes cell survival, tumor angiogenesis, and metastasis [40,41]. Studies have shown that hypoxic tumor cells and spheroid cultures are more resistant to chemotherapy including BRAF inhibitors [31,42,43]. Likewise, we have found that 3D cultures of 1205Lu melanoma cells grown as spheroids on NCPs are more resistant to PLX4720 treatment compared to 1205Lu cells grown in 2D monolayer culture in ambient air. Knowing that polyamine uptake is induced in hypoxic regions of tumor spheroids, we hypothesized that treatment with the PTS ligand AP would increase the sensitivity of 1205Lu spheroid cells to PLX4720. Indeed, the increased resistance of melanoma spheroids to PLX4720 was overcome with AP co-treatment. In contrast, AP co-treatment had no significant effect on the sensitivity of 2D monolayer cultures of 1205Lu cells to PLX4720.

Accumulating literature shows that treatment with a BRAF inhibitor such as PLX4720 enriches a slow-cycling cancer stem cell-like (CSC) subpopulation of melanoma cells that is characterized by stem cell markers such as Lysine-Specific Demethylase 5B (JARID1B) and spheroid formation [44,45]. It is thought that cancer stem cell populations exist in a hypoxic microenvironment [46–48]. Roesch et al. [45] have found that endogenous reactive oxygen species (ROS) levels are increased in slow cycling JARID1B[high] melanoma cells as a result of increased mitochondrial respiration and oxidative phosphorylation. This high oxygen consumption contributes to hypoxic conditions that have been shown to favor the JARID1B[high] slow-cycling CSC-like phenotype [44]. Using 1205Lu melanoma cells stably transduced with a JARID1B-promoter-green fluorescent protein (GFP)-reporter construct [44], we found that a short 2-day exposure to PLX4720 led to a 4-fold enrichment of JARID1B-driven GFP expressing 1205Lu melanoma cells grown as spheroids (data not shown). Our data show that PLX4720-resistant melanoma spheroids are made more sensitive to PLX4720 with AP co-treatment, and it is likely that AP is targeting CSC subpopulations that are enriched in the PLX4720-resistant melanoma spheroids.

Polyamines may also contribute to tumor survival by inducing an autophagic state [32–35]. For instance, spermidine has been shown to induce autophagy in multiple systems including yeast cells, *Caenorhabditis elegans*, *Drosophila melanogaster*, and human tumor cells [35,49] and to increase survival of pluripotent stem cells in culture [50]. Autophagy has recently emerged as a common survival process that tumors undergo when assaulted by chemotherapy and radiation [51]. It is induced by cellular stress such as nutrient deprivation, withdrawal of growth factors, and hypoxia [52]. In established tumors, autophagy is also a resistance mechanism to many therapeutic modalities including BRAF inhibitors [53]. Increased polyamine uptake provides a mechanism for BRAFi-resistant melanoma cells to acquire sufficient polyamines to undergo autophagy to survive treatment with BRAF inhibitors.

Previous clinical trials have tested the anti-tumor efficacy of the ODC inhibitor DFMO. However, treatment with DFMO alone demonstrated only moderate success in treating cancer patients [54]. Subsequent studies discovered that DFMO-inhibition of polyamine biosynthesis leads to upregulation of PTS activity with resulting increased uptake of polyamines from the diet and gut flora into the tumor cells [18]. Our findings show that DFMO induces PTS activity and increases AP sensitivity in melanoma cells that harbor a mutated BRAF protein. In summary, treatment with AP, with or without DFMO, offers an exciting potential as adjunct cancer therapy to overcome drug resistance in mutant BRAF^{V600E} melanoma.

Supplementary Materials: The following are available online at www.mdpi.com/2076-3271/6/1/3/s1, Figure S1: Greater PTS activity and increased sensitivity to AP in BRAF^{V600E} human melanoma cells compared to BRAF^{WT} cells.

Acknowledgments: We thank Meenhard Herlyn (The Wistar Institute, Philadelphia, PA, USA) and Marcus Bosenburg (Yale University, New Haven, CT, USA) for kindly providing melanoma cell lines. This work was supported by the United States Department of Defense grant CA150356 (S.K.G.).

Author Contributions: S.K.G. conceived and designed the experiments; M.C.P. and A.M. performed the experiments; S.K.G., M.C.P. and A.M. analyzed the data; O.P. provided materials and edited the paper; and S.K.G. wrote the paper.

References

1. Davies, H.; Bignell, G.R.; Cox, C.; Stephens, P.; Edkins, S.; Clegg, S.; Teague, J.; Woffendin, H.; Garnett, M.J.; Bottomley, W.; et al. Mutations of the *BRAF* gene in human cancer. *Nature* **2002**, *417*, 949–954. [CrossRef] [PubMed]

2. Solit, D.B.; Rosen, N. Resistance to *BRAF* inhibition in melanomas. *N. Engl. J. Med.* **2011**, *364*, 772–774. [CrossRef] [PubMed]

3. Haq, R.; Fisher, D.E. Targeting melanoma by small molecules: Challenges ahead. *Pigment Cell Melanoma Res.* **2013**, *26*, 464–469. [CrossRef]

4. Flaherty, K.T.; Puzanov, I.; Kim, K.B.; Ribas, A.; McArthur, G.A.; Sosman, J.A.; O'Dwyer, P.J.; Lee, R.J.; Grippo, J.F.; Nolop, K.; et al. Inhibition of mutated, activated *BRAF* in metastatic melanoma. *N. Engl. J. Med.* **2010**, *363*, 809–819. [CrossRef]

5. Sosman, J.A.; Kim, K.B.; Schuchter, L.; Gonzalez, R.; Pavlick, A.C.; Weber, J.S.; McArthur, G.A.; Hutson, T.E.; Moschos, S.J.; Flaherty, K.T.; et al. Survival in *BRAF* V600-mutant advanced melanoma treated with vemurafenib. *N. Engl. J. Med.* **2012**, *366*, 707–714. [CrossRef]

6. Hauschild, A.; Grob, J.J.; Demidov, L.V.; Jouary, T.; Gutzmer, R.; Millward, M.; Rutkowski, P.; Blank, C.U.; Miller, W.H., Jr.; Kaempgen, E.; et al. Dabrafenib in *BRAF*-mutated metastatic melanoma: A multicentre, open-label, phase 3 randomised controlled trial. *Lancet* **2012**, *380*, 358–365. [CrossRef]

7. Bello-Fernandez, C.; Packham, G.; Cleveland, J.L. The ornithine decarboxylase gene is a transcriptional target of c-Myc. *Proc. Natl. Acad. Sci. USA* **1993**, *90*, 7804–7808. [CrossRef]

8. Casero, R.A., Jr.; Marton, L.J. Targeting polyamine metabolism and function in cancer and other hyperproliferative diseases. *Nat. Rev. Drug Discov.* **2007**, *6*, 373–390. [CrossRef]

9. Forshell, T.P.; Rimpi, S.; Nilsson, J.A. Chemoprevention of B-cell lymphomas by inhibition of the Myc target spermidine synthase. *Cancer Prev. Res.* **2010**, *3*, 140–147. [CrossRef]

10. Origanti, S.; Shantz, L.M. Ras transformation of RIE-1 cells activates cap-independent translation of ornithine decarboxylase: Regulation by the Raf/MEK/ERK and phosphatidylinositol 3-kinase pathways. *Cancer Res.* **2007**, *67*, 4834–4842. [CrossRef]

11. Poulin, R.; Casero, R.A.; Soulet, D. Recent advances in the molecular biology of metazoan polyamine transport. *Amino Acids* **2012**, *42*, 711–723. [CrossRef] [PubMed]
12. Bachrach, U.; Seiler, N. Formation of acetylpolyamines and putrescine from spermidine by normal and transformed chick embryo fibroblasts. *Cancer Res.* **1981**, *41*, 1205–1208. [PubMed]
13. Chang, B.K.; Libby, P.R.; Bergeron, R.J.; Porter, C.W. Modulation of polyamine biosynthesis and transport by oncogene transfection. *Biochem. Biophys. Res. Commun.* **1988**, *157*, 264–270. [CrossRef]
14. Roy, U.K.; Rial, N.S.; Kachel, K.L.; Gerner, E.W. Activated *K*-RAS increases polyamine uptake in human colon cancer cells through modulation of caveolar endocytosis. *Mol. Carcinog.* **2008**, *47*, 538–553. [CrossRef] [PubMed]
15. Pegg, A.E. Polyamine metabolism and its importance in neoplastic growth as a target for chemotherapy. *Cancer Res.* **1988**, *48*, 759–774. [PubMed]
16. Tabor, C.W.; Tabor, H. Polyamines. *Ann. Rev. Biochem.* **1984**, *53*, 749–790. [CrossRef] [PubMed]
17. Pegg, A.E. Recent advances in the biochemistry of polyamines in eukaryotes. *Biochem. J.* **1986**, *234*, 249–262. [CrossRef] [PubMed]
18. Gerner, E.W.; Meyskens, F.L., Jr. Polyamines and cancer: Old molecules, new understanding. *Nat. Rev. Cancer* **2004**, *4*, 781–792. [CrossRef] [PubMed]
19. Wallace, H.M.; Fraser, A.V.; Hughes, A. A perspective of polyamine metabolism. *Biochem. J.* **2003**, *376*, 1–14. [CrossRef] [PubMed]
20. Seiler, N.; Delcros, J.G.; Moulinoux, J.P. Polyamine transport in mammalian cells. An update. *Int. J. Biochem. Cell Biol.* **1996**, *28*, 843–861. [CrossRef]
21. Muth, A.; Kamel, J.; Kaur, N.; Shicora, A.C.; Ayene, I.S.; Gilmour, S.K.; Phanstiel, O. Development of polyamine transport ligands with improved metabolic stability and selectivity against specific human cancers. *J. Med. Chem.* **2013**, *56*, 5819–5828. [CrossRef] [PubMed]
22. Phanstiel, O.; Kaur, N.; Delcros, J.G. Structure-activity investigations of polyamine-anthracene conjugates and their uptake via the polyamine transporter. *Amino Acids* **2007**, *33*, 305–313. [CrossRef] [PubMed]
23. Kramer, D.L.; Miller, J.T.; Bergeron, R.J.; Khomutov, R.; Khomutov, A.; Porter, C.W. Regulation of polyamine transport by polyamines and polyamine analogs. *J. Cell. Physiol.* **1993**, *155*, 399–407. [CrossRef] [PubMed]
24. Nilsson, J.A.; Keller, U.B.; Baudino, T.A.; Yang, C.; Norton, S.; Old, J.A.; Nilsson, L.M.; Neale, G.; Kramer, D.L.; Porter, C.W.; et al. Targeting ornithine decarboxylase in Myc-induced lymphomagenesis prevents tumor formation. *Cancer Cell* **2005**, *7*, 433–444. [CrossRef] [PubMed]
25. Schayowitz, A.; Bertenshaw, G.; Jeffries, E.; Schatz, T.; Cotton, J.; Villanueva, J.; Herlyn, M.; Krepler, C.; Vultur, A.; Xu, W. Functional profiling of live melanoma samples using a novel automated platform. *PLoS ONE* **2012**, *7*, e52760. [CrossRef] [PubMed]
26. Alhonen-Hongisto, L.; Seppanen, P.; Janne, J. Intracellular putrescine and spermidine deprivation induces increased uptake of the natural polyamines and methylglyoxal bis(guanylhydrazone). *Biochem. J.* **1980**, *192*, 941–945. [CrossRef] [PubMed]
27. Obenauf, A.C.; Zou, Y.; Ji, A.L.; Vanharanta, S.; Shu, W.; Shi, H.; Kong, X.; Bosenberg, M.C.; Wiesner, T.; Rosen, N.; et al. Therapy-induced tumour secretomes promote resistance and tumour progression. *Nature* **2015**, *520*, 368–372. [CrossRef] [PubMed]
28. Yoshii, Y.; Waki, A.; Yoshida, K.; Kakezuka, A.; Kobayashi, M.; Namiki, H.; Kuroda, Y.; Kiyono, Y.; Yoshii, H.; Furukawa, T. The use of nanoimprinted scaffolds as 3D culture models to facilitate spontaneous tumor cell migration and well-regulated spheroid formation. *Biomaterials* **2011**, *32*, 6052–6058. [CrossRef] [PubMed]
29. Yamada, K.M.; Cukierman, E. Modeling tissue morphogenesis and cancer in 3D. *Cell* **2007**, *130*, 601–610. [CrossRef] [PubMed]
30. Haycock, J.W. 3D cell culture: A review of current approaches and techniques. In *3D Cell Culture: Methods and Protocols*; Springer: Berlin, Germany, 2011; pp. 1–15.
31. Qin, Y.; Roszik, J.; Chattopadhyay, C.; Hashimoto, Y.; Liu, C.; Cooper, Z.A.; Wargo, J.A.; Hwu, P.; Ekmekcioglu, S.; Grimm, E.A. Hypoxia-driven mechanism of vemurafenib resistance in melanoma. *Mol. Cancer Ther.* **2016**, *15*, 2442–2454. [CrossRef] [PubMed]
32. Cufi, S.; Vazquez-Martin, A.; Oliveras-Ferraros, C.; Martin-Castillo, B.; Vellon, L.; Menendez, J.A. Autophagy positively regulates the CD44$^+$ CD24$^{-/\text{low}}$ breast cancer stem-like phenotype. *Cell Cycle* **2011**, *10*, 3871–3885. [CrossRef] [PubMed]

33. Mirzoeva, O.K.; Hann, B.; Hom, Y.K.; Debnath, J.; Aftab, D.; Shokat, K.; Korn, W.M. Autophagy suppression promotes apoptotic cell death in response to inhibition of the PI3K—mTOR pathway in pancreatic adenocarcinoma. *J. Mol. Med.* **2011**, *89*, 877–889. [CrossRef] [PubMed]

34. Morselli, E.; Galluzzi, L.; Kepp, O.; Marino, G.; Michaud, M.; Vitale, I.; Maiuri, M.C.; Kroemer, G. Oncosuppressive functions of autophagy. *Antioxid. Redox Signal.* **2011**, *14*, 2251–2269. [CrossRef] [PubMed]

35. Morselli, E.; Marino, G.; Bennetzen, M.V.; Eisenberg, T.; Megalou, E.; Schroeder, S.; Cabrera, S.; Benit, P.; Rustin, P.; Criollo, A.; et al. Spermidine and resveratrol induce autophagy by distinct pathways converging on the acetylproteome. *J. Cell Biol.* **2011**, *192*, 615–629. [CrossRef] [PubMed]

36. Dallavalle, S.; Giannini, G.; Alloatti, D.; Casati, A.; Marastoni, E.; Musso, L.; Merlini, L.; Morini, G.; Penco, S.; Pisano, C.; et al. Synthesis and cytotoxic activity of polyamine analogues of camptothecin. *J. Med. Chem.* **2006**, *49*, 5177–5186. [CrossRef] [PubMed]

37. Wang, H.; Davis, A.; Yu, S.; Ahmed, K. Response of cancer cells to molecular interruption of the CK2 signal. *Mol. Cell Biochem.* **2001**, *227*, 167–174. [CrossRef] [PubMed]

38. Svensson, K.J.; Welch, J.E.; Kucharzewska, P.; Bengtson, P.; Bjurberg, M.; Pahlman, S.; Ten Dam, G.B.; Persson, L.; Belting, M. Hypoxia-mediated induction of the polyamine system provides opportunities for tumor growth inhibition by combined targeting of vascular endothelial growth factor and ornithine decarboxylase. *Cancer Res.* **2008**, *68*, 9291–9301. [CrossRef] [PubMed]

39. Mozdzan, M.; Szemraj, J.; Rysz, J.; Stolarek, R.A.; Nowak, D. Anti-oxidant activity of spermine and spermidine re-evaluated with oxidizing systems involving iron and copper ions. *Int. J. Biochem. Cell Biol.* **2006**, *38*, 69–81. [CrossRef] [PubMed]

40. Pouyssegur, J.; Dayan, F.; Mazure, N.M. Hypoxia signalling in cancer and approaches to enforce tumour regression. *Nature* **2006**, *441*, 437–443. [CrossRef] [PubMed]

41. Keith, B.; Simon, M.C. Hypoxia-inducible factors, stem cells, and cancer. *Cell* **2007**, *129*, 465–472. [CrossRef] [PubMed]

42. O'Connell, M.P.; Marchbank, K.; Webster, M.R.; Valiga, A.A.; Kaur, A.; Vultur, A.; Li, L.; Herlyn, M.; Villanueva, J.; Liu, Q. Hypoxia induces phenotypic plasticity and therapy resistance in melanoma via the tyrosine kinase receptors ROR1 and ROR2. *Cancer Discov.* **2013**, *3*, 1378–1393. [CrossRef] [PubMed]

43. Pucciarelli, D.; Lengger, N.; Takáčová, M.; Csaderova, L.; Bartosova, M.; Breiteneder, H.; Pastorekova, S.; Hafner, C. Hypoxia increases the heterogeneity of melanoma cell populations and affects the response to vemurafenib. *Mol. Med. Rep.* **2016**, *13*, 3281–3288. [CrossRef] [PubMed]

44. Roesch, A.; Fukunaga-Kalabis, M.; Schmidt, E.C.; Zabierowski, S.E.; Brafford, P.A.; Vultur, A.; Basu, D.; Gimotty, P.; Vogt, T.; Herlyn, M. A temporally distinct subpopulation of slow-cycling melanoma cells is required for continuous tumor growth. *Cell* **2010**, *141*, 583–594. [CrossRef] [PubMed]

45. Roesch, A.; Vultur, A.; Bogeski, I.; Wang, H.; Zimmermann, K.M.; Speicher, D.; Korbel, C.; Laschke, M.W.; Gimotty, P.A.; Philipp, S.E.; et al. Overcoming intrinsic multidrug resistance in melanoma by blocking the mitochondrial respiratory chain of slow-cycling JARID1B^{high} cells. *Cancer Cell* **2013**, *23*, 811–825. [CrossRef] [PubMed]

46. Schwab, L.P.; Peacock, D.L.; Majumdar, D.; Ingels, J.F.; Jensen, L.C.; Smith, K.D.; Cushing, R.C.; Seagroves, T.N. Hypoxia inducible factor-1α promotes primary tumor growth and tumor-initiating cell activity in breast cancer. *Breast Cancer Res.* **2012**, *14*, R6. [CrossRef] [PubMed]

47. Mathieu, J.; Zhang, Z.; Zhou, W.; Wang, A.J.; Heddleston, J.M.; Pinna, C.M.; Hubaud, A.; Stadler, B.; Choi, M.; Bar, M. HIF induces human embryonic stem cell markers in cancer cells. *Cancer Res.* **2011**, *71*, 4640–4652. [CrossRef] [PubMed]

48. Mohyeldin, A.; Garzon-Muvdi, T.; Quinones-Hinojosa, A. Oxygen in stem cell biology: A critical component of the stem cell niche. *Cell Stem Cell* **2010**, *7*, 150–161. [CrossRef] [PubMed]

49. Eisenberg, T.; Knauer, H.; Schauer, A.; Buttner, S.; Ruckenstuhl, C.; Carmona-Gutierrez, D.; Ring, J.; Schroeder, S.; Magnes, C.; Antonacci, L.; et al. Induction of autophagy by spermidine promotes longevity. *Nat. Cell Biol.* **2009**, *11*, 1305–1314. [CrossRef] [PubMed]

50. Chen, T.; Shen, L.; Yu, J.; Wan, H.; Guo, A.; Chen, J.; Long, Y.; Zhao, J.; Pei, G. Rapamycin and other longevity-promoting compounds enhance the generation of mouse induced pluripotent stem cells. *Aging Cell* **2011**, *10*, 908–911. [CrossRef] [PubMed]

51. Strohecker, A.M.; White, E. Targeting mitochondrial metabolism by inhibiting autophagy in *BRAF*-driven cancers. *Cancer Discov.* **2014**, *4*, 766–772. [CrossRef] [PubMed]

52. Kroemer, G.; Marino, G.; Levine, B. Autophagy and the integrated stress response. *Mol. Cell* **2010**, *40*, 280–293. [CrossRef] [PubMed]
53. Ma, X.H.; Piao, S.F.; Dey, S.; McAfee, Q.; Karakousis, G.; Villanueva, J.; Hart, L.S.; Levi, S.; Hu, J.; Zhang, G.; et al. Targeting ER stress-induced autophagy overcomes *BRAF* inhibitor resistance in melanoma. *J. Clin. Investig.* **2014**, *124*, 1406–1417. [CrossRef] [PubMed]
54. Seiler, N. Thirty years of polyamine-related approaches to cancer therapy. Retrospect and prospect. Part 1. Selective enzyme inhibitors. *Curr. Drug Targets* **2003**, *4*, 537–564. [CrossRef] [PubMed]

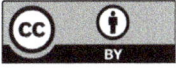

medical
sciences

MDPI

Article

The ODC 3′-Untranslated Region and 5′-Untranslated Region Contain *cis*-Regulatory Elements: Implications for Carcinogenesis

Shannon L. Nowotarski [1,*] and Lisa M. Shantz [2]

[1] Division of Science, The Pennsylvania State University Berks Campus, Reading, PA 19610, USA; sln167@psu.edu

[2] Department of Cellular and Molecular Physiology, The Pennsylvania State University College of Medicine, Hershey, PA 17033, USA; lms17@psu.edu

* Correspondence: sln167@psu.edu; Tel.: +1-610-396-6005

Received: 16 November 2017; Accepted: 15 December 2017; Published: 22 December 2017

Abstract: It has been hypothesized that both the 3′-untranslated region (3′UTR) and the 5′-untranslated region (5′UTR) of the ornithine decarboxylase (ODC) mRNA influence the expression of the ODC protein. Here, we use luciferase expression constructs to examine the influence of both UTRs in keratinocyte derived cell lines. The ODC 5′UTR or 3′UTR was cloned into the pGL3 control vector upstream or downstream of the luciferase reporter gene, respectively, and luciferase activity was measured in both non-tumorigenic and tumorigenic mouse keratinocyte cell lines. Further analysis of the influence of the 3′UTR on luciferase activity was accomplished through site-directed mutagenesis and distal deletion analysis within this region. Insertion of either the 5′UTR or 3′UTR into a luciferase vector resulted in a decrease in luciferase activity when compared to the control vector. Deletion analysis of the 3′UTR revealed a region between bases 1969 and 2141 that was inhibitory, and mutating residues within that region increased luciferase activity. These data suggest that both the 5′UTR and 3′UTR of ODC contain *cis*-acting regulatory elements that control intracellular ODC protein levels.

Keywords: ornithine decarboxylase; polyamines; untranslated region

---

1. Introduction

Polyamines are small ubiquitously expressed polycations that are essential for normal cell growth and development [1,2]. Their positive charge allows them to bind to DNA, RNA, proteins, and acidic phospholipids [3]. Under normal physiological conditions, the concentration of polyamines is tightly regulated by biosynthetic, catabolic and poorly understood transport mechanisms [4,5]. Ornithine decarboxylase (ODC) is the first and usually rate-limiting enzyme in the polyamine biosynthetic pathway and converts the amino acid ornithine into the diamine putrescine, which is subsequently converted to the higher polyamines spermidine and spermine [6]. Changes in transcription, translation and protein degradation have all been shown to maintain ODC intracellular levels under normal physiological conditions [7–10]. Recently, our group described the post-transcriptional regulation of ODC through the RNA binding proteins (RBPs) HuR and TTP [11,12].

Post-transcriptional regulation occurs in a variety of transcripts and encompasses mRNA stability and mRNA translation efficiency with both the 3′ and 5′ untranslated regions (UTRs) playing a role in these processes [13,14]. Along with regulation due to secondary structure, 5′UTRs may contain a collection of regulatory elements such as upstream start codons (AUGs), internal open reading frames, and internal ribosome entry sites that affect translation initiation [15]. In addition, the 3′UTR regulates processes such as transcript cleavage, mRNA stability, mRNA localization, and translation [15].

Post-transcriptional regulation can be carried out by RNA binding proteins (RBPs), which can bind to adenosine- and uracil-rich elements (AREs) within either the 5′UTR or 3′UTR [16,17]. Classically, this sequence has been denoted as AUUUA. These sequences behave as *cis*-acting elements and are located in numerous proto-oncogene, cytokine, and transcription factor mRNAs as binding sites for RBPs [18–20].

Studies investigating ODC regulation have shown that both the 5′UTR and 3′UTR control ODC mRNA translation. The mammalian ODC 5′UTR is long, consisting of over 300 bases [21,22]. The size of this region, in conjunction with a high G-C content on the 5′ distal end, promotes the formation of secondary structure within the ODC 5′UTR [21–23]. In addition, the 5′UTR contains a short internal open reading frame that is located 150 bases upstream of the translational start site [21,22,24]. These features have been found to inhibit ODC translation [8,25]. In studies conducted in ODC-deficient Chinese hamster ovary (CHO) cells expressing a firefly luciferase reporter gene, the inhibitory nature of the ODC 5′UTR was partially released by the addition of the ODC 3′UTR [26]. The goal of the studies described here is to further these previous findings by investigating the influence of the ODC UTRs on luciferase activity in normal keratinocytes, which contain low levels of endogenous ODC, and keratinocyte-derived spindle carcinoma cells with high ODC activity.

Insertion of either the entire mouse ODC 3′UTR or 5′UTR into a luciferase control plasmid resulted in decreased luciferase activity in both C5N keratinocytes and A5 spindle carcinoma cells when compared to cells that contained only the luciferase open reading frame. Deletion analysis identified the region between bases 1969 and 2141 in the ODC 3′UTR as inhibitory. Moreover, mutation of the classical AUUUA sequence within the ODC 3′UTR significantly increased luciferase activity. Overall, these studies identify this AUUUA sequence as a negative regulatory element within the ODC 3′UTR.

2. Materials and Methods

2.1. Cell Culture

The C5N and A5 mouse keratinocytes (a generous gift from Dr. Allan Balmain, UCSF, San Francisco, CA, USA) were cultured in Dulbecco's Modified Eagle's Medium (DMEM) (Life Technologies, Carlsbad, CA, USA) supplemented with 10% fetal bovine serum (Atlanta Biologicals, Lawrenceville, GA, USA), 1% penicillin streptomycin, and 1% glutamine (Life Technologies). These cells have been described previously [27]. Passages 5–20 were used in the experiments, and experimental results were consistent regardless of passage number. Stock flasks were incubated at 37 °C in a humidified atmosphere of 95% air/5% CO_2. Cells were passaged one time after thawing and before use.

2.2. ODC 3′UTR and 5′UTR Luciferase Assays

The mouse ODC 3′UTR (NM_013614) was cloned into the pGL3 control vector (Promega, Madison, WI, USA) and placed downstream of the firefly luciferase reporter gene while the ODC 5′UTR (NM_013614) was cloned into the pGL3 control vector (Promega) and placed upstream of the firefly luciferase reporter gene. These plasmids are denoted pODC3′UTRLuc and pODC5′UTRLuc respectively. Additional changes in the ODC 3′UTR include a 381 base pair truncation of the distal ODC 3′UTR denoted herein as ARE03 and a 553 base pair truncation of the distal ODC 3′UTR denoted ARE02, cloned into the pGL3 control vector downstream of the firefly luciferase reporter gene (Figure 1). Site-directed mutagenesis was conducted to change the AUUUA sequence present in the ARE02 vector to GGGUA using the Stratagene Quikchange Site-directed Mutagenesis Kit as per the manufacturer's instructions (Stratagene, La Jolla, CA, USA) (Figure 1). This mutation was introduced to determine whether the putative RBP binding site influenced luciferase activity. The primers used to create the AUUUA to GGGUA mutation were:

5′-GGCATTTGGGGGGACCGGGTAACTTAATTACTGCTAGTTTGG-3′ (sense);
5′-CCAAACTAGCAGTAATTAAGTTACCCGGTCCCCCCAAATGCC-3′ (antisense).

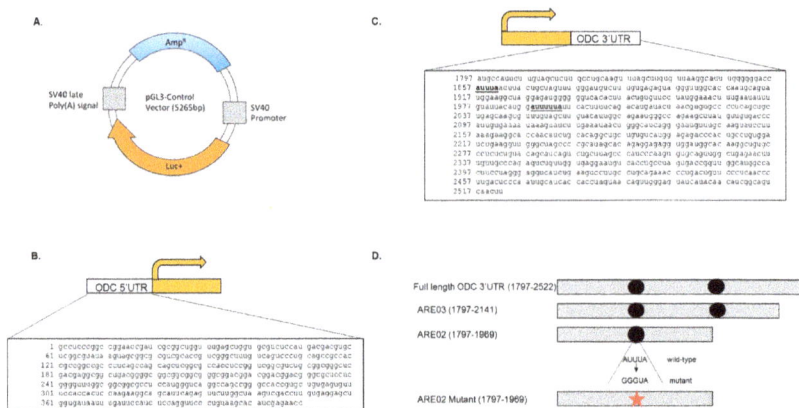

Figure 1. Luciferase constructs used to measure the influence of the ornithine decarboxylase (ODC) untranslated regions (UTRs) on luciferase activity. (**A**) Schematic of the base pGL3 control vector (adapted from Promega). The ODC 5′UTR and 3′UTR sequences were ligated immediately upstream or downstream of the luciferase open reading frame (Luc+), respectively. (**B**) Schematic of the ODC 5′UTR luciferase plasmid (pODC5′UTRLuc) and sequence of the complete ODC 5′UTR that was cloned upstream of the luciferase reporter gene. (**C**) Schematic of the full length ODC 3′UTR luciferase plasmid (pODC3′UTRLuc) and the complete ODC 3′UTR sequence that was cloned downstream of the luciferase reporter gene. Putative adenosine- and uracil-rich elements (AREs) are in bold and underlined. (**D**) Schematic of the full length ODC 3′UTR and distal end truncation constructs used in the luciferase experiments. The full length ODC 3′UTR vector was comprised of bases 1797–2522, the ARE03 vector was comprised of bases 1797–2141, and the ARE02 vectors were comprised of bases 1797–1969. Black circles indicate the location of putative ARE sequences. Star denotes the location of the AUUUA to GGGUA mutation in the ARE02 vector.

The AUUUA to GGGUA mutation was validated by sequencing. For all luciferase assays, cells were transfected at 70% confluence with 2 µg per plate of vector using the Lipofectamine 2000 transfection reagent as per the manufacturer's protocol (Life Technologies). Mock transfected cells were treated with Lipofectamine 2000 only. The pRL-SV40 renilla reporter plasmid (Promega) was transfected at 0.2 µg per plate in order to act as a transfection efficiency control. Forty-eight h post-transfection, cells were harvested in $1\times$ Passive Lysis Buffer and assayed using the Dual-Luciferase Kit as per manufacturer's instructions (Promega). For each sample, the firefly luciferase activity was normalized to the renilla luciferase activity, and the data were expressed as the firefly/renilla ratio. The data were normalized to the pGL3 control.

2.3. Statistics

Results are expressed as means $\pm$ standard errors (SE) from three to nine samples. Statistical analysis was performed using Student's unpaired *t*-test on the Graphpad webtool. *p*-values of <0.05 were considered significant.

3. Results

3.1. The ODC 5′UTR and 3′UTR Decrease Expression of the Luciferase Reporter Gene

To study the influence of the ODC UTRs on luciferase activity, we transfected both normal keratinocyte C5N cells and spindle carcinoma A5 cells with either the parental pGL3 control vector, pODC3′UTRLuc, or pODC5′UTRLuc. We hypothesized that both the 3′UTR and 5′UTR of ODC would greatly reduce the luciferase activity in both cell lines. The insertion of the full-length ODC 3′UTR resulted in an approximate 65% decrease in luciferase activity in both non-tumorigenic C5N and tumorigenic A5 cells (Figure 2).

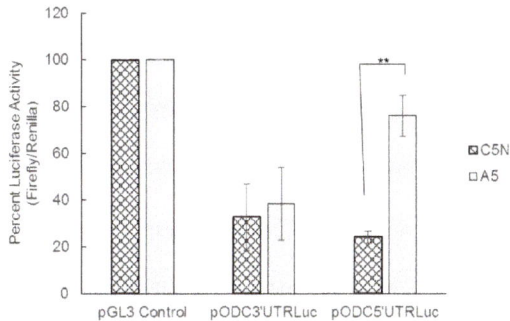

Figure 2. Insertion of the ODC UTRs in the pGL3 control vector causes a decrease in luciferase activity. C5N and A5 cells were transfected with either the pGL3 control vector, pODC3′UTRLuc, or pODC5′UTRLuc. A plasmid containing the renilla luciferase gene was co-transfected into these cells and used as a transfection efficiency control. Luciferase activity was measured 48 h after transfection. Firefly luciferase activity was normalized to renilla luciferase for each sample. The luciferase activity of pGL3 control was set to 100% and the samples from cells transfected with pODC3′UTRLuc and pODC5′UTRLuc are shown as a percentage of the pGL3 control luciferase activity. Values are means ± S.E. ($n = 9$). The difference in luciferase activity between the pGL3 control and both pODC3′UTRLuc and pODC5′UTRLuc were statistically significant for each cell line. ** $p < 0.005$.

Insertion of the full-length 5′UTR of ODC resulted in a more dramatic inhibition of luciferase activity in C5N cells when compared to A5 cells. C5N cells showed a reduction in luciferase activity of approximately 80% compared to control, which is similar to previous results in CHO cells [26]. Unexpectedly, luciferase activity in A5 cells transfected with the ODC 5′UTR was reduced by only 20% when compared to the pGL3 control vector (Figure 2). Because the reporter constructs contain identical promoter and enhancing elements, these data suggest the presence of *cis*-acting negative regulatory elements within both the ODC 3′UTR and 5′UTR. Furthermore, the results are consistent with the presence of *trans*-acting factors for the ODC 5′UTR that either enhance expression in the A5 cells or inhibit it in C5N cells, since the ODC 5′UTR sequence is identical in the two cell lines (data not shown). This is in keeping with the higher ODC protein levels observed in A5 cells and would be consistent with an increase in ODC protein synthesis in these cells [12].

3.2. The ODC 3′UTR Contains a Negative Regulatory Element between Bases 1969 and 2141

We decided to focus on the influence of the ODC 3′UTR because the effects of the ODC 5′UTR on ODC expression are well-described and because our previous results show that the ODC 3′UTR is important for ODC post-transcriptional regulation through both *cis* and *trans*-acting factors [11,12,26,28]. Given that the luciferase activity was similar between both cell lines, we decided to focus our 3′UTR studies on the C5N cells. To fully understand the impact of the ODC 3′UTR, we measured the luciferase activity in cells that had been transfected with either the full length ODC

3′UTR or two distal end truncation constructs, ARE02 (bases 1797–1969) and ARE03 (bases 1797–2141), both of which contain the classical putative ARE AUUUA (Figures 1 and 3).

While the presence of the full length ODC 3′UTR caused an 80% decrease in luciferase activity compared to control, the ARE02 deletion mutant attenuated this inhibition to 50%. Interestingly, luciferase activity was undetectable in cells transfected with ARE03 (Figure 3), which is 172 bases longer than ARE02. These results indicate that a negative *cis*-regulatory element resides between bases 1969 and 2141. Moreover, the data suggest that the most distal region of the ODC 3′UTR (bases 2141–2522) contains positive regulatory elements, since we see a rescue in luciferase activity when we compare the full length ODC 3′UTR and ARE03.

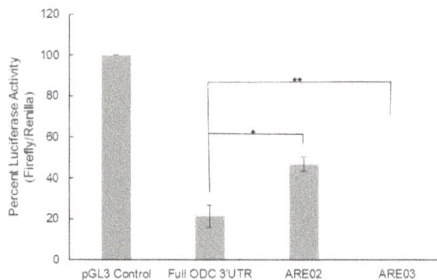

Figure 3. The ODC 3′UTR contains a negative regulatory element between bases 1969 and 2141. The full length ODC 3′UTR as well as two distal end truncations of the ODC 3′UTR were inserted into the pGL3 control vector in order to show the influence of the 3′UTR of ODC on luciferase activity. C5N cells were transfected with either the pGL3 control vector, pODC3′UTRLuc (Full ODC 3′UTR), ARE02 or ARE03. A plasmid containing the renilla luciferase gene was co-transfected into these cells and used as a transfection efficiency control. Luciferase activity was measured 48 h after transfection. Firefly luciferase activity was normalized to renilla luciferase for each sample. The luciferase activity of pGL3 control was set to 100% and the samples from cells transfected with pODC3′UTRLuc (Full ODC 3′UTR), ARE02 and ARE03 are shown as a percentage of the pGL3 control luciferase activity. Values are means $\pm$ S.E. (n = 6). Differences in luciferase activity between pGL3 control and Full ODC 3′UTR, ARE02 and ARE03 were all statistically significant. Statistics were performed to compare the luciferase activity between Full ODC 3′UTR, ARE02, and ARE03. * $p < 0.05$ and ** $p < 0.005$.

3.3. Mutation of the AUUUA Classical ARE Dramatically Increases Luciferase Activity

To further elucidate the influence of the AUUUA sequence on luciferase activity we investigated the effect of mutating the AUUUA ARE site on the ODC 3′UTR (Figure 1). These experiments were conducted to verify that the AUUUA site was indeed a *cis*-acting regulatory element on the ODC 3′UTR. Using site-directed mutagenesis we mutated the AUUUA in the ARE02 vector to GGGUA. In C5N cells, the full length ODC 3′UTR exhibited an 85% reduction in luciferase activity compared to pGL3 control, while expression of the ARE02 construct containing wild-type AUUUA resulted in a 45% reduction in luciferase activity. Interestingly, the GGGUA mutant ARE02 construct produced higher luciferase activity than the pGL3 control. These results suggest that the AUUUA sequence acts as an inhibitory *cis*-acting regulatory element within the ODC 3′UTR (Figure 4).

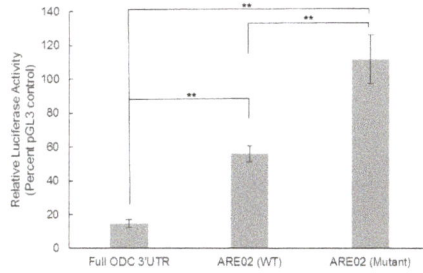

Figure 4. Mutating the AUUUA ARE to GGGUA increases the luciferase activity in non-tumorigenic C5N keratinocytes. Site directed mutagenesis was used to change the AUUUA classical ARE sequence to GGGUA in the ARE02 vector. Sequencing confirmed this 3 base mutation. C5N cells were transfected with either the pGL3 control vector, pODC3'UTRLuc (Full ODC 3'UTR), the ARE02 wild-type plasmid (WT), or the ARE02 mutant plasmid. A plasmid containing the renilla luciferase gene was co-transfected into these cells and used as a transfection efficiency control. Luciferase activity was measured 48 h after transfection. Firefly luciferase activity was normalized to renilla luciferase for each sample. The luciferase activity of pGL3 control was set to 100% and the samples from cells transfected with pODC3'UTRLuc (Full ODC 3'UTR), ARE02 WT, and ARE02 mutant constructs are shown as a percentage of the pGL3 control luciferase activity. Values are means $\pm$ S.E. ($n = 9$). ** $p < 0.005$ for all comparisons.

4. Discussion

We have previously determined that A5 spindle carcinoma cells have much higher levels of ODC activity and protein than C5N keratinocytes [12]. We therefore decided to investigate the influence of the ODC UTRs in these cell lines using luciferase constructs containing either the wild-type ODC UTR sequences, or a variety of mutations and deletions within the UTRs. In agreement with previous studies, we show that the insertion of the 5'UTR or 3'UTR of ODC into the pGL3 luciferase control vector causes a reduction in the luciferase activity (Figure 2) [26,28]. The level of 3'UTR-influenced luciferase activity inhibition is similar in both C5N keratinocytes and A5 spindle carcinoma cells; however the effect of the 5'UTR on luciferase activity is markedly different between the two cell lines, suggesting that different *trans*-acting factors within these cell lines play a role in ODC regulation. These data fit with our previous work showing that ODC enzyme activity is higher in A5 cells when compared to C5N cells and suggest that a possible mechanism for this difference is altered post-transcriptional control of the ODC mRNA [12].

We observed that insertion of the full length ODC 3'UTR dramatically reduced luciferase activity. Moreover, an ODC 3'UTR distal end truncation of 553 bases to create ARE02 mitigates the repression caused by the full length ODC 3'UTR. In contrast, the ARE03 construct (1797–2141) completely ablated the luciferase activity, and addition of the most distal end of the ODC 3'UTR (2141–2522) relieves some of this repression. This is observed when we compare the luciferase activities between ARE03 and the full length ODC 3'UTR (Figure 3). These data can be interpreted two ways, which are not mutually exclusive. First, there are multiple positive and negative regulatory elements along the ODC 3'UTR that may work in concert to regulate the ODC mRNA transcript. Alternately, the secondary structure of the ODC 3'UTR is altered in the truncation constructs, which affects the binding of *trans*-acting factors. We are currently investigating these two possible mechanisms of regulation. Moreover, we are currently investigating the influence of the AUUUUA sequence on ODC regulation. This non-classical ARE is part of both the full ODC 3'UTR and ARE03 constructs (Figure 1) and may contribute to the inhibitory effect of these two constructs by regulating ODC mRNA stability.

Mutation of the ARE sequence AUUUA to GGGUA within the ODC 3'UTR resulted in a significant induction of luciferase activity. In fact, the luciferase activity observed in C5N cells that had been transfected with the ARE02 mutant was higher than the luciferase activity in cells transfected with the control vector (Figure 4). These data are in agreement with previous studies using the mouse COX-2 3′UTR, which showed that the removal of the first 60 nucleotides, which contained 7 out of 12 AUUUA sequences, caused an increase in luciferase activity, demonstrating that the AUUUA consensus sequence was inhibitory [29]. Similarly, our results suggest that the AUUUA sequence is an inhibitory element on the ODC 3′UTR.

We previously used Ras-transformed RIE-1 cells, which are characterized by high levels of ODC [30,31] to demonstrate that the RBP HuR bound more strongly to the ARE02 mutant than the wild-type sequence. These findings support our current luciferase activity data as the binding of HuR to the ARE02 mutant would stabilize the mRNA and lead to a higher luciferase activity.

MicroRNAs (MiRs) behave as *trans*-acting factors that are involved in the post-transcriptional regulation of mRNAs by imperfectly binding to a target mRNAs 3′UTR. This binding typically results in accelerated mRNA turnover and a decrease in mRNA translation [32]. MiRNA databases predict that numerous miRs can bind to the ODC 3′UTR, and it has been found that ultraviolet-B (UVB), the primary carcinogen in non-melanoma skin cancer (NMSC), alters the expression of numerous miRs in mice [33]. This has also been demonstrated in human patients who suffer from NMSC [34]. We are currently using our mouse keratinocyte cell models used here as well as a human keratinocyte cell model (HaCaT cells) treated with apoptotic doses of UVB to address the influence of predicted miRs on the post-transcriptional regulation of ODC. We believe these studies will give us better insight into the complex regulation of ODC during the process of skin carcinogenesis.

In conclusion, the studies described here compliment other work investigating regulation by the ODC UTRs [26,30]. We demonstrate using a series of luciferase reporter plasmids that both the ODC 3′UTR and 5′UTR influence luciferase activity, and establish that both positive and negative regulatory elements are contained within the UTRs of ODC. Moreover, we show that a specific site on the ODC 3′UTR, the putative ARE sequence AUUUA, is inhibitory in a keratinocyte model. These studies further our understanding of ODC regulation at the post-transcriptional level. We are currently investigating the mechanisms of ODC 3′UTR and 5′UTR mediated regulation, and how a specific RBP, HuR, is involved in these processes, both in normal keratinocytes and cutaneous carcinomas. We hope to elucidate the binding sequence for this RBP and further investigate the interplay between HuR and other *trans*-acting factors.

Acknowledgments: These studies were funded by NIH grants ES19242, CA142051 and the Pennsylvania Department of Health Tobacco CURE funds (to LMS).

Author Contributions: S.L.N. and L.M.S. conceived and designed the experiments; S.L.N. performed the experiments; S.L.N. and L.M.S. analyzed the data; S.L.N. and L.M.S. contributed reagents/materials/analysis tools; S.L.N. wrote the manuscript, which was reviewed and edited by L.M.S.

Conflicts of Interest: The authors declare no conflict of interest. The founding sponsors had no role in the design of the study; in the collection, analyses, or interpretation of data; in the writing of the manuscript, and in the decision to publish the results.

References

1. Fozard, J.R.; Part, M.L.; Prakash, N.J.; Grove, J. Inhibition of murine embryonic development by alpha-difluoromethylornithine, an irreversible inhibitor of ornithine decarboxylase. *Eur. J. Pharmacol.* **1980**, *65*, 379–391. [CrossRef]
2. Pendeville, H.; Carpino, N.; Marine, J.C.; Takahashi, Y.; Muller, M.; Martial, J.A.; Cleveland, J.L. The ornithine decarboxylase gene is essential for cell survival during early murine development. *Mol. Cell. Biol.* **2001**, *21*, 6549–6558. [CrossRef] [PubMed]
3. Wallace, H.M.; Fraser, A.V.; Hughes, A. A perspective of polyamine metabolism. *Biochem. J.* **2003**, *376*, 1–14. [CrossRef] [PubMed]
4. Nowotarski, S.L.; Woster, P.M.; Casero, R.A., Jr. Polyamines and cancer: Implications for chemotherapy and chemoprevention. *Expert Rev. Mol. Med.* **2013**, *15*. [CrossRef] [PubMed]

5. Pegg, A.E. Mammalian polyamine metabolism and function. *IUBMB Life* **2009**, *61*, 880–894. [CrossRef] [PubMed]

6. Pegg, A.E. Regulation of ornithine decarboxylase. *J. Biol. Chem.* **2006**, *281*, 14529–14532. [CrossRef] [PubMed]

7. Bello-Fernandez, C.; Packham, G.; Cleveland, J.L. The ornithine decarboxylase gene is a transcriptional target of c-Myc. *Proc. Natl. Acad. Sci. USA* **1993**, *90*, 7804–7808. [CrossRef] [PubMed]

8. Shantz, L.M.; Pegg, A.E. Translational regulation of ornithine decarboxylase and other enzymes of the polyamine pathway. *Int. J. Biochem. Cell Biol.* **1999**, *31*, 107–122. [CrossRef]

9. Wallon, U.M.; Persson, L.; Heby, O. Regulation of ornithine decarboxylase during cell growth. Changes in the stability and translatability of the mRNA, and in the turnover of the protein. *Mol. Cell. Biochem.* **1995**, *146*, 39–44. [CrossRef] [PubMed]

10. Zhao, B.; Butler, A.P. Core promoter involvement in the induction of rat ornithine decarboxylase by phorbol esters. *Mol. Carcinogenes.* **2001**, *32*, 92–99. [CrossRef] [PubMed]

11. Nowotarski, S.L.; Origanti, S.; Sass-Kuhn, S.; Shantz, L.M. Destabilization of the ornithine decarboxylase mRNA transcript by the RNA-binding protein tristetraprolin. *Amino Acids* **2016**, *48*, 2303–2311. [CrossRef] [PubMed]

12. Nowotarski, S.L.; Shantz, L.M. Cytoplasmic accumulation of the RNA-binding protein HuR stabilizes the ornithine decarboxylase transcript in a murine nonmelanoma skin cancer model. *J. Biol. Chem.* **2010**, *285*, 31885–31894. [CrossRef] [PubMed]

13. Brooks, S.A.; Blackshear, P.J. Tristetraprolin (TTP): Interactions with mRNA and proteins, and current thoughts on mechanisms of action. *Biochim. Biophys. Acta* **2013**, *1829*, 666–679. [CrossRef] [PubMed]

14. Hinman, M.N.; Lou, H. Diverse molecular functions of Hu proteins. *Cell. Mol. Life Sci.* **2008**, *65*, 3168–3181. [CrossRef] [PubMed]

15. Barrett, L.W.; Fletcher, S.; Wilton, S.D. Regulation of eukaryotic gene expression by the untranslated gene regions and other non-coding elements. *Cell. Mol. Life Sci.* **2012**, *69*, 3613–3634. [CrossRef] [PubMed]

16. Parker, R.; Sheth, U. P bodies and the control of mRNA translation and degradation. *Mol. Cell* **2007**, *25*, 635–646. [CrossRef] [PubMed]

17. Griseri, P.; Pages, G. Control of pro-angiogenic cytokine mRNA half-life in cancer: The role of AU-rich elements and associated proteins. *J. Interferon Cytokine Res.* **2014**, *34*, 242–254. [CrossRef] [PubMed]

18. Audic, Y.; Hartley, R.S. Post-transcriptional regulation in cancer. *Biol. Cell* **2004**, *96*, 479–498. [CrossRef] [PubMed]

19. Liao, W.L.; Wang, W.-C.; Chang, W.-C.; Tseng, J.T. The RNA-binding protein HuR stabilizes cytosolic phospholipase A2α mRNA under interleukin-1β treatment in non-small cell lung cancer A549 Cells. *J. Biol. Chem.* **2011**, *286*, 35499–35508. [CrossRef] [PubMed]

20. Datta, S.; Novotny, M.; Pavicic, P.G., Jr.; Zhao, C.; Herjan, T.; Hartupee, J.; Hamilton, T. IL-17 regulates CXCL1 mRNA stability via an AUUUA/tristetraprolin-independent sequence. *J. Immunol.* **2010**, *184*, 1484–1491. [CrossRef] [PubMed]

21. Gupta, M.; Coffino, P. Mouse ornithine decarboxylase. Complete amino acid sequence deduced from cDNA. *J. Biol. Chem.* **1985**, *260*, 2941–2944. [PubMed]

22. Kahana, C.; Nathans, D. Translational regulation of mammalian ornithine decarboxylase by polyamines. *J. Biol. Chem.* **1985**, *260*, 15390–15393. [PubMed]

23. Pegg, A.E.; Shantz, L.M.; Coleman, C.S. Ornithine decarboxylase: Structure, function and translational regulation. *Biochem. Soc. Trans.* **1994**, *22*, 846–852. [CrossRef] [PubMed]

24. Wen, L.; Huang, J.K.; Blackshear, P.J. Rat ornithine decarboxylase gene. Nucleotide sequence, potential regulatory elements, and comparison to the mouse gene. *J. Biol. Chem.* **1989**, *264*, 9016–9021. [PubMed]

25. Van Steeg, H.; Van Oostrom, C.T.; Hodemaekers, H.M.; Peters, L.; Thomas, A.A. The translation in vitro of rat ornithine decarboxylase mRNA is blocked by its 5′ untranslated region in a polyamine-independent way. *Biochem. J.* **1991**, *274*, 521–526. [CrossRef] [PubMed]

26. Lorenzini, E.C.; Scheffler, I.E. Co-operation of the 5′ and 3′ untranslated regions of ornithine decarboxylase mRNA and inhibitory role of its 3′ untranslated region in regulating the translational efficiency of hybrid RNA species via cellular factor. *Biochem. J.* **1997**, *326*, 361–367. [CrossRef] [PubMed]

27. Zoumpourlis, V.; Solakidi, S.; Papathoma, A.; Papaevangeliou, D. Alterations in signal transduction pathways implicated in tumour progression during multistage mouse skin carcinogenesis. *Carcinogenesis* **2003**, *24*, 1159–1165. [CrossRef] [PubMed]

28. Manzella, J.M.; Blackshear, P.J. Regulation of rat ornithine decarboxylase mRNA translation by its 5′-untranslated region. *J. Biol. Chem.* **1990**, *265*, 11817–11822. [PubMed]

29. Cok, S.J.; Morrison, A.R. The 3′-untranslated region of murine cyclooxygenase-2 contains multiple regulatory elements that alter message stability and translational efficiency. *J. Biol. Chem.* **2001**, *276*, 23179–23185. [CrossRef] [PubMed]

30. Origanti, S.; Nowotarski, S.L.; Carr, T.D.; Sass-Kuhn, S.; Xiao, L.; Wang, J.Y.; Shantz, L.M. Ornithine decarboxylase mRNA is stabilized in an mTORC1-dependent manner in Ras-transformed cells. *Biochem. J.* **2012**, *442*, 199–207. [CrossRef] [PubMed]

31. Origanti, S.; Shantz, L.M. Ras transformation of RIE-1 cells activates cap-independent translation of ornithine decarboxylase: Regulation by the Raf/MEK/ERK and phosphatidylinositol 3-kinase pathways. *Cancer Res.* **2007**, *67*, 834–842. [CrossRef] [PubMed]

32. Liu, Q.; Paroo, Z. Biochemical principles of small RNA pathways. *Annu. Rev. Biochem.* **2010**, *79*, 295–319. [CrossRef] [PubMed]

33. Zhou, B.R.; Xu, Y.; Luo, D. Effect of UVB irradiation on microRNA expression in mouse epidermis. *Oncol. Lett.* **2012**, *3*, 560–564. [PubMed]

34. Sonkoly, E.; Lovén, J.; Xu, N.; Meisgen, F.; Wei, T.; Brodin, P.; Jaks, V.; Kasper, M.; Shimokawa, T.; Harada, M.; et al. MicroRNA-203 functions as a tumor suppressor in basal cell carcinoma. *Oncogenesis* **2012**, *1*. [CrossRef] [PubMed]

medical sciences

MDPI

Article

Investigation of Polyamine Metabolism and Homeostasis in Pancreatic Cancers

Chelsea Massaro, Jenna Thomas and Otto Phanstiel IV *

Department of Medical Education, College of Medicine, University of Central Florida, Orlando, FL 32826-3227, USA; clm12d@my.fsu.edu (C.M.); jenna.thomas97@Knights.ucf.edu (J.T.)
* Correspondence: otto.phanstiel@ucf.edu; Tel.: +1-407-823-6545

Received: 7 November 2017; Accepted: 5 December 2017; Published: 7 December 2017

Abstract: Pancreatic cancers are currently the fourth leading cause of cancer-related death and new therapies are desperately needed. The most common pancreatic cancer is pancreatic ductal adenocarcinoma (PDAC). This report describes the development of therapies, which effectively deplete PDAC cells of their required polyamine growth factors. Of all human tissues, the pancreas has the highest level of the native polyamine spermidine. To sustain their high growth rates, PDACs have altered polyamine metabolism, which is reflected in their high intracellular polyamine levels and their upregulated import of exogenous polyamines. To understand how these cancers respond to interventions that target their specific polyamine pools, L3.6pl human pancreatic cancer cells were challenged with specific inhibitors of polyamine biosynthesis. We found that pancreatic cell lines have excess polyamine pools, which they rebalance to address deficiencies induced by inhibitors of specific steps in polyamine biosynthesis (e.g., ornithine decarboxylase (ODC), spermidine synthase (SRM), and spermine synthase (SMS)). We also discovered that combination therapies targeting ODC, SMS, and polyamine import were the most effective in reducing intracellular polyamine pools and reducing PDAC cell growth. A combination therapy containing difluoromethylornithine (DFMO, an ODC inhibitor) and a polyamine transport inhibitor (PTI) were shown to significantly deplete intracellular polyamine pools. The additional presence of an SMS inhibitor as low as 100 nM was sufficient to further potentiate the DFMO + PTI treatment.

Keywords: polyamine; cancer; metabolism; difluoromethylornithine; polyamine transport inhibitor; pancreatic ductal adenocarcinoma

1. Introduction

Pancreatic ductal adenocarcinoma (PDAC) typically develops slowly, over many years, from precursor pancreatic intraepithelial neoplasias (PanINs) [1,2]. With a five-year survival rate of less than 8%, new therapies are desperately needed to address this fourth leading cause of cancer-related death. The most frequently mutated gene in human PDAC is KRAS (95%) and human PDAC cell lines often (50%) have a copy number gain in MYC (c-myc) [3]. Since the MYC gene family is activated in nearly 70% of all human cancers, targeting myc-driven processes could have broad applications. Both KRAS and MYC are upstream activators of polyamine metabolism, and their mutations are known to increase intracellular polyamine levels to presumably drive tumor growth [4].

At physiological pH, the native polyamines exist as polycations and can interact with anionic biomolecules like RNA, DNA, and proteins. As a result, polyamines have pleiotropic effects in cells and play critical roles in chromatin remodeling, transcription, translation, eIF5A activation, potassium channel regulation, and in the immune response [4]. The normal pancreas has the highest level of the native polyamine spermidine (Spd) of any mammalian tissue [5–7]. Our preliminary investigations revealed that PDAC cells are addicted to polyamines and have high intracellular polyamine levels and upregulated polyamine import activity [8,9]. Polyamines are charged and exist in millimolar levels inside cells, and maintaining polyamine homeostasis is critical to cell viability. As shown in Figure 1, homeostasis is tightly regulated via a balance between polyamine biosynthesis, catabolism, and transport.

This report investigated how PDAC cells respond to inhibition of discrete steps in polyamine biosynthesis and importation. While homeostatic control was expected where polyamine pools could be interconverted to replenish a specific polyamine deficiency, it was unclear as to which specific enzyme (or combination of enzymes) needed to be inhibited to affect PDAC cell growth. Understanding this paradigm is important as it provides an opportunity to design new therapies to treat these deadly cancers by effectively targeting their enhanced reliance upon polyamine growth factors. Specific inhibitors for the polyamine biosynthetic enzymes ornithine decarboxylase (ODC), spermidine synthase (SRM), and spermine synthase (SMS) are known: these are α-difluoromethylornithine (DFMO) [2], trans-4-methylcyclohexylamine (MCHA) [10], and N-cyclohexyl-1,3-diaminopropane (CDAP) [11], respectively (Figure 1). Our group also developed a polyamine transport inhibitor (PTI, trimer44, Figure 1), which competitively inhibits import of the native polyamines and works synergistically with DFMO to inhibit cell growth in vitro [9,12]. In this manner, we were able to block both polyamine biosynthesis and importation, and decrease specific intracellular polyamine pools.

By using single agents and combinations of these compounds, we investigated how these interventions affected pancreatic cancer cell growth and intracellular polyamine pools.

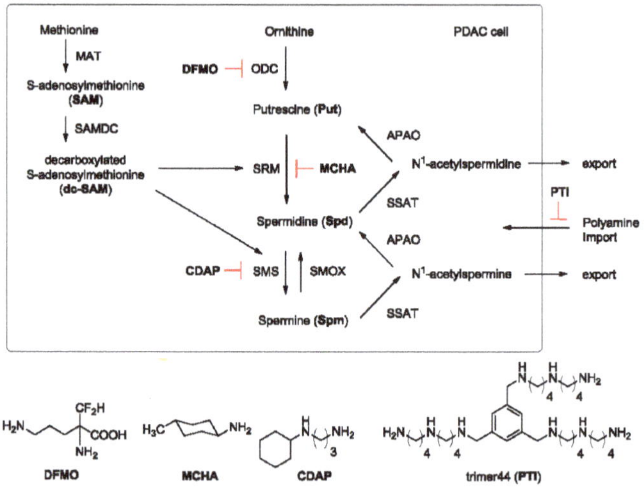

Figure 1. Polyamine Metabolism and Transport. Ornithine is converted to putrescine via ornithine decarboxylase (ODC). Methionine is converted to S-adenosylmethionine (SAM) via methionine adenosyltransferase (MAT) and SAM is converted to decarboxylated S-adenosylmethionine (dc-SAM) via the action of S-adenosylmethionine decarboxylase (SAMDC). Putrescine is converted to spermidine via spermidine synthase (SRM) and an aminopropyl fragment derived from dc-SAM. Similarly, spermidine is converted to spermine via spermine synthase (SMS) and dc-SAM. Back conversion can occur via N-acetylation using spermidine/spermine N^1-acetyltransferase (SSAT) to form N^1-acetyl derivatives, which can be oxidized by acetylpolyamine oxidase (APAO) to generate the respective polyamine. N^1-acetylpolyamines can also be excreted by cells to maintain intracellular polyamine levels and exogenous polyamines can be imported to increase intracellular polyamine pools via the polyamine transport system. Spermine oxidase (SMOX) allows direct conversion of spermine to spermidine. PDAC: pancreatic ductal adenocarcinoma; DFMO: difluoromethylornithine; MCHA: *trans*-4-methyl-cyclohexylamine; CDAP: N-cyclohexyl-1,3-diaminopropane; PTI: polyamine transport inhibitor.

2. Materials and Methods

2.1. Materials

DFMO was obtained as a gift from Dr. Patrick Woster at the Medical University of South Carolina. The inhibitors MCHA and CDAP were obtained from Acros Organics (Fair Lawn, NJ, USA) and Alfa Aesar (Ward Hill, MA, USA), respectively. The trimer44NMe PTI was prepared as previously described [9]. The compounds were readily soluble in water and stock solutions were made in phosphate buffered saline (PBS).

2.2. Biological Studies

L3.6pl cells were grown in Roswell Park Memorial Institute (RPMI) 1640 medium supplemented with 10% fetal bovine serum (FBS) and 1% penicillin/streptomycin. The cells were grown at 37 °C under a humidified 5% CO_2 atmosphere. Cells in early- to mid-log phase were used.

2.3. IC_{50} Determinations and Cell Growth Studies

Cell growth assays were performed in sterile 96-well microtiter plates (Costar 3599, Corning, NY, USA). L3.6pl cells were seeded at 500 cells/70 μL with 250 μM aminoguanidine in each well. Aminoguanidine was added to prevent the oxidation of polyamine compounds by the bovine serum amine oxidase present in the FBS. Drug solutions (10 μL per well) of appropriate concentration in PBS were added after overnight incubation of each cell line. When necessary, PBS was added to make the total volume 100 μL in each well. After drug exposure for 72 h at 37 °C, cell growth was determined by measuring formazan formation from the 3-(4,5-dimethylthiazol-2-yl)-5-(3-carboxymethoxyphenyl)-2-(4-sulfenyl)-2*H*-tetrazolium inner salt (MTS) using a SynergyMx Biotek microplate reader measuring absorbance at 490 nm [13]. All experiments were run in triplicate and compared to untreated controls.

2.4. HPLC and Polyamine Level Determination

L3.6pl cells (500,000 cells/10 mL media) were incubated with aminoguanidine (250 μM) at 37 °C for 24 h. Each compound (dissolved in PBS) was then added either alone or in combination with other agents, and the cells were incubated for 72 h at 37 °C. The cells were then washed extensively with ice cold PBS (once with 5 mL and twice with 2 mL). Each PBS wash was removed by suction. To the washed cells, an additional 2 mL of ice cold PBS was added and the cells were scraped off the dish and collected in a centrifuge tube. This scraping step was done twice for each experimental condition. The cell suspensions were then centrifuged at 1000 rpm for 4 min to provide a cell pellet. The supernatant was carefully removed by suction. The cell pellet was then lysed using a 0.2 M perchloric acid/1 M NaCl solution (200 μL), sonicated, and centrifuged. The resultant supernatant and pellet were separated. To 100 μL of the polyamine-containing supernatant 1,7-diaminoheptane (30 μL of 1.5×10^{-4} M) was added as an internal standard. The supernatant was then treated on the rotary shaker at 200 rpm for 1 h at 65 °C with 1 M Na_2CO_3 (200 μL) and dansyl chloride (5 mg/mL, 400 μL) to generate the respective *N*-dansylated polyamines. Proline (1 M, 100 μL) was added and the reaction mixture shaken on the rotary shaker at 200 rpm for 20 min at 65 °C. The dansylated polyamines were then extracted into chloroform (1 mL) and the organic layer separated and concentrated. The resulting residue was redissolved in MeOH (1 mL) and the solution filtered through hydrophobic reversed phase C_{18} cartridges (50 mg bed weight). Each polyamine was then quantified via HPLC analysis [14] using authentic standards. The protein content of the pellet was quantified using the Pierce BCA Protein Assay Kit from ThermoFisher Scientific (Waltham, MA, USA). Final results were expressed as nmol polyamine/mg protein. Each condition was performed in duplicate.

2.5. Statistical Analysis

All cell growth studies were performed in triplicate. Polyamine level determination studies were performed in duplicate. Analyses were done using an unpaired *t*-test on GraphPad Prism 7 (GraphPad Software, San Diego, CA, USA). In all cases, values of $p < 0.05$ were regarded as being statistically significant.

3. Results

Bioevaluation

A previous investigation of a series of pancreatic cancer cell lines identified the human L3.6pl cell line as an excellent model to look at polyamine metabolism and import due to its high polyamine transport activity [8]. A dose–response curve was obtained for each compound tested as a single agent to determine the IC_{50} value: the dose at which the growth of cells was inhibited 50% compared to the untreated control. We observed that L3.6pl cells (500 cells/well with 250 μM aminoguanidine) became more sensitive to DFMO over time (i.e., after 48 h, 72 h, and 96 h of incubation at 37 °C). The 72 h incubation time was selected to balance the cells' DFMO sensitivity and the ability of these DFMO-treated cells to be rescued back with exogenous spermidine (1 μM) to ≥90% of the growth observed with the untreated control. The IC_{50} value of DFMO was 4.2 mM after 72 h incubation and these cells could be rescued back to ≥90% of the growth observed with the untreated control by Spd (1 μM). [8] We also screened MCHA, CDAP and the trimer44 PTI for their ability to affect L3.6pl cell growth as single agents after 72 h of incubation (Figure 2). As shown in Figure 2, both MCHA and CDAP were relatively non-toxic and required high concentrations to affect L3.6pl cell growth. The L3.6pl 72 h IC_{50} value of PTI trimer44 was 69.6 ± 1.8 μM and typically the trimer44 alone could be dosed at 4 μM with no effect on cell growth.

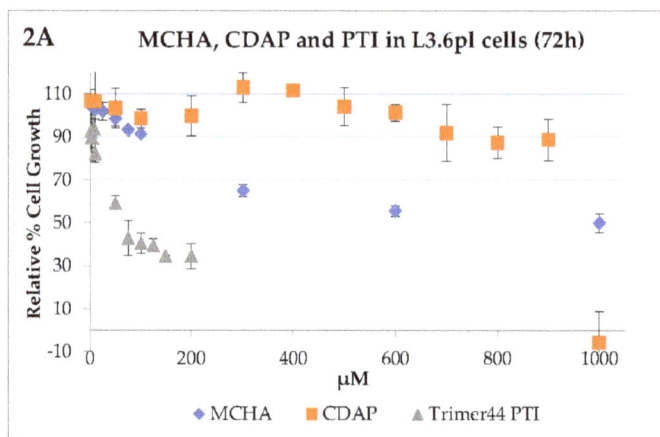

Figure 2. *Cont.*

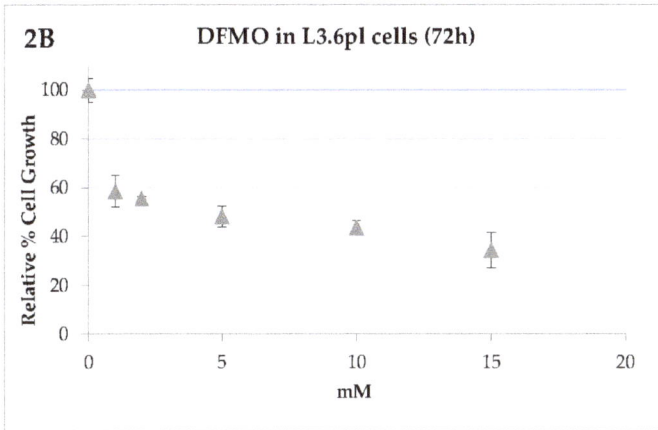

Figure 2. Influence of trans-4-methylcyclohexylamine (MCHA), *N*-cyclohexyl-1,3-propanediamine (CDAP), polyamine transport inhibitor (PTI, trimer 44), and difluoromethylornithine (DFMO) on L3.6pl human pancreatic cancer cell growth. Panel A: Growth inhibition by MCHA, CDAP, and the trimer44 PTI in L3.6pl cells after 72 h incubation at 37 °C in a 5% CO_2 atmosphere. The 72 h IC_{50} values of MCHA, CDAP, and trimer44 PTI in L3.6pl human pancreatic cancer cells were 1000 ± 4.4 μM, 941 ± 12.1 μM, and 69.6 ± 1.8 μM, respectively. In general, the PTI was well tolerated at low doses (e.g., 4 μM with no reduction in growth) and the other inhibitors could be dosed at high concentrations with little effect on cell growth. For example, after 72 h, MCHA and CDAP at 100 μM gave 91% and 99% relative growth, respectively, versus the untreated control. Panel B: Relative cell growth in the presence of DFMO in L3.6pl cells after 72 h incubation at 37 °C in a 5% CO_2 atmosphere. The 72 h IC_{50} value of DFMO in L3.6pl cells was 4.24 ± 0.11 mM.

Armed with this knowledge of how L3.6pl cells responded to these compounds as single agents, we explored how MCHA and CDAP affected intracellular polyamine pools in a concentration-dependent manner. These results are shown in Figure 3.

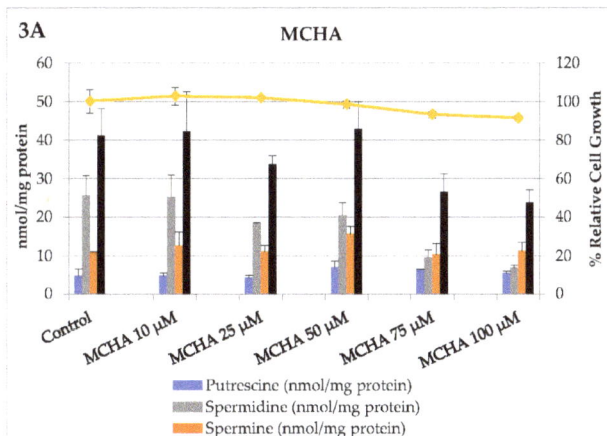

Figure 3. *Cont.*

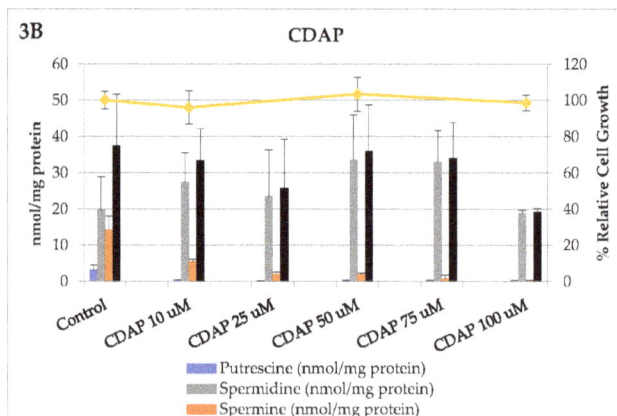

Figure 3. The effects of MCHA (Panel A) and CDAP (Panel B) on intracellular polyamine pools of L3.6pl cells after 72 h incubation. As shown in the top panel, the SRM inhibitor, MCHA, reduced intracellular spermidine pools in a dose-dependent manner and at 100 µM resulted in a 73% reduction of intracellular spermidine pools (gray bars, $p < 0.05$) and provided an overall 42% reduction of total intracellular polyamine pools (black bars). In the bottom panel, the SMS inhibitor, CDAP, reduced intracellular spermine pools in a dose-dependent manner and at 100 µM resulted in a 99.9% reduction in intracellular spermine levels and provided a 48% reduction of total intracellular polyamine pools. The reduction in spermine levels was statistically significant ($p < 0.05$) for concentrations of CDAP greater than 10 µM compared to untreated controls. Neither intervention significantly reduced relative cell growth (vs. untreated controls), consistent with these cells having excess polyamine pools.

Next, we measured how DFMO and the PTI (trimer44) modulated polyamine pools (Tables 1 and 2), and then tested how CDAP and MCHA, when tested individually in combination with DFMO or DFMO + PTI, affected intracellular polyamine levels and % relative cell growth. These 72 h experiments were conducted in L3.6pl cells and the results are shown in Figures 4 and 5.

Table 1. Polyamine levels as a function of DFMO concentration in L3.6 pl cells incubated at 37 °C for 72 h [a].

Experiment	DFMO (mM)	Spermidine nmoles/mg Protein	Spermine nmoles/mg Protein	Total Polyamines nmoles/mg Protein	Spd/Spm Ratio
Control	0	13.0 ± 1.6	10.1 ± 0.9	23.1	1.29
DFMO	0.1	12.0 ± 2.8	13.4 ± 4.3	25.3	0.90
DFMO	0.5	0.7 ± 0.1	15.3 ± 2.3	16.1	0.05
DFMO	1	0.2 ± 0.0	14.9 ± 0.8	15.2	0.01
DFMO	2	0.6 ± 0.7	9.2 ± 3.9	9.8	0.06
DFMO	3	0.1 ± 0.0	6.1 ± 0.2	6.2	0.01
DFMO	4.2	0.2 ± 0.0	7.7 ± 1.0	7.9	0.02

[a] error is listed as standard deviation around the mean. Note the putrescine level for the control was 0.8 ± 0.6 nmoles/mg protein and was undetectable after DFMO addition.

Table 2. Polyamine levels as a function of trimer44 concentration in L3.6pl cells incubated at 37 °C for 72 h [a].

Experiment	trimer44 (µM)	Spermidine nmoles/mg Protein	Spermine nmoles/mg Protein	Total Polyamines nmoles/mg Protein	Spd/Spm Ratio
Control	0	14.0 ± 3.6	8.5 ± 1.8	22.5	1.65
trimer 44	1	17.0 ± 6.3	12.9 ± 3.4	29.9	1.31
trimer 44	2	12.2 ± 2.8	8.9 ± 1.2	21.1	1.36
trimer 44	4	12.0 ± 5.3	10.4 ± 2.1	22.4	1.15
trimer 44	5	11.4 ± 2.5	13.5 ± 0.8	24.9	0.84
trimer 44	10	5.5 ± 1.5	11.2 ± 2.1	16.7	0.49

[a] error is listed as standard deviation around the mean.

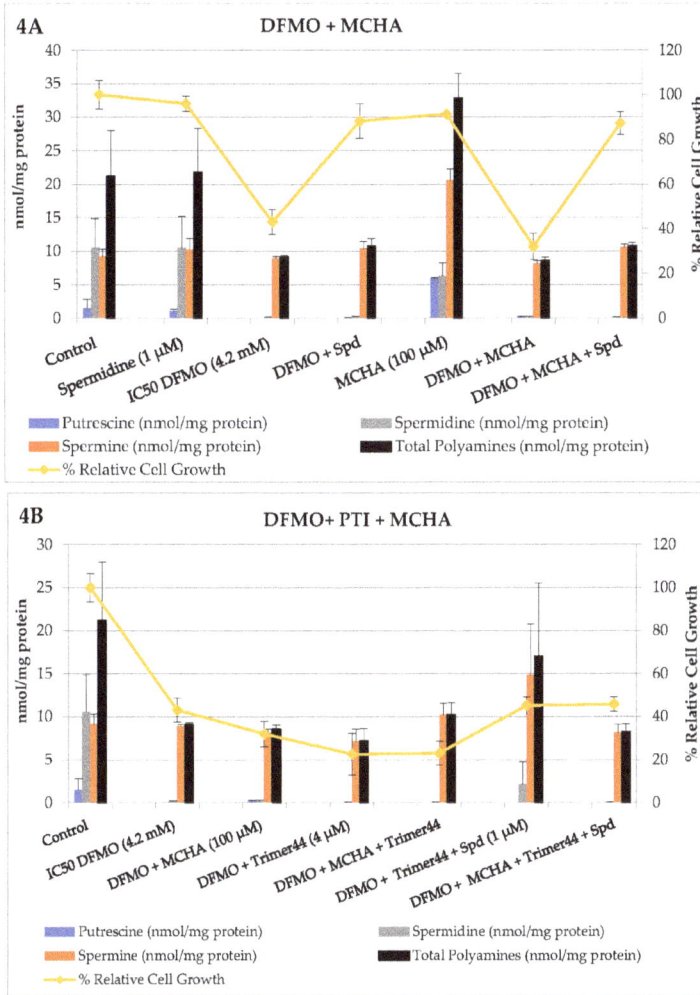

Figure 4. Single and combination therapies in L3.6pl cells with the spermidine synthase inhibitor, MCHA, at 100 μM. Altered polyamine pools (expressed as nmoles polyamine/mg protein) and L3.6pl relative % cell growth (vs. an untreated control) were observed after 72 h incubation. Controls were run in parallel, polyamine levels were determined in duplicate, and % cell growth determined in triplicate. The concentrations of compounds were: DFMO (4.2 mM), MCHA (100 μM), trimer44 PTI (4 μM), and spermidine (Spd, 1 μM). At these doses only DFMO gave significant reductions in cell growth when tested alone. In Panel A, the reduction in putrescine and spermine was statistically significant ($p < 0.05$) compared to untreated controls. Relative cell growth tracked fairly well with total intracellular polyamine pools.

As shown in Figure 4 Panel A, only DFMO alone (43% cell growth, 4.2 mM) and in combination with MCHA (32%, 100 μM) resulted in significantly reduced cell growth. Exogenous spermidine (1 μM) was shown to rescue the growth of DFMO and DFMO + MCHA treated L3.6pl cells. As shown in Figure 4 Panel B, the presence of either MCHA, the trimer44 PTI, or the combination of MCHA + PTI with DFMO decreased cell growth. The combination of DFMO + MCHA + PTI + Spd resembled the DFMO + MCHA control, which is consistent with the trimer44 PTI inhibiting spermidine import. Indeed, the presence of the PTI prevented spermidine rescue because L3.6pl cells treated with DFMO + MCHA + Spd gave 87% relative cell growth, whereas cells treated with DFMO + MCHA + PTI + Spd gave 46% relative cell growth and gave nearly identical intracellular polyamine pools to the DFMO + MCHA entry.

Having surveyed MCHA and its effectiveness in combination with DFMO and the trimer44 PTI, we next evaluated CDAP and the results are shown in Figure 5.

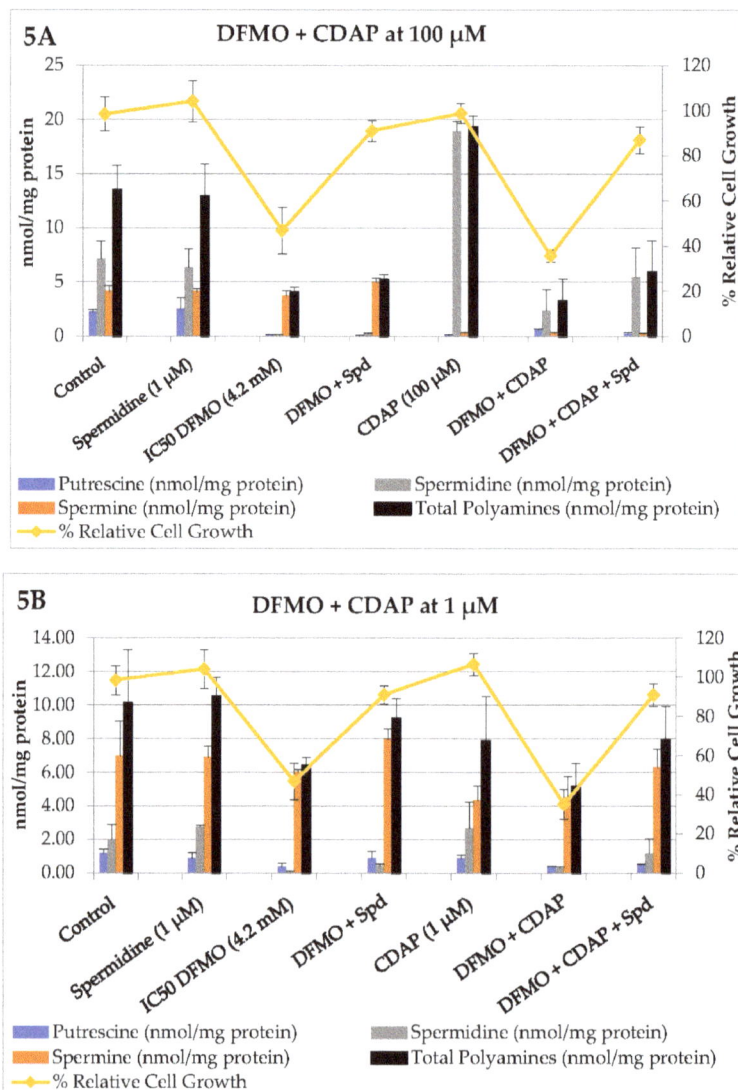

Figure 5. *Cont.*

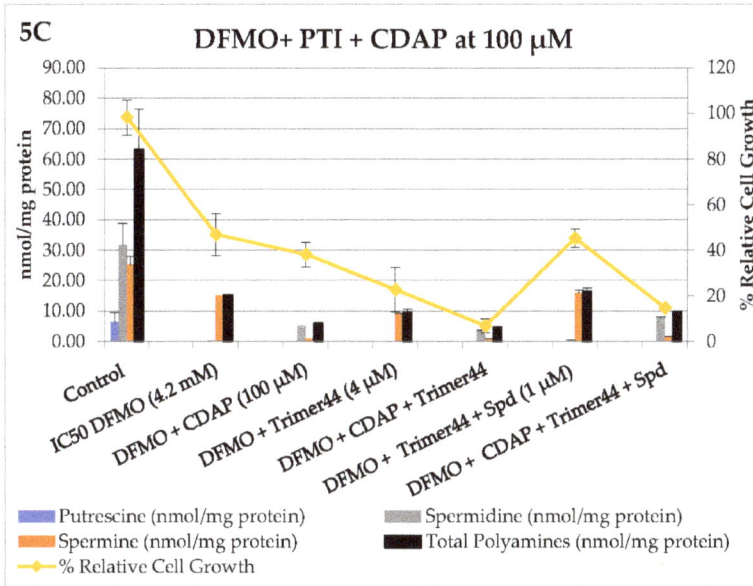

Figure 5. Single and combination therapies in L3.6pl cells with the spermine synthase inhibitor, CDAP. Altered polyamine pools (expressed as nmoles polyamine/mg protein) and L3.6pl relative % cell growth (vs. an untreated control) were observed after 72 h incubation. Controls were run in parallel, polyamine levels were determined in duplicate, and % cell growth in triplicate. The concentrations of compounds were: DFMO (4.2 mM), CDAP (100 μM in Panels A and C and 1 μM in Panel B), trimer44 PTI (4 μM), and spermidine (Spd, 1 μM). Note: in Panel A, we also tested CDAP (100 μM) + Spd (1 μM), which gave 99% relative cell growth and 0.08 ± 0.20, 13.72 ± 0.85, 0.20 ± 0.09, and 14.00 ± 0.97 nmol/mg protein for putrescine, spermidine, spermine, and total polyamines, respectively (not shown). This revealed that exogenous spermidine (1 μM) did not further enhance intracellular spermidine levels in the presence of CDAP (100 μM). This observation is consistent with homeostatic control of polyamine levels. In contrast, in the presence of DFMO (4.2 mM), CDAP (100 μM), and exogenous spermidine (1 μM), significant rescue as evidenced by increased cell growth (from 35% to 86%) was observed, which is consistent with the observed increase in intracellular spermidine pools.

As shown in Figure 5 Panel A, exogenous spermidine (1 μM) alone had little effect on intracellular polyamine pools and the growth of L3.6pl cells, presumably due to homeostatic controls which maintain preferred polyamine levels. In contrast, as seen in Panel A, DFMO (4.2 mM) resulted in 47% relative growth and a significant loss of intracellular putrescine and spermidine pools. DFMO-treated cells were rescued with exogenous spermidine (1 μM), which resulted in 90% relative growth and increased spermine pools. Interestingly, as will be illustrated again later in the discussion, these cells had a potential choice between living on spermidine or spermine, and elected to maintain spermine levels by shunting the imported spermidine to spermine. This commitment to spermine maintenance is consistent with previous findings [4,11].

CDAP at 100 μM was very effective in depleting ~90% and 95% of intracellular putrescine and spermine pools in L3.6pl cells after 72 h. As shown in Figure 5 Panel A, CDAP (100 μM)-treated L3.6pl cells increased their total intracellular polyamine pools via a 2.6-fold increase in intracellular spermidine levels. Interestingly, these high spermidine levels could not be driven higher in the presence of exogenous spermidine (1 μM), suggesting a possible upper limit under these conditions (see legend of Figure 5). The combination of CDAP (100 μM) and DFMO (4.2 mM) resulted in decreased cell growth and decreased intracellular spermidine pools. Both cell growth and spermidine pools could be rescued from DFMO + CDAP treatment by the addition of exogenous spermidine (1 µM). These results illustrate the dynamic exchange between polyamine pools and the plasticity of the homeostatic system to inhibitors of polyamine biosynthesis.

Interestingly, these same changes in relative cell growth were also observed when CDAP was reduced to 1 µM (see purple lines in Panels A and B). The reduced ability of CDAP (at 1 µM) to inhibit SMS significantly affected the composition of the polyamine pools. Indeed, the commitment to maintain spermine pools in the presence of DFMO (4.2 mM), DFMO + Spd, and DFMO + CDAP (1 µM) becomes obvious in Panel B. Both DFMO and DFMO + CDAP (1 µM) caused a significant decrease in relative cell growth (46.7% and 35%, respectively), which could be rescued to >90% relative cell growth by the availability of exogenous spermidine (1 µM). We noted that the presence of exogenous spermidine resulted in an approximate 2 nmol/mg protein increase in spermine levels (see red bars in Figure 5 Panel B). Interestingly, as seen in Panel A, when there is efficient inhibition of spermine biosynthesis, the intracellular spermidine pools were increased by 3 nmol/mg protein in the presence of exogenous spermidine (1 µM). These results suggest that polyamine import supplies at least 2–3 nmol/mg of spermidine to these cells, and when possible these spermidine resources are preferentially shunted to spermine pools. This preference for spermine is interesting because, as shown in the far right entries in Panels A and B (DFMO + CDAP + Spd), these cells can maintain high (>90%) relative growth in the presence of sufficient spermidine or spermine pools.

Given the impact of polyamine import on these results, we evaluated how the presence of a PTI (trimer44) affected inhibitor performance. The trimer44 PTI was shown to be a competitive inhibitor of polyamine transport via ^{3}H-Spd uptake assays. Typically, as shown in Panel A, exogenous spermidine (Spd, 1 µM) was shown to rescue DFMO-treated L3.6pl cells from 46.7% relative growth with DFMO only (4.2 mM) to over 90% with DFMO + Spd. As shown in Figure 5 Panel C, the DFMO, DFMO + PTI, and DFMO + PTI + Spd treatments provided 46.7, 22.6, and 45.2% relative cell growth, respectively. In addition, the DFMO-only and DFMO + PTI + Spd entries had nearly identical polyamine pool distributions and % relative growth (46.7% and 45.2% in Figure 5, Panel C), illustrating the ability of the PTI to block both spermidine uptake and the rescue of DFMO-treated L3.6pl cells. The presence of the trimer44 PTI (4 µM) with DFMO + CDAP (100 µM) significantly lowered the cell growth to 6.8%. We also noted that the distribution of intracellular polyamine pools of DFMO + CDAP (100 µM) was similar to DFMO + CDAP (100 µM) + PTI + Spd, consistent with inhibited Spd import. In summary, only the therapies containing DFMO reduced cell growth at the concentrations used. This reduction in cell growth was augmented to some degree by the presence of CDAP and the PTI as single additives, and more so when they were used in combination with DFMO to provide extensive inhibition of polyamine metabolism.

In our titrations of CDAP with a fixed dose of DFMO (4.2 mM), we noticed a consistently lower relative cell growth (10%) in the presence of CDAP, even at concentrations where CDAP was not toxic. Indeed, this modest potentiation of DFMO was noted at as low as 100 nM CDAP, suggesting that even mild perturbation of spermine biosynthesis via SMS inhibition can augment the effectiveness of DFMO. Since DFMO + CDAP could be readily rescued by exogenous polyamines, the combination of these two agents may not be effective in vivo and may require the additional presence of the PTI for effective anti-proliferative activity.

4. Discussion

Of all three native polyamines, spermine has been shown to be the most effective immune suppressant, with inhibitory activity noted in T-cells, monocytes, and macrophages [14–19]. Compared to other human tissues, the human pancreas has the highest amount of spermidine. Armed with significant stores of spermidine, we hypothesized that PDAC tumors with upregulated SMS can convert spermidine to spermine (Figure 1) for immune suppression. Indeed, spermine is naturally present in amniotic fluid to suppress the maternal immune response and spermine has been shown to inhibit virtually all immune cells [14–19]. We speculated that PDAC uses this 'fetal strategy' to create a spermine-rich zone of immune privilege via spermine production and secretion. Rewardingly, a search of six existing pancreatic databases found that SMS mRNA is universally upregulated in PDAC, which is consistent with our hypothesis. This insight is potentially paradigm-shifting because it suggests that, unless spermine is downregulated in the PDAC tumor microenvironment, immunotherapies will continue to fail [20,21].

The results reported here suggest that even though PDAC cells can survive on either spermidine or spermine, they prefer spermine when given the choice (e.g., see DFMO results in Figure 5). This preference is consistent with the apparent high SMS expression in PDAC cells and may in part be critical for tumor survival by establishing immune privilege via the excretion of spermine or its metabolites.

Prior in vivo studies in healthy rats using MCHA showed that delivery was more effective via MCHA (0.1%, pH 6) in the drinking water than by a single intraperitoneal (i.p.) injection. A significant dose-dependent decrease in spermidine content was observed [10]. For example, using a 0.1% solution of MCHA for 10 days, Shirahata et al. observed a decrease in spermidine levels in prostate (28%), liver (21%), and kidney (33%) tissue with a concomitant increase in spermine levels. The total polyamine levels in these tissues did not fluctuate strikingly with the treatment, presumably due to polyamine import [10]. These results are very similar to what we observed with MCHA in culture with PDAC cells, where a shift in the intracellular polyamine distribution from spermidine to spermine occurs (Figure 4). The authors also reported the low toxicity of MCHA in rats with a lethal dose needed to kill 50% of the rats (LD_{50}) of 250 mg/kg [10]. In summary, MCHA was effective in vivo in modulating spermidine pools in rat tissues.

Related studies in healthy rats with CDAP showed similar results to what we observed in cell culture, including a significant decrease in both spermine and putrescine pools, and a buildup of spermidine in the presence of CDAP [10]. For example, using a 0.1% CDAP solution in the drinking water for 10 days, the authors observed a 95% decrease in spermine content in the prostate and ~50% reduction in the liver and kidney tissues. Exposure to 0.1% CDAP in the drinking water showed a 30% decrease in spermine content in the liver of rats after 10 days and a 79% decrease in liver spermine levels and 211% increase in liver spermidine levels after 120 days of CDAP exposure. The authors also reported the low toxicity of CDAP in rats with a LD_{50} of 500 mg/kg [10]. In summary, CDAP was effective in vivo in modulating spermine pools in rat tissues and our in vitro results here are consistent with these prior findings in healthy rats.

While this is the first report of CDAP and MCHA assessment in pancreatic cancers, these compounds have been previously evaluated by He et al. in the treatment of P388 leukemia cells in mice [11]. Specifically, DFMO (1.5 g/kg) + CDAP (25 mg/kg) gave a 1.3-fold increase in survival [11]. The authors concluded that the anti-proliferative effect of DFMO was strengthened by CDAP due to a reduction in spermine content. When DFMO + CDAP was administered, a compensatory increase in spermidine was observed in P388 cells, but remained significantly lower than untreated P388 cells. Notably, the authors suggested that polyamine import may explain why, when CDAP was increased to 50 mg/kg, it failed to further decrease spermine levels [11]. In addition, when MCHA and CDAP were added individually with DFMO, only CDAP + DFMO significantly strengthened the antiproliferative effect of DFMO against leukemia cells [11]. This is interesting because we found in L3.6pl cells in vitro that both MCHA and CDAP can potentiate DFMO (i.e., further decrease cell growth) when used in combination with DFMO (Figures 4 and 5).

In summary, many inhibitors of SRM and SMS have been developed, but few provide reduced cell viability [22–25]. One possible explanation is the observed increase in *S*-adenosylmethionine decarboxylase (SAMDC) activity when either spermidine or spermine intracellular stores are reduced [10]. Shirahata et al. demonstrated (albeit in the prostate) that either MCHA or CDAP, when dosed at 0.1% in the drinking water for 10 days, leads to the expected reduced levels of spermidine and spermine, respectively. MCHA (0.1%) and CDAP (0.1%) individually induced a 2.2-fold and 2.7-fold increase in SAMDC activity, respectively (see Figure 1) [10]. Indeed, depending upon the inhibitor used, the higher activity of SAMDC would lead to increased amounts of dc-SAM (i.e., *S*-adenosylmethioninamine) to facilitate the respective conversions of either putrescine to spermidine or spermidine to spermine as needed (Figure 1).

Therefore, in the presence of either MCHA or CDAP, the cells are likely primed for polyamine biosynthesis with high dc-SAM levels, but lack the specific polyamine substrates and non-inhibited enzymes needed to replenish their deficient pool(s). This explanation is consistent with the classic expectation that substrates buildup behind an enzymatic block. However, our results suggest that polyamine import plays a key role for how PDAC cells circumvent these inhibitors (by importing polyamine substrates like Spd) and suggest that the development of PTIs like trimer44 is warranted for future use in combination with DFMO or DFMO + CDAP. With the ability to decrease spermine and total polyamine pools in the presence of exogenous polyamines, the combination therapy of DFMO + CDAP + PTI provides a new way to target PDAC cells via their altered polyamine metabolism.

5. Conclusions

Understanding how PDAC cells respond to inhibitors of polyamine metabolism and import could lead to new approaches to treat PDAC. For example, combination therapies like DFMO + CDAP + PTI, which reduce intratumoral spermine pools, even in the presence of exogenous polyamines, may provide a new approach to PDAC tumors by targeting their high reliance upon spermine. Reduced spermine production by the tumor would result in less spermine available for secretion to establish immune privilege. Indeed, others have shown that DFMO + PTI therapies provide an enhanced immune response in other cancers [26–28]. While the precise genes and proteins involved in polyamine transport are only now coming to light [29], this area is ripe for future drug discovery. The findings that DFMO and CDAP are both orally bioavailable, have low toxicity, and CDAP levels even as low as 100 nM in culture provide beneficial potentiation of DFMO, provide additional rationale for continued studies of these interventions in PDAC.

Acknowledgments: The authors thank the Department of Defense Idea Award with Special Focus program for partial financial support of this work via award CA # 150356. The authors also appreciate the assistance of Kelvin Chaplin in the culture, isolation, and processing of L3.6pl cell pellets.

Author Contributions: O.P. conceived and designed the experiments; C.M. and J.T. performed the experiments; C.M. analyzed the data; C.M. and O.P. wrote the paper.

Conflicts of Interest: The authors declare no conflict of interest.

References

1.   American Cancer Society. *Cancer Facts & Figures 2016*; American Cancer Society: Atlanta, GA, USA, 2016.
2.   Mohammed, A.; Janakiram, N.B.; Madka, V.; Ritchie, R.L.; Brewer, M.; Biddick, L.; Patlolla, J.M.; Sadeghi, M.; Lightfoot, S.; Steele, V.E.; et al. Eflornithine (DFMO) prevents progression of pancreatic cancer by modulating ornithine decarboxylase signaling. *Cancer Prev. Res. (Phila)* **2014**, *7*, 1198–1209. [CrossRef] [PubMed]
3.   Gysin, S.; Rickert, P.; Kastury, K.; McMahon, M. Analysis of genomic DNA alterations and mRNA expression patterns in a panel of human pancreatic cancer cell lines. *Genes Chromosom. Cancer* **2005**, *44*, 37–51. [CrossRef] [PubMed]
4.   Casero, R.A., Jr.; Marton, L.J. Targeting polyamine metabolism and function in cancer and other hyperproliferative diseases. *Nat. Rev. Drug Discov.* **2007**, *6*, 373–390. [CrossRef] [PubMed]
5.   Loser, C.; Folsch, U.R.; Paprotny, C.; Creutzfeldt, W. Polyamine concentrations in pancreatic tissue, serum, and urine of patients with pancreatic cancer. *Pancreas* **1990**, *5*, 119–127. [CrossRef] [PubMed]
6.   Hyvonen, M.T.; Merentie, M.; Uimari, A.; Keinanen, T.A.; Janne, J.; Alhonen, L. Mechanisms of polyamine catabolism-induced acute pancreatitis. *Biochem. Soc. Trans.* **2007**, *35*, 326–330. [CrossRef] [PubMed]
7.   Loser, C.; Folsch, U.R.; Cleffmann, U.; Nustede, R.; Creutzfeldt, W. Role of ornithine decarboxylase and polyamines in camostate (Foy-305)-induced pancreatic growth in rats. *Digestion* **1989**, *43*, 98–112. [CrossRef] [PubMed]

8. Madan, M.; Patel, A.; Skruber, K.; Geerts, D.; Altomare, D.A.; Phanstiel, O., IV. ATP13A3 and Caveolin-1 as Potential Biomarkers for Difluoromethylornithine-based therapies in Pancreatic Cancers. *Am. J. Cancer Res.* **2016**, *6*, 1231–1252. [PubMed]
9. Muth, A.; Madan, M.; Archer, J.J.; Ocampo, N.; Rodriguez, L.; Phanstiel, O. Polyamine transport inhibitors: Design, synthesis, and combination therapies with difluoromethylornithine. *J. Med. Chem.* **2014**, *57*, 348–363. [CrossRef] [PubMed]
10. Shirahata, A.; Takahashi, N.; Beppu, T.; Hosoda, H.; Samejima, K. Effects of inhibitors of spermidine synthase and spermine synthase on polyamine synthesis in rat tissues. *Biochem. Pharmacol.* **1993**, *45*, 1897–1903. [CrossRef]
11. He, Y.; Shimogori, T.; Kashiwagi, K.; Shirahata, A.; Igarashi, K. Inhibition of cell growth by combination of alpha-difluoromethylornithine and an inhibitor of spermine synthase. *J. Biochem.* **1995**, *117*, 824–829. [CrossRef] [PubMed]
12. Muth, A.; Kamel, J.; Kaur, N.; Shicora, A.C.; Ayene, I.S.; Gilmour, S.K.; Phanstiel, O. Development of polyamine transport ligands with improved metabolic stability and selectivity against specific human cancers. *J. Med. Chem.* **2013**, *56*, 5819–5828. [CrossRef] [PubMed]
13. Mosmann, T. Rapid colorimetric assay for cellular growth and survival: Application to proliferation and cytotoxicity assays. *J. Immunol. Methods* **1983**, *65*, 55–63. [CrossRef]
14. Minocha, S.C.; Minocha, R.; Robie, C.A. High-performance liquid chromatographic method for the determination of dansyl-polyamines. *J. Chromatogr.* **1990**, *511*, 177–183. [CrossRef]
15. Evans, C.H.; Lee, T.S.; Flugelman, A.A. Spermine-directed immunosuppression of cervical carcinoma cell sensitivity to a majority of lymphokine-activated killer lymphocyte cytotoxicity. *Nat. Immun.* **1995**, *14*, 157–163. [PubMed]
16. Kano, Y.; Soda, K.; Nakamura, T.; Saitoh, M.; Kawakami, M.; Konishi, F. Increased blood spermine levels decrease the cytotoxic activity of lymphokine-activated killer cells: A novel mechanism of cancer evasion. *Cancer Immunol. Immunother.* **2007**, *56*, 771–781. [CrossRef] [PubMed]
17. Ogata, K.; Nishimoto, N.; Uhlinger, D.J.; Igarashi, K.; Takeshita, M.; Tamura, M. Spermine suppresses the activation of human neutrophil NADPH oxidase in cell-free and semi-recombinant systems. *Biochem. J.* **1996**, *313*, 549–554. [CrossRef] [PubMed]
18. Soda, K.; Kano, Y.; Nakamura, T.; Kasono, K.; Kawakami, M.; Konishi, F. Spermine, a natural polyamine, suppresses LFA-1 expression on human lymphocyte. *J. Immunol.* **2005**, *175*, 237–245. [CrossRef] [PubMed]
19. Zhang, M.; Borovikova, L.V.; Wang, H.; Metz, C.; Tracey, K.J. Spermine inhibition of monocyte activation and inflammation. *Mol. Med.* **1999**, *5*, 595–605. [PubMed]
20. Zhang, M.; Wang, H.; Tracey, K.J. Regulation of macrophage activation and inflammation by spermine: A new chapter in an old story. *Crit. Care Med.* **2000**, *28*, N60–N66. [CrossRef] [PubMed]
21. Paniccia, A.; Merkow, J.; Edil, B.H.; Zhu, Y. Immunotherapy for pancreatic ductal adenocarcinoma: An overview of clinical trials. *Chin. J. Cancer Res.* **2015**, *27*, 376–391. [PubMed]
22. Kunk, P.R.; Bauer, T.W.; Slingluff, C.L.; Rahma, O.E. From bench to bedside a comprehensive review of pancreatic cancer immunotherapy. *J. Immunother. Cancer* **2016**, *4*, 14. [CrossRef] [PubMed]
23. Pegg, A.E.; Coward, J.K.; Talekar, R.R.; Secrist, J.A., 3rd. Effects of certain 5′-substituted adenosines on polyamine synthesis: Selective inhibitors of spermine synthase. *Biochemistry* **1986**, *25*, 4091–4097. [CrossRef] [PubMed]
24. Pegg, A.E.; Coward, J.K. Effect of *N*-(n-butyl)-1,3-diaminopropane on polyamine metabolism, cell growth and sensitivity to chloroethylating agents. *Biochem. Pharmacol.* **1993**, *46*, 717–724. [CrossRef]
25. Pegg, A.E.; Wechter, R.; Poulin, R.; Woster, P.M.; Coward, J.K. Effect of S-adenosyl-1,12-diamino-3-thio-9-azadodecane, a multisubstrate adduct inhibitor of spermine synthase, on polyamine metabolism in mammalian cells. *Biochemistry* **1989**, *28*, 8446–8453. [CrossRef] [PubMed]
26. Hayes, C.S.; Shicora, A.C.; Keough, M.P.; Snook, A.E.; Burns, M.R.; Gilmour, S.K. Polyamine-blocking therapy reverses immunosuppression in the tumor microenvironment. *Cancer Immunol. Res.* **2014**, *2*, 274–285. [CrossRef] [PubMed]
27. Hayes, C.S.; Burns, M.R.; Gilmour, S.K. Polyamine blockade promotes antitumor immunity. *Oncoimmunology* **2014**, *3*, e27360. [CrossRef] [PubMed]

28. Alexander, E.; Minton, A.; Peters, M.; Phanstiel, O.; Gilmour, S. A Novel Polyamine Blockade Therapy Activates an Anti-Tumor Immune Response. *Oncotarget* **2017**, *8*, 84140–84152. [CrossRef] [PubMed]
29. Poulin, R.; Casero, R.A.; Soulet, D. Recent advances in the molecular biology of metazoan polyamine transport. *Amino Acids* **2012**, *42*, 711–723. [CrossRef] [PubMed]

![medical sciences logo]

MDPI

Article

Evaluation of Polyamine Transport Inhibitors in a *Drosophila* Epithelial Model Suggests the Existence of Multiple Transport Systems

Minpei Wang [1], Otto Phanstiel IV [2] and Laurence von Kalm [1,*]

[1] Department of Biology, University of Central Florida, Orlando, FL 32816, USA;
minpei.wang@knights.ucf.edu

[2] Department of Medical Education, College of Medicine, University of Central Florida,
Orlando, FL 32827, USA; otto.phanstiel@ucf.edu

* Correspondence: lvonkalm@ucf.edu; Tel.: +1-407-823-6684

Received: 17 October 2017; Accepted: 9 November 2017; Published: 14 November 2017

Abstract: Increased polyamine biosynthesis activity and an active polyamine transport system are characteristics of many cancer cell lines and polyamine depletion has been shown to be a viable anticancer strategy. Polyamine levels can be depleted by difluoromethylornithine (DFMO), an inhibitor of the key polyamine biosynthesis enzyme ornithine decarboxylase (ODC). However, malignant cells frequently circumvent DFMO therapy by up-regulating polyamine import. Therefore, there is a need to develop compounds that inhibit polyamine transport. Collectively, DFMO and a polyamine transport inhibitor (PTI) provide the basis for a combination therapy leading to effective intracellular polyamine depletion. We have previously shown that the pattern of uptake of a series of polyamine analogues in a *Drosophila* model epithelium shares many characteristics with mammalian cells, indicating a high degree of similarity between the mammalian and *Drosophila* polyamine transport systems. In this report, we focused on the utility of the *Drosophila* epithelial model to identify and characterize polyamine transport inhibitors. We show that a previously identified inhibitor of transport in mammalian cells has a similar activity profile in *Drosophila*. The *Drosophila* model was also used to evaluate two additional transport inhibitors. We further demonstrate that a cocktail of polyamine transport inhibitors is more effective than individual inhibitors, suggesting the existence of multiple transport systems in *Drosophila*. Our findings reinforce the similarity between the *Drosophila* and mammalian transport systems and the value of the *Drosophila* model to provide inexpensive early screening of molecules targeting the transport system.

Keywords: polyamine transport inhibitor; *Drosophila* imaginal discs; difluoromethylorthinine; DFMO

1. Introduction

The common native polyamines (putrescine **1**, spermidine **2** and spermine **3**; Figure 1) are a family of ubiquitous low molecular weight organic polycations containing two to four amine moieties separated by methylene groups. In eukaryotes, polyamines are essential for a variety of cellular processes including cell proliferation, transcription, translation, apoptosis and cytoskeletal dynamics [1–4]. Polyamines can also bind to intracellular polyanions including nucleic acids and ATP, as well as specific proteins such as N-methyl-D-aspartate receptors and inward rectifier potassium ion channels to regulate their functions [5–8].

A balance between biosynthesis, degradation and transport of polyamines is required to maintain polyamine homeostasis [9–12] and an increased intracellular polyamine content due to increased biosynthesis and transport activity is a hallmark of many types of malignant cells [13–15]. Difluoromethylornithine (DFMO **4**; Figure 1) is an inhibitor of polyamine biosynthesis and has been used in the treatment of several cancers [13,14]. DFMO binds irreversibly to ornithine decarboxylase (ODC), the rate limiting enzyme of the polyamine biosynthetic pathway, resulting in the proteasomal degradation of ODC [15]. The clinical effectiveness of DFMO, however, is often limited due to the up-regulation of the polyamine transport system (PTS) to access polyamines from the extracellular milieu [16,17]. To this end, there is a need to develop compounds that inhibit polyamine import. Use of polyamine transport inhibitor compounds with DFMO should simultaneously inhibit biosynthesis and transport, and efficiently deplete polyamine pools in malignant cells.

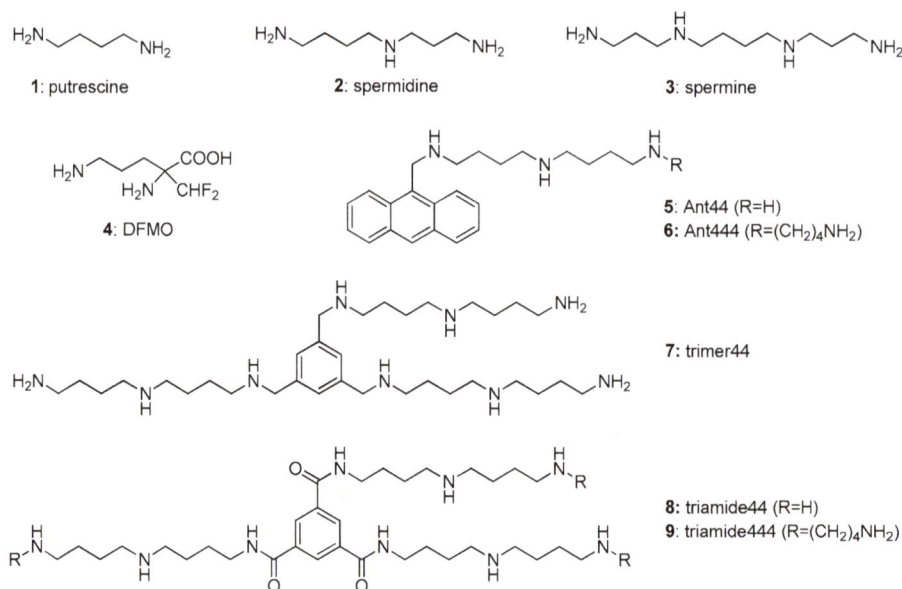

Figure 1. Structures of the native polyamines (**1–3**), difluoromethylornithine (DFMO) (**4**), polyamine analogue (**5**) and candidate polyamine transport inhibitors (PTIs; **6–9**).

The mechanism of polyamine transport has been well characterized in unicellular organisms, such as *Escherichia coli* [18,19], yeast [20,21], *Leishmania* [22] and *Treponema* [23]. In contrast, in multicellular animals only a few PTS components have been identified [24–33] and it is not understood how these components interact, or whether they comprise one or more transport systems. The current understanding has been reviewed by Poulin et al., where evidence for three models is presented [34]. In one model, cell surface glypican-1-anchored heparan sulfate proteoglycans capture extracellular polyamines and these complexes are then endocytosed into endosomes [24]. A second model involves caveolin-mediated endocytosis of polyamines via an unknown receptor [35]. In both the glypican-1 and caveolin-mediated models the sequestration of polyamines into endosomes is followed by nitric oxide-mediated release of polyamines from these vesicles. A third model proposes that an energy-dependent cell-surface transporter/channel allows entry of free polyamines into the cytosol and that these are rapidly sequestered into the endosomal sorting pathway, where they are stored or trafficked to specific cellular locations as needed [36]. In reality, none of these models are mutually exclusive and the PTS may well be a combination of all three.

In previous work, we reported a novel assay to study polyamine transport in *Drosophila* leg imaginal discs [37]. Leg imaginal discs are the embryonic and larval precursors of adult legs. In the larval stage prior to adult development, imaginal discs exist as a single-cell-thick folded epithelium. In response to exposure to the steroid hormone ecdysone at the onset of metamorphosis, they rapidly develop into rudimentary legs (see Figure 2) [38]. Using the *Drosophila* assay we directly compared a series of toxic polyamine ligands for their PTS selectivity in *Drosophila* and mammalian cells. The behavior of the polyamine compounds in imaginal discs was very similar to their behavior in mammalian cell culture, suggesting broad similarities between the PTS of *Drosophila* and mammals. A major advantage of the leg imaginal disc assay is that compounds that access cells through the PTS or inhibit transport can be studied in an environment where cells exhibit normal adhesion properties and are surrounded by extracellular matrix. Thus, the *Drosophila* assay potentially provides an inexpensive animal model for early testing of compounds targeting the PTS.

In this study, we identified and characterized two compounds that act as polyamine transport inhibitors in *Drosophila*. We also demonstrated that a cocktail of polyamine transport inhibitors was more effective than individual inhibitors, suggesting the existence of multiple transport systems in *Drosophila*.

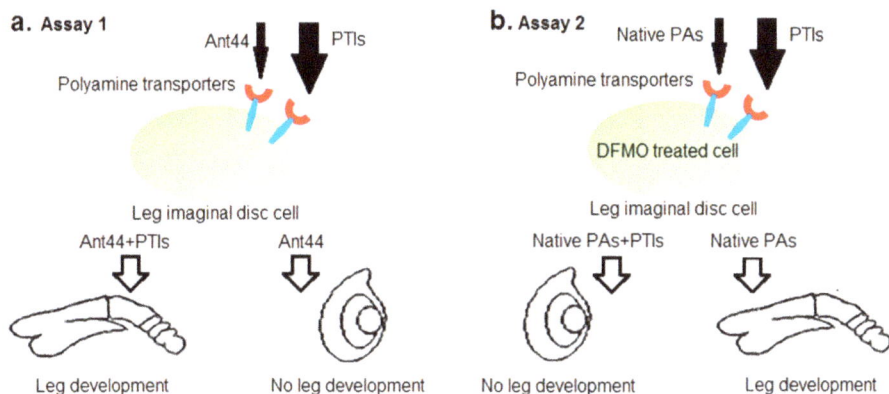

Figure 2. *Drosophila* assays used to characterize polyamine transport inhibitors. Native PAs: native polyamines; PTIs: polyamine transport inhibitors. (**a**) Assay 1: Undeveloped leg imaginal discs were incubated with ecdysone to promote development. In the presence of Ant44 (**5**), a toxic polyamine analog that targets the transport system, leg imaginal discs will not develop. The ability of candidate PTIs to rescue development of imaginal discs treated with Ant44 (**5**) was then assayed by monitoring and scoring the leg development process. (**b**) Assay 2: Leg imaginal discs treated with DFMO fail to develop in the presence of ecdysone. Uptake of exogenous native polyamines can rescue DFMO inhibition of disc development. The ability of candidate PTIs to block rescue of disc development in the presence of DFMO and native polyamines was tested.

2. Materials and Methods

2.1. Synthesis

The synthesis of the anthracene-polyamine conjugates (**5** and **6**) and the aryl-polyamine conjugates (**7–9**) have been described [39,40].

2.2. Drosophila Strains and Larval Collections

The Oregon-R variant of *Drosophila melanogaster* was used in all experiments. Larval preparation and staging were performed as previously described [37,41]. All larvae used in the experiments were synchronized to within 7 h of pupariation, immediately prior to the pulse of 20-hydroxyecdysone that triggers imaginal disc development. Imaginal discs dissected from larvae at this developmental stage develop into rudimentary legs when exposed to 20-hydroxyecdysone in in vitro culture.

2.3. Imaginal Disc Culture and Scoring

Leg imaginal discs were dissected at room temperature in Ringer's solution (130 mM NaCl, 5 mM KCl, 15 mM CaCl$_2$·2H$_2$O) containing 0.1% bovine serum albumin (BSA, w/v), which was added to the Ringer's solution immediately prior to use. Up to 150 discs were dissected in less than 1 h to avoid prolonged storage in Ringer's solution. After dissection, discs were transferred to 12-well plastic culture plates containing Ringer's solution (1 mL). Before the disc culture medium was added, dissected imaginal discs were washed once with 1× minimal Robb's medium (see Section 2.4). To begin a culture, a solution of 1 mL of 1× minimal Robb's medium (final concentration) containing 20-hydroxyecdysone (1 µg/mL) and each of the compounds to be tested was added to each well. Control experiments lacking polyamine transport inhibitor (PTI) were run in parallel. Imaginal discs were incubated for 18 h at 25 °C. After 18 h, the discs were scored as developed or non-developed. Fully developed discs (the leg is fully extended from the epithelium) and partially developed discs (the leg protrudes from the epithelium but is not fully extended) were scored as developed. Non-developed discs showed no sign of development. For each experiment, the percent development was determined by ([(number of developed discs)/(total number of discs)] × 100.

2.4. Robb's Minimal Medium

2× Minimal Robb's medium consisting of 80 mM KCl, 0.8 mM KH$_2$PO$_4$, 80 mM NaCl, 0.8 mM NaH$_2$PO$_4$·7H$_2$O, 2.4 mM MgSO$_4$·7H$_2$O, 2.4 mM MgCl$_2$·6H$_2$O, 2 mM CaCl$_2$·2H$_2$O, 20 mM glucose, 8.0 mM L-glutamine, 0.32 mM glycine, 1.28 mM L-leucine, 0.64 mM L-proline, 0.32 mM L-serine and 1.28 mM L-valine, pH 7.2 was prepared and stored at −20 °C. Immediately prior to use, 20 µL of 10% BSA (w/v) was added to 1 mL of medium [42].

2.5. Statistical Analysis

Statistical analysis was performed using IBM SPSS Statistics 19 with one-way ANOVA.

3. Results

In order to identify PTIs using the *Drosophila* assay, we selected four compounds for study. Ant444 (**6** in Figure 1) is a N^1-anthracenylmethyl substituted polyamine that binds tightly to the surface of mammalian A375 cells with high affinity for the PTS, which suggests that it could be an effective transport inhibitor [39]. However, the ability of this compound to inhibit polyamine transport has never been directly demonstrated. We also tested Triamide444 (**9** in Figure 1), a compound with relatively high toxicity in Chinese Hamster Ovary (CHO) and human pancreatic cancer L3.6pl cells, which precluded an analysis of its transport inhibitory properties in these cell lines. Trimer44 (**7** in Figure 1) has been previously shown to be an effective inhibitor of spermidine uptake in the presence of DFMO in mammalian L3.6pl cells. [40,43] Triamide44 (**8** in Figure 1) was previously shown to be a poor transport inhibitor [40]. We, therefore, used the transport inhibition properties of Trimer44 (**7**) and Triamide44 (**8**) as a baseline for comparison to Ant444 (**6**) and Triamide444 (**9**). Armed with these molecular tools, we assessed their ability to perform as PTIs in the *Drosophila* model.

3.1. Compounds Ant444 (**6**) and Triamide444 (**9**) Block the Toxicity of the Polyamine Analog Ant44 (**5**) that Gains Entry to Cells via the PTS

In the first experiments, all compounds were tested in two different *Drosophila* assays. In Assay 1, these compounds were tested for their ability to block toxicity of a polyamine analogue, Ant44 (**5**, Figure 1), which gains access to leg imaginal disc cells via the polyamine transport system (Figure 2a) [37,39]. At the concentrations of Ant44 (**5**) used in our experiments (40–50 µM), fewer than 10% of imaginal discs develop. We hypothesized that an effective PTI would inhibit Ant44 uptake or release, and thus reduce the toxicity of Ant44 (**5**) and permit development of leg imaginal discs. A potential caveat of this approach is that a toxic PTI compound would generate a false negative result in this assay. Therefore, it was critical that we first determine the highest dose of PTI compound that could be used without toxicity to avoid biasing the results.

Addition of Ant444 (**6**) and Triamide444 (**9**) at non-toxic concentrations to the assay showed significant rescue of imaginal disc development in the presence of Ant44 (**5**) (Figure 3a,b). Their effectiveness as PTIs was ranked via determination of EC_{50} values. The EC_{50} value was defined as the effective concentration of the compound which decreased the inhibition of disc development by Ant44 (**5**) to 50% of the untreated control value (i.e., 50% inhibited). For both Ant444 (**6**) and Triamide444 (**9**) the EC_{50} values (3.6 and 2.8 µM, respectively) were 10 to 15-fold lower than the concentration of Ant44 (**5**, e.g., 40–50 µM) used in the assays. Maximum protection from Ant44 was observed at 10 µM **6** and 5 µM **9**, respectively. These activity profiles are similar to Trimer44 (**7**) which is an effective transport inhibitor in mammalian Chinese Hamster Ovary (CHO) and L3.6pl cells (Figure 3c) [40]. In contrast, Triamide44 (**8**) was a less effective PTI in the *Drosophila* model with an EC_{50} of 144 µM and gave maximum protection at 300 µM (Figure 3d). These observations are consistent with similar findings in mammalian L3.6pl cells [40].

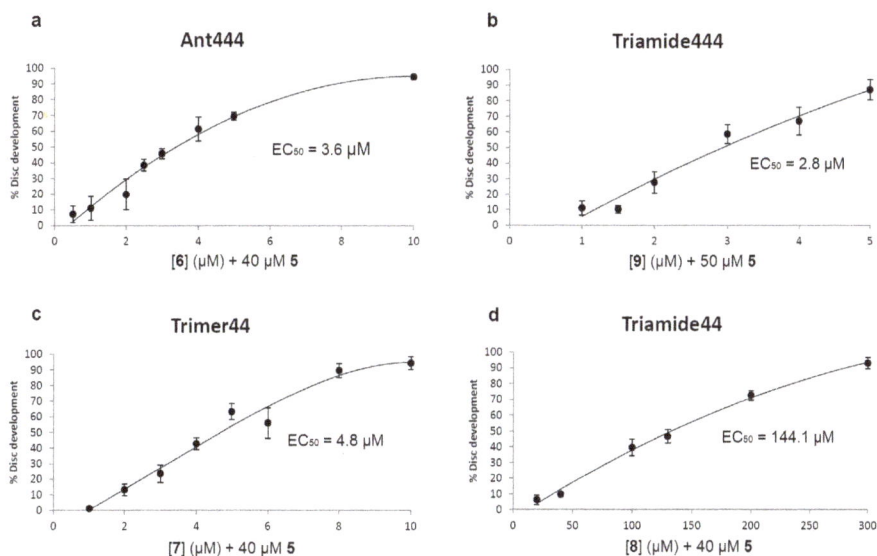

Figure 3. Compounds Ant444 (**6**) and Triamide444 (**9**) are effective PTIs. Candidate PTIs Ant444 (**6**) and Triamide444 (**9**) were tested in the presence of a toxic concentration of Ant44 (**5**) that by itself permitted the development of fewer than 10% of imaginal discs. The percentage of imaginal discs that developed was determined for each PTI concentration tested (see Section 2.3 for details). All assays were repeated at least in triplicate. Error bars reflect the standard error of the mean (SEM). (**a–d**) Respective dose-response curves of (**a**) Ant444 (**6**), (**b**) Triamide444 (**9**), (**c**) Trimer44 (**7**) and (**d**) Triamide44 (**8**) in blocking the inhibitory effect of Ant44 (**5**) on imaginal disc development. Note: the EC_{50} value is the concentration of the compound needed to block 50% of the inhibitory effect of Ant44 (**5**) on imaginal disc development.

3.2. Ant444 (6) and Triamide444 (9) Are More Effective than the Native Polyamines in Inhibiting the Toxicity of Ant44 (5) in Imaginal Discs

Compounds containing recognizable polyamine sequences should be able to compete for access to the polyamine receptor on the cell surface. Our previous work has shown that spermidine is able to inhibit the toxicity of Ant44 (**5**) on mammalian cells and *Drosophila* leg imaginal discs by competing for binding and transport via the PTS [37]. In the present study, the efficiencies of the native polyamines (spermidine and spermine) in rescuing disc development from a toxic concentration of Ant44 (**5**) were evaluated in Assay 1 (Figures 2a and 4a,b). As shown in Figure 4a, the EC_{50} of spermidine was 43.6 μM and complete rescue of imaginal disc development was observed at 80 μM. In contrast, the EC_{50} values of Ant444 (**6**) and Triamide444 (**9**) are 3.6 μM and 2.8 μM, respectively (Figure 3a,b). In short, compounds **6** and **9** were approximately 12–15 times better than spermidine in inhibiting the toxicity of Ant44 (**5**).

Spermine—a native tetraamine—was more effective than spermidine in blocking Ant44 (**5**) inhibition of imaginal disc development with an EC_{50} value of 19.7 μM and afforded complete protection at 40 μM (Figure 4b). The EC_{50} values of Ant444 (**6**) and Triamide444 (**9**) were 5-fold and 7-fold lower than spermine respectively, demonstrating that these compounds are more efficient at competing for access to the PTS than either of the native polyamines spermidine or spermine. The data for Ant444 (**6**) and Triamide444 (**9**) are similar to Trimer44 (**7**). In contrast, Triamide44 (**8**) (EC_{50} 144 μM; Figure 3d) was 3-fold *less* effective than spermidine and 7-fold *less* effective than spermine in inhibiting the toxicity of Ant44 (**5**).

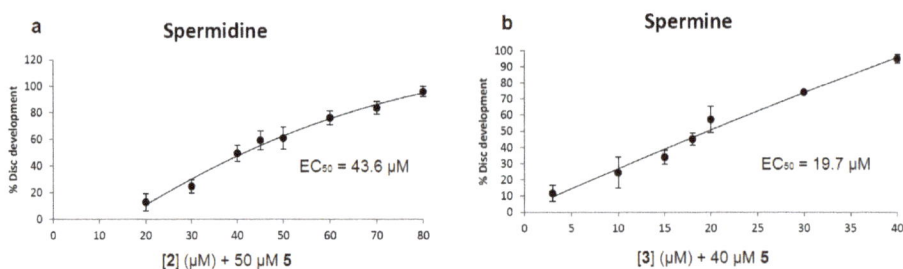

Figure 4. Spermidine and spermine block the inhibitory effect of Ant44 (**5**) on imaginal disc development. Spermidine (**2**) and spermine (**3**) were tested at different concentrations in the presence of Ant44 (**5**) and the percentage of imaginal disc development was recorded for each concentration. (**a**) Effective concentration of spermidine in blocking the inhibitory effect of Ant44 (**5**) on imaginal disc development; (**b**) Effective concentration of spermine in blocking the inhibitory effect of Ant44 (**5**) on imaginal disc development. Every data point was repeated at least in triplicate. Error bars reflect the standard error of the mean (SEM). Note: the EC_{50} value is the concentration of the polyamine needed to block 50% of the inhibitory effect of Ant44 (**5**) on imaginal disc development.

In contrast to spermidine and spermine, the native diamine, putrescine, was unable to rescue the inhibition of Ant44 (**5**) in imaginal discs. Concentrations of up to 1 mM putrescine had no effect on the inhibition of imaginal disc development by Ant44 (**5**) (Figure S1). One interpretation of these observations is that the diamine putrescine presents fewer charges to the cell surface receptors than Ant44 (**5**), which is a triamine analogue. Therefore, the inability of putrescine to rescue cells from Ant44 (**5**) could be due to differences in relative binding affinity. An alternative interpretation is that Ant44 (**5**) is imported into the cell via a polyamine transporter which does not recognize putrescine. Indeed, the existence of multiple polyamine transporters with different affinities and selectivity for the native polyamines has been suggested in mammalian cells [44] and also in this study (see Section 3.4).

In conclusion, Ant444 (**6**), Trimer44 (**7**) and Triamide444 (**9**) are all considerably more effective than either of the native polyamines spermidine or spermine in competing with Ant44 (**5**) for access to the PTS. Because putrescine could not rescue Ant44 (**5**) toxicity in disc development, no comparisons can be made for this native diamine.

*3.3. Ant444 (**6**) and Triamide444 (**9**) Effectively Prevent Rescue by Native Polyamines of DFMO-Treated Imaginal Discs*

Assay 1 tested the ability of candidate PTIs to block the toxicity of Ant44 (**5**), which accessed cells via the PTS (Figure 2a). Assay 2 tests the ability of each candidate PTI to block the uptake of exogenous polyamines into DFMO-treated imaginal discs (Figure 2b). Since DFMO inhibits polyamine biosynthesis [45], intracellular polyamine levels are depleted and cell viability is decreased. The effect of DFMO in mammalian cell culture is dose-dependent and typically cytostatic and this inhibition can be reversed by the addition of native polyamines to the cell culture medium [14,16]. Therefore, we investigated if DFMO inhibits imaginal disc development and if the compounds Ant444 (**6**), Trimer44 (**7**), Triamide44 (**8**) and Triamide444 (**9**) could prevent the rescue of DFMO-treated imaginal disc development by exogenous native polyamines. Essentially, we asked if these compounds could effectively compete with native polyamines for access to the PTS in DFMO-treated imaginal discs.

When imaginal discs are cultured in the presence of 10 mM DFMO greater than 95% of the discs fail to develop (Figure 5). As in mammalian cell culture, DFMO inhibition of disc development was dose-dependent. As shown in Figure 5, the 18 h IC_{50} value of DFMO on imaginal disc development was 4.4 mM, a value similar to that reported for CHO cells at 48 h and L3.6pl cells at 72 h [40]. Here the 18 h IC_{50} value is defined as the concentration of DFMO required to inhibit 50% of leg development after 18 h of incubation.

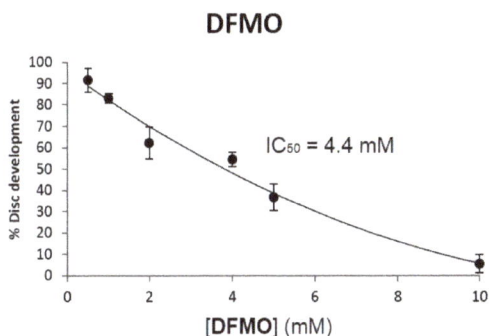

Figure 5. DFMO inhibits imaginal disc development. DFMO (**4**) was tested at different concentrations and the percentage of disc development was determined for each concentration after 18 h of incubation. All data points were repeated at least in triplicate and error bars reflect the standard error of the mean (SEM). The IC_{50} value corresponds to the concentration of DFMO needed to inhibit 50% of discs from developing into rudimentary legs.

In the presence of exogenous polyamines, polyamines from outside the cell should enter into imaginal disc cells to rescue inhibition of development by DFMO. In contrast, in the presence of DFMO and an effective PTI, exogenous polyamines are expected to be unable to gain access to the cell resulting in inhibition of development. DFMO was used at 10 mM in all experiments because at this dose imaginal discs showed little development and retained the same shape as controls treated with culture medium only (i.e., with no steroid hormone to stimulate development). Data for these experiments are shown in Figures 6 and 7.

Each of the three native polyamines were evaluated for their ability to rescue the development of leg discs treated with DFMO. Addition of 500 μM putrescine to the culture medium resulted in a significant increase (5% to 59%, Figure 6a; 4% to 66%, Figure 7a) in imaginal disc development compared to DFMO alone (compare blue and green columns in both Figures). Similarly, addition of 200 μM spermidine or spermine to DFMO-treated leg discs also significantly increased imaginal disc development (see Figures 6 and 7). Thus, each of the native polyamines was able to rescue imaginal disc development in the presence of DFMO (10 mM). These results mirror the ability of DFMO-treated mammalian cells to be rescued by each of the native polyamines. We note that the concentrations of native polyamines needed to rescue inhibition by DFMO in the *Drosophila* model assay were much higher (200–500 μM) than those observed in mammalian cells (around 1 μM). The higher doses are likely due to the fact that unlike cell culture, imaginal discs are an intact epithelial tissue surrounded by extracellular matrix, which may impede polyamine access to the PTS.

As with Assay 1, it was important to use a non-toxic dose of each PTI compound because in Assay 2 a toxic PTI would generate a false positive. To avoid introducing this bias, non-toxic concentrations of the PTI compounds were determined and used in both assays. In a series of control experiments, Ant444 (**6**), Trimer44 (**7**) and Triamide444 (**9**) were each found to be non-toxic to imaginal disc development at 100 µM, whereas Triamide 44 (**8**) was non-toxic at 300 µM.

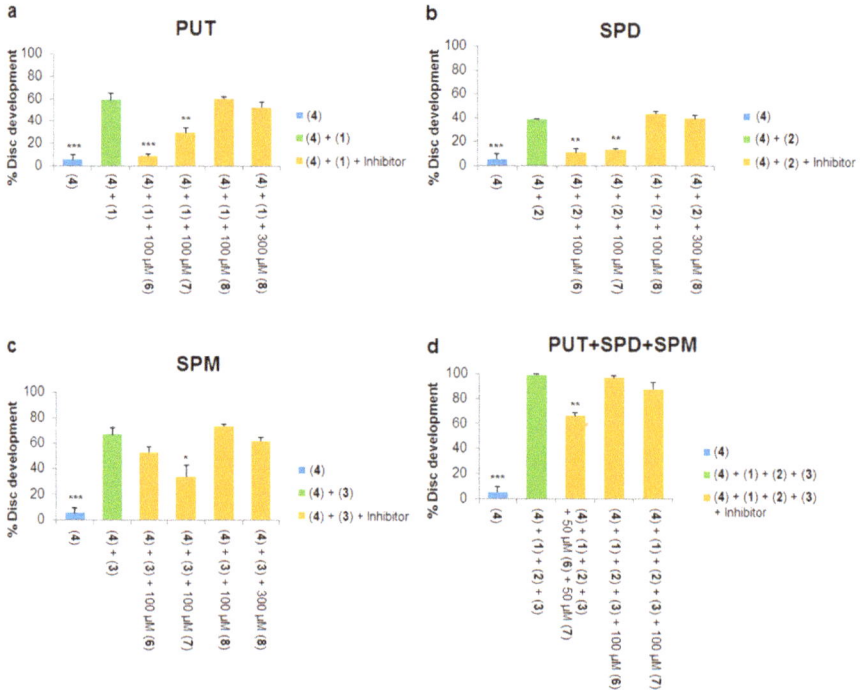

Figure 6. Candidate PTIs prevent native polyamine rescue of imaginal discs treated with DFMO. Polyamine transport inhibitors at the indicated concentrations were used to block the rescue of DFMO treated imaginal discs (10 mM) by the native polyamines putrescine (PUT; **1**), spermidine (SPD; **2**) and spermine (SPM; **3**). DFMO alone (10 mM) results in approximately 5% imaginal disc development. Native polyamines were tested at the following concentrations (putrescine **1**: 500 µM; spermidine **2**: 200 µM; spermine **3**: 200 µM). Polyamines and PTIs were individually tested in the absence of DFMO to ensure there was no inhibition of imaginal disc development at the concentrations used. In addition, polyamines and PTIs were tested in combination for possible negative synergy on imaginal disc development and none was observed at the concentrations used. Compounds are numbered as described in Figure 1. All data points were repeated at least in triplicate and error bars reflect the SEM. Significant differences * $p < 0.05$; ** $p < 0.01$; *** $p < 0.001$ from treatment with DFMO and native polyamine alone are indicated. (**a**) Ability of PTIs to prevent rescue of DFMO treated imaginal discs with putrescine; (**b**) ability of PTIs to prevent rescue of DFMO treated imaginal discs with spermidine; (**c**) ability of PTIs to prevent rescue of DFMO treated imaginal discs with spermine; and (**d**) ability of PTIs to prevent rescue of DFMO treated imaginal discs with a cocktail containing all three native polyamines (putrescine, spermidine and spermine).

We next asked if non-toxic concentrations of PTIs could block the developmental rescue of DFMO-treated imaginal discs by native polyamines. Rescue of DFMO-treated imaginal discs by putrescine was significantly reduced by addition of 100 µM of Ant444 (**6**), Trimer44 (**7**) and Triamide444 (**9**) (Figures 6a and 7a). Imaginal disc development was reduced from 59% (500 µM putrescine, 10 mM DFMO) to 9% in the presence of 100 µM Ant444, 500 µM putrescine and 10 mM DFMO, a result which was similar to the control with DFMO alone (Figure 6a). Likewise, addition of Trimer44 (100 µM) in the presence of putrescine and DFMO reduces imaginal disc development from 59% to 29% (Figure 6a). Addition of 100 µM Triamide444 reduced imaginal disc development from 66% to 18% (Figure 7a). While the decrease in imaginal disc development in the presence of Ant444 (**6**), Trimer44 (**7**) or Triamide444 (**9**) is significant, our data suggest that Trimer44 (**7**) is less effective than Ant444 (**6**) or Triamide444 (**9**) in inhibiting the uptake of putrescine. Consistent with earlier studies, 100 µM or 300 µM Triamide44 (**8**) was unable to compete with putrescine for access to the *Drosophila* PTS (Figure 6a).

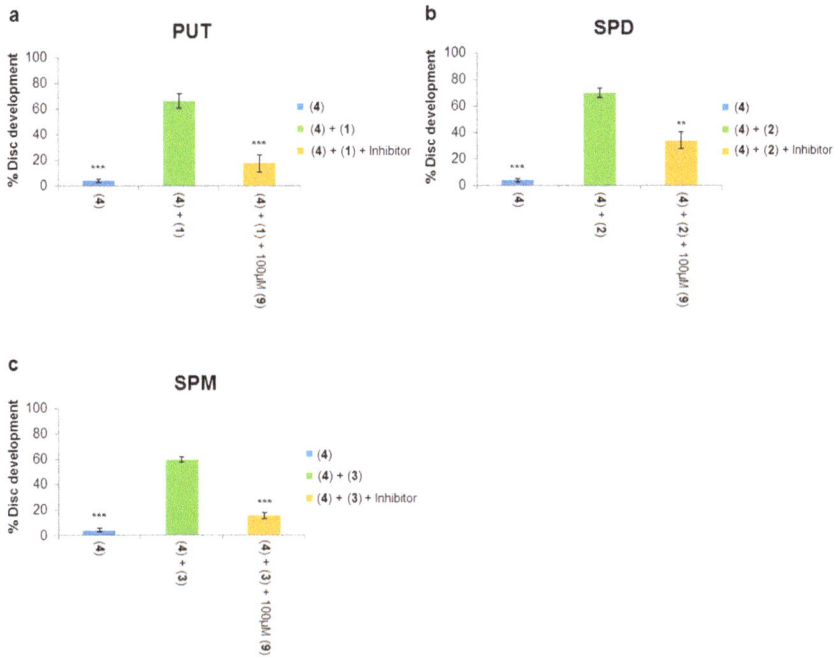

Figure 7. Triamide444 (**9**) is an effective inhibitor of native polyamine uptake. The ability of 100 µM Triamide444 (**9**) to block the rescue of DFMO **4** treated imaginal discs (10 mM) by native polyamines putrescine (PUT; **1**), spermidine (SPD; **2**) and spermine (SPM; **3**). 10 mM DFMO alone (10 mM) results in approximately 5% disc development. Native polyamines were tested at the following concentrations (putrescine **1**: 500 µM; spermidine **2**: 200 µM; spermine **3**: 200 µM). Triamide444 (**9**) and individual polyamines were tested in the absence of DFMO to ensure there was no inhibition of imaginal disc development at the concentrations used. In addition, Triamide444 and individual polyamines were tested in combination for possible negative synergy on imaginal disc development and none was observed at the concentrations used. All data points were repeated at least in triplicate and error bars reflect the SEM. Significant differences * $p < 0.05$; ** $p < 0.01$; *** $p < 0.001$ from treatment with DFMO and native polyamine alone are indicated. Ability of Triamide444 (**9**) to prevent rescue of DFMO treated imaginal discs with (**a**) putrescine; (**b**) spermidine and (**c**) spermine.

Similar results were observed with spermidine. At 100 µM, Ant444 (**6**), Trimer44 (**7**) and Triamide444 (**9**) were all able to significantly inhibit import of spermidine. In the presence of 10 mM DFMO and 200 µM spermidine imaginal disc development decreased from 39% to 11% in the presence of 100 μM Ant444 and to 13% in the presence of 100 μM Trimer44 (Figure 6b). In the presence of 100 μM Triamide444 imaginal disc development decreased from 70% to 34% (Figure 7b). In contrast, Triamide44 (**8**) failed to inhibit import of spermidine even at 300 μM (Figure 6b).

Finally, we tested the ability of the PTIs to inhibit import of spermine in the presence of 10 mM DFMO and 200 μM spermine. As shown in Figure 6c, 100 μM Ant444 (**6**) did not reduce uptake of spermine, whereas 100 μM Trimer44 (**7**) significantly reduced imaginal disc development from 67% to 34%. Triamide444 (**9**) showed even greater ability to reduce spermine uptake reducing imaginal disc development from 60% to 15% (Figure 7c). Thus, the PTIs can be ranked Triamide444 > Trimer44 > Ant444 with respect to their relative abilities to inhibit spermine uptake. As with the previous assays, Triamide44 (**8**) was unable to inhibit import of spermine even at 300 μM concentration.

In summary, even though Ant444 (**6**), Trimer44 (**7**) and Triamide444 (**9**) have similar EC_{50} values for protection against toxicity of Ant44 (**5**) and a similar concentration of full protection against Ant44 (**5**) (Figure 3a–c), they show different specificities in blocking the uptake of native polyamines into imaginal discs treated with DFMO (Figures 6a–c and 7). Ant444 (**6**) is better at blocking uptake of putrescine, Ant444 (**6**) and Trimer44 (**7**) show similar abilities to block uptake of spermidine and Triamide444 (**9**) is the most potent of the PTIs at blocking spermine uptake. These findings suggest that the PTIs have different specificities for the polyamine transport systems active in the presence of DFMO. In this regard, there may be a basal and DFMO-stimulated PTS in *Drosophila*. The basal PTS is assessed via the Ant44 assay (Assay 1), whereas the DFMO-stimulated PTS is assessed via Assay 2 (Figure 2). The poor performance of triamide44 (**8**) in these assays is consistent with the inability of this compound to block the toxicity of Ant44 (**5**) (Figure 3d) and suggests that presenting polyamine chains containing only two charges per polyamine arm limits interactions with the putative PTS extracellular receptor (e.g., glypican-1 anchored heparan sulfate proteoglycans [24]).

3.4. A Cocktail of Ant444 (**6**) and Trimer44 (**7**) Is More Potent than Either Compound Alone at Inhibiting the Import of Native Polyamines into DFMO-Treated Imaginal Discs, Suggesting the Existence of Multiple Transport Systems

In the next experiments, we further examined our finding that the PTIs have different specificities for the PTS. Specifically, we asked if a cocktail of PTIs was more effective than individual PTIs in inhibiting rescue of DFMO-treated imaginal discs in the presence of all three native polyamines. In our prior experiments, we studied the effects of individual native polyamines, however, all three polyamines are present in vivo. For example in circulating red blood cells, the levels of putrescine, spermidine and spermine were found to be 3, 55 and 35 pmol/mg protein respectively [46]. Because Ant444 (**6**) and Trimer44 (**7**) are effective PTIs and showed different specificities towards putrescine, spermidine and spermine respectively, a combination of these inhibitors was used to block the rescue of DFMO treated leg discs by a mixture of all of the native polyamines. As shown in Figure 6d, a cocktail of native polyamines (500 μM putrescine, 200 μM spermidine and 200 μM spermine) was able to fully rescue inhibition of leg disc development by DFMO (compare blue and green columns). In contrast to experiments using just one PTI, 100 μM of either Ant444 (**6**) or Trimer44 (**7**) alone was unable to significantly inhibit the rescue of DFMO-treated discs by the exogenous native polyamine cocktail. In contrast, a combination of 50 μM Ant444 (**6**) and 50 μM Trimer44 (**7**) significantly inhibited rescue by native polyamines even though the amount of each PTI was reduced by half compared to experiments when only one PTI was used. This result suggests that a combination of polyamine transport inhibitors will be more effective in inhibiting the import of all three native polyamines than individual inhibitors dosed alone.

The different selectivity of Ant444 (**6**), Trimer44 (**7**) and Triamide444 (**9**) towards native polyamines and the ability of a cocktail of PTIs to inhibit transport more effectively than individual PTI's suggests the existence of multiple polyamine transport systems in *Drosophila* as has been observed in unicellular organisms [44]. Ant444 (**6**) shows a greater ability to inhibit uptake of putrescine, whereas Trimer44 (**7**) is more effective in inhibiting uptake of spermine (Figure 6a,c) which may be the underlying basis for the improved ability of a cocktail of these compounds to inhibit rescue in the presence of all three native polyamines. Further support for multiple transporters with different specificities for the native polyamines comes from our observation that 500 µM putrescine can rescue inhibition by DFMO (Figure 6a) whereas 1 mM putrescine is unable to rescue the toxicity of 40 µM Ant44 (**5**) (Figure S1), consistent with the notion that putrescine is imported into cells through a transport system different from Ant44. The underlying transport pathway selection may be charge-dependent because unlike the diamine putrescine, Ant44 is a triamine and presents three positive charges to the putative cell surface receptor. In addition, Ant44 is a homospermidine analogue and its toxicity can be rescued by the higher polyamines, spermidine and spermine. An alternative explanation for our observations is that Ant44 bears both a hydrophobic anthryl substituent along with the hydrophilic polyamine head group and thus its amphiphilic properties may facilitate its uptake via a specific transport system.

4. Discussion

Our work reinforces the value of the *Drosophila* imaginal disc assay as an early and inexpensive system in which to evaluate compounds targeting the mammalian PTS. There are several advantages to our approach. First, mammalian cell culture is not a natural cellular environment because cells lack cell-cell contacts and extracellular matrix, both of which are factors influencing drug accessibility to cells in vivo. In contrast, the imaginal disc assay tests the effects of medicinal compounds on cells in a more natural environment. Second, inexpensive early animal model testing of promising compounds can reduce the time it takes successful compounds to reach the clinic by up to fifty percent. Mice are more expensive to use in the early stages of drug development where most compounds will fail, therefore a cheaper system such as our *Drosophila* assay is useful. Third, experiments in mice can only be performed on a small scale, whereas we can assay relatively large numbers of imaginal discs, typically more than 100 per assay.

Of course, the imaginal disc assay is only useful to understand mammalian transport if the polyamine transport system is similar in *Drosophila* and mammals. Our work suggests that this is the case. In previous work, we compared the uptake of nine polyamine analogs in mammalian CHO and L1210 cells and *Drosophila* imaginal discs [37]. Two of the compounds tested in those experiments, Ant44 (**5**) and N^1-(3-aminopropyl)-N^4-(anthracen-9-ylmethyl)butane-1,4-diamine (Ant43) gain entry to mammalian cells via the polyamine transport system as evidenced by spermidine competition experiments and greatly reduced uptake in CHO-MG cells, which lack a functional transport system [47]. In imaginal discs, uptake of Ant44 and Ant43 is also greatly reduced in spermidine competition experiments. In contrast, uptake of the other seven polyamine analogs cannot be competed with spermidine in mammalian cells or *Drosophila* imaginal discs, suggesting that they do not utilize the transport system to gain access to cells in either system. In addition, Trimer44 (**7**) has previously been shown to be an effective inhibitor in mammalian cells, whereas Triamide44 (**8**) was not [40] and these results are mirrored in the *Drosophila* assay.

Use of the *Drosophila* imaginal disc assay has added to our knowledge of polyamine transport inhibitors. We show that two compounds that exhibit toxicity in mammalian cell culture, Ant444 (**6**) and Triamide444 (**9**), are non-toxic in the *Drosophila* assay and are effective PTIs with activity profiles similar to that of Trimer44 (**7**). The reduced toxicity of Ant444 and Triamide444 in *Drosophila* may due to a lower effective concentration of these compounds reaching the cell surface due to the presence of intact cell-cell adhesions and extracellular matrix. We also provide activity data for the PTIs against all three native polyamines (putrescine, spermidine, spermine) whereas most mammalian cell culture studies focus on spermidine uptake. This approach revealed differences in the ability of each PTI to inhibit uptake of individual polyamines suggesting the existence of multiple transport systems. This view is further reinforced by our finding that a mixture of two PTIs is more effective than either PTI alone at inhibiting uptake of a cocktail of all three native polyamines.

In this study, we assayed the ability of PTIs to inhibit the rescue of DFMO treated imaginal discs in the presence of exogenous polyamines. This approach is clinically relevant in that many tumors circumvent DFMO treatment via upregulation of their polyamine transport systems. Our previous work indicates that the PTIs inhibit polyamine uptake. Our data are consistent with the reported K_i values for several of these compounds in terms of competing with ^{3}H-radiolabeled spermidine for the putative cell surface receptors in L1210 murine leukemia cells. The L1210 K_i values for putrescine, spermidine and spermine are 208.2, 2.46 and 1.34 µM, respectively [39]. The L1210 cell K_i values for Ant44 (**5**), Ant444 (**6**) and Trimer44 (**7**) are 1.8 µM, 0.05 µM and 0.49 µM, respectively [39,43]. Although the K_i value of Triamide44 (**8**) was not determined in L1210 cells, a comparative study of the Trimer44 and Triamide44 compounds in human L3.6pl pancreatic cancer cells revealed K_i values of 36 nM and 398 nM, respectively [40], suggesting a significantly lower affinity of Triamide44 for the putative cell surface receptors of the polyamine transport system.

The low K_i values of Ant44, Ant444 and Trimer44 suggest that these compounds compete with the native polyamines for uptake. For example, Ant44 (a triamine) has a L1210 K_i value of 1.8 µM and provides a fluorescent molecule with similar affinity for the polyamine transport system as the native polyamines spermidine (L1210 K_i = 2.46 µM) and spermine (L1210 K_i = 1.34 µM). We speculate that in order to be successfully imported, compounds must bind and release from the cell surface receptors. The K_i values of the native polyamines (spermidine and spermine) suggest that K_i values in the low µM range are optimal for these binding and releasing properties. The related Ant444 compound **6** (a tetraamine) has a significantly lower L1210 K_i value (51 nM) indicating high affinity for the cell surface receptors. Using confocal microscopy, we have demonstrated that this higher affinity of Ant444 was observed as a compound which could not be washed off the surface of L1210 cells by phosphate buffered saline (PBS). In contrast, the triamine Ant44 could be readily washed off the surface of L1210 cells by PBS and appeared to have improved uptake past the cell membrane [39]. This data is consistent with the higher toxicity of Ant44 (5:48 h L1210 IC_{50} = 0.3 µM) compared to Ant444 (6:48 h L1210 IC_{50} = 7.5 µM) [39]. In summary, highly charged lipophilic tetraamines like Ant444 tend to stick and not enter, which likely contributes to their ability to act as less toxic PTIs.

Our finding that Ant444 (**6**) and Triamide444 (**9**) are effective PTIs expands our understanding of the chemical rules governing an effective PTI design. Inhibitors presenting diamine arms, like Triamide44 (**8**), are ineffective transport inhibitors. In contrast, compounds containing a higher number of charges in their polyamine arms such as Trimer44 (**7**) and Triamide444 (**9**) are effective PTIs. In this regard, N^1-substituted triamine and tetraamine analogues can be used to design efficient ligands and inhibitors of polyamine transport. Our work and previous studies suggest that presentation of at least three or more positive charges is necessary for efficient competitive binding to the PTS.

A combination therapy using DFMO and a PTI has shown promise in cancer growth inhibition [48,49]. While the lack of knowledge of the genes and proteins involved in polyamine transport has hampered the development of PTIs, structure-activity relationship studies have nevertheless resulted in the development of effective PTIs. One effective PTI is AMXT-1501 (**11**, Figure 8) [48]. In combination with DFMO, AMXT-1501 inhibits cancer cell growth in several cancer cell lines and mouse models [49]. Recently this compound was also found to reverse immunosuppression in the tumor microenvironment [50]. Structurally the compounds we tested here are different from AMXT-1501, which is a lipophilic palmitic acid–lysine spermine conjugate. Indeed, the hydrophilic compound **12** (Figure 8), which is a N-methylated derivative of Trimer44 (**7**), was recently shown to behave in a similar manner as AMXT-1501 (**11**) both in its ability to shrink tumors in vivo as well as to beneficially modulate the immune response [51]. Thus, this report provides alternative three-arm PTI designs and new insights as to how combinations of PTIs can be used to effectively inhibit the import of all three native polyamines. Going forward the *Drosophila* model can be used to pre-screen PTIs prior to more expensive testing in mouse models. Having a cheap model system for early animal testing will reduce the time from conceptual PTI design to future validation in clinical trials.

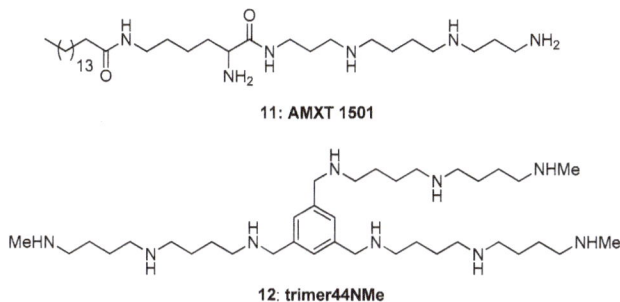

Figure 8. Structures of PTI compounds **11** and **12**.

Supplementary Materials: Supplementary Materials are available online at www.mdpi.com/2076-3271/5/4/27/s1.

Acknowledgments: The authors thank David Jenkins of the Department of Biology for the help with statistical analysis. L.v.K. thanks the UCF Colleges of Sciences for support for this work. The authors thank the 2011 Department of Defense Congressionally Directed Medical Research Program (CDMRP) Peer Review Cancer Research Program (PRCRP) Discovery Award CA110724 for financial support of the synthesis of the PTI compounds used herein. The funding sources had no involvement in study design, collection, analysis or interpretation of data, writing of the manuscript or the decision to submit for publication.

Author Contributions: M.W. and L.v.K. conceived and designed the experiments; M.W. performed the experiments; L.v.K. and M.W. analyzed the data; O.P. contributed reagents/materials/analysis tools. All authors contributed to the writing of the paper.

Conflicts of Interest: The authors declare no competing financial interests.

References

1. Cohen, S.S. *A Guide to the Polyamines*; Oxford University Press: Oxford, UK, 1998; pp. 1–543.

2. Igarashi, K.; Kashiwagi, K. Polyamines: Mysterious modulators of cellular functions. *Biochem. Biophys. Res. Commun.* **2000**, *271*, 559–564. [CrossRef] [PubMed]

3. Packham, G.; Cleveland, J.L. Ornithine decarboxylase is a mediator of c-Myc-induced apoptosis. *Mol. Cell. Biol.* **1994**, *14*, 5741–5747. [CrossRef] [PubMed]

4. Oriol-Audit, C. Polyamine-induced actin polymerization. *Eur. J. Biochem.* **1978**, *87*, 371–376. [CrossRef] [PubMed]

5. Watanabe, S.; Kusama-Eguchi, K.; Kobayashi, H.; Igarashi, K. Estimation of polyamine binding to macromolecules and ATP in bovine lymphocytes and rat livers. *J. Biol. Chem.* **1991**, *266*, 20803–20809. [PubMed]

6. Lopatin, A.N.; Makhina, E.N.; Nichols, C.G. Potassium channel block by cytoplasmic polyamines as the mechanism of intrinsic rectification. *Nature* **1994**, *372*, 366–369. [CrossRef] [PubMed]

7. Williams, K.; Dawson, V.L.; Romano, C.; Dichter, M.A.; Molinoff, P.B. Characterization of polyamines having agonist, antagonist and inverse agonist effects at the polyamine recognition site of the NMDA receptor. *Neuron* **1990**, *5*, 199–208. [CrossRef]

8. Matsufuji, S.; Miyazaki, Y.; Kanamoto, R.; Kameji, T.; Murakami, Y.; Baby, T.G.; Fujita, K.; Ohno, T.; Hayashi, S. Analyses of ornithine decarboxylase antizyme mRNA with a cDNA cloned from rat liver. *J. Biochem.* **1990**, *108*, 365–371. [CrossRef] [PubMed]

9. Wallace, H.M.; Fraser, A.V.; Hughes, A. A perspective of polyamine metabolism. *Biochem. J.* **2003**, *376*, 1–14. [CrossRef] [PubMed]

10. Igarashi, K.; Kashiwagi, K. Polyamine transport in bacteria and yeast. *Biochem. J.* **1999**, *344*, 633–642. [CrossRef] [PubMed]

11. Seiler, N.; Delcros, J-G.; Moulinoux, J.P. Polyamine transport in mammalian cells. An update. *Int. J. Biochem. Cell Biol.* **1996**, *28*, 843–861. [CrossRef]

12. Igarashi, K.; Ito, K.; Kashiwagi, K. Polyamine uptake systems in *Escherichia coli*. *Res. Microbiol.* **2001**, *152*, 271–278. [CrossRef]

13. Koomoa, D.L.; Yco, L.P.; Borsics, T.; Wallick, C.J.; Bachmann, A.S. Ornithine decarboxylase inhibition by α-difluoromethylornithine activates opposing signaling pathways via phosphorylation of both Akt/protein kinase B and p27Kip1 in neuroblastoma. *Cancer Res.* **2008**, *68*, 9825–9831. [CrossRef] [PubMed]

14. Meyskens, F.L., Jr.; Gerner, E.W. Development of difluoromethylornithine (DFMO) as a chemoprevention agent. *Clin. Cancer Res.* **1999**, *5*, 945–951. [PubMed]

15. Poulin, R.; Lu, L.; Ackermann, B.; Bey, P.; Pegg, A.E. Mechanism of the irreversible inactivation of mouse ornithine decarboxylase by α-difluoromethylornithine. Characterization of sequences at the inhibitor and coenzyme binding sites. *J. Biol. Chem.* **1992**, *267*, 150–158. [PubMed]

16. Ask, A.; Persson, L.; Heby, O. Increased survival of L1210 leukemic mice by prevention of the utilization of extracellular polyamines. Studies using a polyamineuptake mutant, antibiotics and a polyamine-deficient diet. *Cancer Lett.* **1992**, *66*, 29–34. [CrossRef]

17. Stabellini, G.; Calastrini, C.; Gagliano, N.; Dellavia, C.; Moscheni, C.; Vizzotto, L.; Occhionorelli, S.; Gioia, M. Polyamine levels and ornithine decarboxylase activity in blood and erythrocytes in human diseases. *Int. J. Clin. Pharmacol. Res.* **2003**, *23*, 17–22. [PubMed]

18. Kashiwagi, K.; Miyamoto, S.; Nukui, E.; Kobayashi, H.; Igarashi, K. Functions of potA and potD proteins in spermidine-preferential uptake system in *Escherichia coli*. *J. Biol. Chem.* **1993**, *268*, 19358–19363. [PubMed]

19. Kashiwagi, K.; Endo, H.; Kobayashi, H.; Takio, K.; Igarashi, K. Spermidine-preferential uptake system in *Escherichia coli*. ATP hydrolysis by PotA protein and its association with membrane. *J. Biol. Chem.* **1995**, *270*, 25377–25382. [CrossRef] [PubMed]

20. Kakinuma, Y.; Maruyama, T.; Nozaki, T.; Wada, Y.; Ohsumi, Y.; Igarashi, K. Cloning of the gene encoding a putative serine/threonine protein kinase which enhances spermine uptake in *Saccharomyces cerevisiae*. *Biochem. Biophys. Res. Commun.* **1995**, *216*, 985–992. [CrossRef] [PubMed]

21. Uemura, T.; Kashiwagi, K.; Igarashi, K. Uptake of putrescine and spermidine by Gap1p on the plasma membrane in *Saccharomyces cerevisias*. *Biochem. Biophys. Res. Commun.* **2005**, *328*, 1028–1033. [CrossRef] [PubMed]

22. Marie-Pierre, H.; Buddy, U. Identification and characterization of polyamine permease from the protozoan parasite *Leishmania major*. *J. Biol. Chem.* **2005**, *208*, 15188–15194.

23. Machius, M.; Brautigam, C.A.; Tomchick, D.R.; Ward, P.; Otwinowski, Z.; Blevins, J.S.; Deka, R.K.; Norgard, M.V. Structural and biochemical basis for polyamine binding to the Tp0655 lipoprotein of *Treponema pallidum*: Putative role for Tp0655 (TpPotD) as a polyamine receptor. *J. Mol. Biol.* **2007**, *373*, 681–694. [CrossRef] [PubMed]

24. Belting, M.; Mani, K.; Jonsson, M.; Cheng, F.; Sandgren, S.; Jonsson, S.; Ding, K.; Delcros, J.-G.; Fransson, L.A. Glypican-1 is a vehicle for polyamine uptake in mammalian cells: A pivotal role for nitrosothiol-derived nitric oxide. *J. Biol. Chem.* **2003**, *278*, 47181–47189. [CrossRef] [PubMed]

25. Roy, U.K.; Rial, N.S.; Kachel, K.L.; Gerner, E.W. Activated K-RAS increases polyamine uptake in human colon cancer cells through modulation of caveolar endocytosis. *Mol. Carcinog.* **2008**, *47*, 538–553. [CrossRef] [PubMed]

26. Sala, R.; Rotoli, B.M.; Colla, E.; Visigalli, R.; Parolari, A.; Bussolati, O.; Gazzola, G.C.; Dall'Asta, V. Two-way arginine transport in human endothelial cells: TNF-α stimulation is restricted to system y+. *Am. J. Physiol. Cell Physiol.* **2002**, *282*, C134–C143. [PubMed]

27. Uemura, T.; Yerushalmi, H.F.; Tsaprailis, G.; Stringer, D.E.; Pastorian, K.E.; Hawel, L.; Byus, C.V.; Gerner, E.W. Identification and characterization of a diamine exporter in colon epithelial cells. *J. Biol. Chem.* **2008**, *283*, 26428–26435. [CrossRef] [PubMed]

28. Heinick, A.; Urban, K.; Roth, S.; Spies, D.; Nunes, F.; Phanstiel, O., IV; Liebau, E.; Lüersen, K. Caenorhabditis elegans P5B-type ATPase CATP-5 operates in polyamine transport and is crucial for norspermidine-mediated suppression of RNA interference. *FASEB J.* **2010**, *24*, 206–217. [CrossRef] [PubMed]

29. Hiasa, M.; Miyaji, T.; Haruna, Y.; Takeuchi, T.; Harada, Y.; Moriyama, S.; Yamamoto, A.; Omote, H.; Moriyama, Y. Identification of a mammalian vesicular polyamine transporter. *Sci. Rep.* **2014**, *4*, 6836. [CrossRef] [PubMed]

30. Isenring, P.; Daigle, N.D.; Carpentier, G.A.; Frenette-Cotton, R.; Simard, M.G.; Lefoll, M.H.; Noel, M.; Caron, L.; Noel, J. Molecular characterization of a human cation-Cl⁻ cotransporter (SLC12A8A, CCC9A) that promotes polyamine and amino acid transport. *J. Cell. Physiol.* **2009**, *220*, 680–689. [CrossRef]

31. Sharpe, J.G.; Seidel, E.R. Polyamines are absorbed through a y+ amino acid carrier in rat intestinal epithelial cells. *Amino Acids* **2005**, *29*, 245–253. [CrossRef] [PubMed]

32. Aouida, M.; Poulin, R.; Ramotar, D. The human carnitine transporter SLC22A16 mediates high affinity uptake of the anticancer polyamine analogue bleomycin-A5. *J. Biol. Chem.* **2010**, *285*, 6275–6284. [CrossRef] [PubMed]

33. Fairbanks, C.A.; Winter, T.N.; Elmquist, W.F. OCT2 and MATE1 provide bidirectional agmatine transport. *Mol. Pharm.* **2011**, *8*, 133–142. [CrossRef]

34. Poulin, R.; Casero, R.A.; Soulet, D. Recent advances in the molecular biology of metazoan polyamine transport. *Amino Acids* **2012**, *42*, 711–723. [CrossRef] [PubMed]

35. Uemura, T.; Stringer, D.E.; Blohm-Mangone, K.A.; Gerner, E.W. Polyamine transport is mediated by both endocytic and solute carrier transport mechanisms in the gastrointestinal tract. *Am. J. Physiol. Gastrointest. Liver Physiol.* **2010**, *299*, G517–G522. [CrossRef] [PubMed]

36. Soulet, D.; Gagnon, B.; Rivest, S.; Audette, M.; Poulin, R. A fluorescent probe of polyamine transport accumulates into intracellular acidic vesicles via a two-step mechanism. *J. Biol. Chem.* **2004**, *279*, 49355–49366. [CrossRef] [PubMed]

37. Tsen, C.; Iltis, M.; Kaur, N.; Bayer, C.; Delcros, J.G.; von Kalm, L.; Phanstiel, O., IV. A *Drosophila* model to identify polyamine–drug conjugates that target the polyamine transporter in an intact epithelium. *J. Med. Chem.* **2008**, *51*, 324–330. [CrossRef] [PubMed]

38. Von Kalm, L.; Fristrom, D.; Fristrom, J. The making of a fly leg: A model for epithelial morphogenesis. *Bioessays* **1995**, *17*, 693–702. [CrossRef] [PubMed]

39. Wang, C.; Delcros, J.G.; Biggerstaff, J.; Phanstiel, O., IV. Molecular requirements for targeting the polyamine transport system. Synthesis and biological evaluation of polyamine-anthracene conjugates. *J. Med. Chem.* **2003**, *46*, 2672–2682. [CrossRef] [PubMed]

40. Muth, A.; Madan, M.; Archer, J.J.; Ocampo, N.; Rodriguez, L.; Phanstiel, O., IV. Polyamine transport inhibitors: Design, synthesis and combination therapies with difluoromethylornithine. *J. Med. Chem.* **2014**, *57*, 348–363. [CrossRef] [PubMed]

41. Maroni, G.; Stamey, S.C. Use of blue food to select synchronous, late third instar larvae. *Drosoph. Inf. Serv.* **1983**, *59*, 142–143.

42. Fristrom, J.W.; Logan, W.R.; Murphy, C. The synthetic and minimal culture requirements for evagination of imaginal discs of *Drosophila melanogaster* in vitro. *Dev. Biol.* **1973**, *33*, 441–456. [CrossRef]

43. Kaur, N.; Delcros, J.G.; Imran, J.; Khaled, A.; Chehtane, M.; Tschammer, N.; Martin, B.; Phanstiel, O., IV. A comparison of chloroambucil- and xylene-containing polyamines leads to improved ligands for accessing the polyamine transport system. *J. Med. Chem.* **2008**, *51*, 1393–1401. [CrossRef] [PubMed]

44. Igarashi, K.; Kashiwagi, K. Bacterial and eukaryotic transport systems. In *Polyamine Cell Signaling*; Humana Press: New York, NY, USA, 2006; pp. 433–448. ISBN 978-1-58829-625-2.

45. Metcalf, B.W.; Bey, P.; Danzin, C.; Jung, M.J.; Casara, P.; Vevert, J.P. Catalytic irreversible inhibition of mammalian ornithine decarboxylase (EC41117) by substrate and product analogs. *J. Am. Chem. Soc.* **1978**, *100*, 2551–2553. [CrossRef]

46. Jin, H.-T.; Räty, S.; Minkkinen, M.; Järvinen, S.; Sand, J.; Alhonen, L.; Nordback, I. Changes in blood polyamine levels in human acute pancreatitis. *Scand. J. Gastroenterol.* **2009**, *44*, 1004–1011. [CrossRef] [PubMed]

47. Mandel, J.L.; Flintoff, W.F. Isolation of mutant mammalian cells altered in polyamine transport. *J. Cell. Physiol.* **1978**, *97*, 335–343. [CrossRef] [PubMed]

48. Burns, M.R.; Graminski, G.F.; Weeks, R.S.; Chen, Y.; O'Brien, T.G. Lipophilic lysine-spermine conjugates are potent polyamine transport inhibitors for use in combination with a polyamine biosynthesis inhibitor. *J. Med. Chem.* **2009**, *52*, 1983–1993. [CrossRef] [PubMed]

49. Chen, Y.; Weeks, R.S.; Burns, M.R.; Boorman, D.W.; Klein-Szanto, A.; O'Brien, T.G. Combination therapy with α-difluoromethylornithine and a polyamine transport inhibitor against murine squamous cell carcinoma. *Int. J. Cancer Prev.* **2006**, *118*, 2344–2349. [CrossRef] [PubMed]

50. Hayes, C.S.; Shicora, A.C.; Keough, M.P.; Snook, A.E.; Burns, M.R.; Gilmour, S.K. Polyamine-blocking therapy reverses immunosuppression in the tumor microenvironment. *Cancer Immunol. Res.* **2014**, *2*, 274–285. [CrossRef] [PubMed]

51. Alexander, E.T.; Minton, A.; Peters, M.C.; Phanstiel, O., IV; Gilmour, S.K. A novel polyamine blockade therapy activates an anti-tumor immune response. *Oncotarget* **2017**. [CrossRef]

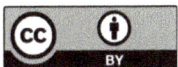

![medical sciences logo] *medical sciences*

MDPI

Review

Antizyme Inhibitors in Polyamine Metabolism and Beyond: Physiopathological Implications

Bruno Ramos-Molina [1,2,3,†], **Ana Lambertos** [1,4,†] and **Rafael Peñafiel** [1,4,*]

1 Department of Biochemistry and Molecular Biology B and Immunology, Faculty of Medicine, University of Murcia, 30100 Murcia, Spain; bruno.ramos@ibima.eu (B.R.-M.); ana.lambertos@um.es (A.L.)
2 Laboratory of Cellular and Molecular Endocrinology, Institute of Biomedical Research in Malaga (IBIMA), Virgen de la Victoria University Hospital, 29010 Málaga, Spain
3 CIBER Physiopathology of Obesity and Nutrition (CIBERobn), Institute of Health Carlos III (ISCIII), 28029 Madrid, Spain
4 Biomedical Research Institute of Murcia (IMIB), 30120 Murcia, Spain
* Correspondence: rapegar@um.es; Tel.: +34-868-887-174
† These authors contributed equally.

Received: 16 September 2018; Accepted: 4 October 2018; Published: 9 October 2018

Abstract: The intracellular levels of polyamines, cationic molecules involved in a myriad of cellular functions ranging from cellular growth, differentiation and apoptosis, is precisely regulated by antizymes and antizyme inhibitors via the modulation of the polyamine biosynthetic and transport systems. Antizymes, which are mainly activated upon high polyamine levels, inhibit ornithine decarboxylase (ODC), the key enzyme of the polyamine biosynthetic route, and exert a negative control of polyamine intake. Antizyme inhibitors (AZINs), which are proteins highly homologous to ODC, selectively interact with antizymes, preventing their action on ODC and the polyamine transport system. In this review, we will update the recent advances on the structural, cellular and physiological functions of AZINs, with particular emphasis on the action of these proteins in the regulation of polyamine metabolism. In addition, we will describe emerging evidence that suggests that AZINs may also have polyamine-independent effects on cells. Finally, we will discuss how the dysregulation of AZIN activity has been implicated in certain human pathologies such as cancer, fibrosis or neurodegenerative diseases.

Keywords: polyamines; polyamine metabolism; antizyme; antizyme inhibitors; ornithine decarboxylase

1. Introduction

In mammalian cells, the control of the polyamine homeostasis is critical for the maintenance of cellular functions, since these molecules participate and modulate cellular processes such as cell growth, proliferation, differentiation and apoptosis [1–3]. Indeed, dysregulation of the intracellular polyamine levels has been observed in pathological conditions, ranging from cancer to inflammation, including neurological disorders [4–11]. The intracellular polyamine pool can be regulated by different mechanisms, including biosynthesis, catabolism and transport [12]. Ornithine decarboxylase (ODC), the rate-limiting enzyme of the polyamine biosynthetic pathway, catalyses the conversion of ornithine into the diamine putrescine by decarboxylation. Once putrescine is generated, the rest of the polyamines, spermidine and spermine, are produced as a consequence of the addition of aminopropyl groups from decarboxylated S-adenosylmethionine by the action of spermidine synthase and spermine synthase, respectively.

Ornithine decarboxylase activity is highly regulated in mammalian cells, being induced by different stimuli including oncogenes, hypoxic conditions or hormones [13]. Ornithine decarboxylase regulation is mediated by transcriptional, post-transcriptional, translational and post-translational mechanisms [3,13–16]. The post-translational control of ODC is mediated by a series of antagonistic proteins, antizymes (AZs) and antizyme inhibitors (AZINs) that down-regulate or up-regulate, respectively, the activity of ODC and the levels of polyamines [17] (Figure 1). In mammals, three different forms of ornithine decarboxylase antizymes (AZ1, AZ2 and AZ3) have been described [18]. Most studies on AZs (or OAZs) have been focused on AZ1, the first discovered antizyme. Although AZ2 and AZ3 share many functional properties with AZ1, they differ in their tissue and cellular distribution [19–25]. The synthesis of functional AZs is mediated by polyamines at the translational level. AZ is encoded by two partially overlapping open reading frames (ORFs), ORF1 and ORF2. At the end of the ORF1 there is a premature stop codon responsible for the synthesis of an incomplete form of AZ that cannot bind to ODC. Under high polyamine levels, a proportion of ribosomes that initiate at the start of ORF1 switch to the +1 reading frame at its last codon, skipping the stop codon, and proceed to decode the ORF2 and synthetize functional AZ [26,27] (Figure 1). The proportion of ribosomes that switch to the ORF2 frame depends on the intracellular polyamine concentration. Antizyme 1 inhibits ODC by interacting with the ODC monomer and therefore preventing the formation of active ODC homodimers, and induces the proteasomal degradation of ODC without ubiquitination [28,29]. Antizymes also affect the polyamine homeostasis by inhibiting the polyamine transport system at the plasma membrane [30–32], although the mechanism by which AZs inhibit the polyamine transporter is still unknown. Antizyme inhibitors (AZIN1 and AZIN2) are proteins homologous to ODC lacking enzymatic activity, that are able to interact with AZs even more efficiently than ODC, counteracting the negative effect of AZs on the biosynthesis of intracellular polyamines [17,33,34] (Figure 2). In addition, AZINs are able to positively modulate the uptake of extracellular polyamines, likely by preventing the inhibitory action of AZs on the polyamine transport system [35]. Interestingly, AZs and AZINs also regulate the uptake of other biogenic amines like agmatine, which could be also transported into the intracellular space by the canonical polyamine transporter [32]. Remarkably, the AZ/AZIN axis is not relevant for the regulation of the cellular uptake of cationic amino acids such as arginine, lysine or ornithine (Ramos-Molina et al., unpublished results), suggesting that both AZs and AZINs act specifically on the polyamine carrier and not on other organic cation transporters. In the subsequent sections, we will summarize the knowledge about these two proteins related with ODC regulation, as well as the most recent advances on our understanding of their pathophysiological implications.

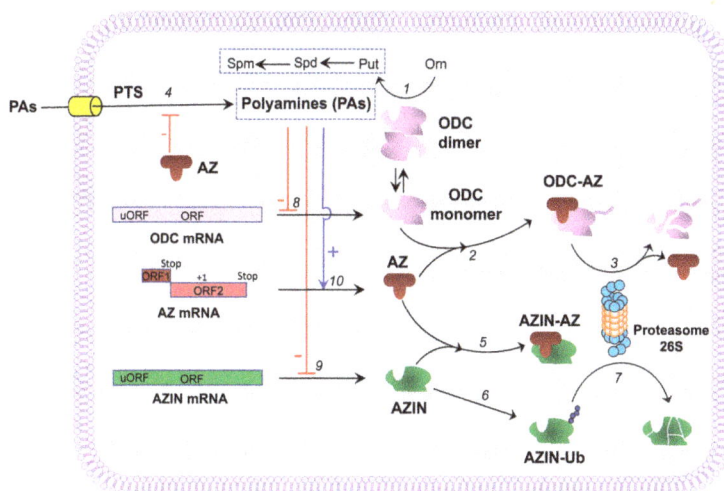

Figure 1. Antizymes and antizyme inhibitors in the control of polyamine homeostasis in mammalian cells. (**1**) The active ornithine decarboxylase (ODC) dimer catalyses the formation of putrescine (Put)

from L-ornithine (Orn). Put serves as precursor in the synthesis of the major polyamines spermidine (Spd) and spermine (Spm). (**2**) Antizyme (AZ) inhibits ODC activity by binding to the ODC monomer. (**3**) The ODC-AZ complex interacts with the 26S proteasome where ODC is degraded and AZ is recycled. (**4**) AZ inhibits the polyamine transport system (PTS) at the plasma membrane by an unknown mechanism. (**5**) Antizyme inhibitor (AZIN) binds with higher affinity to AZ decreasing the negative effect of AZ on ODC and polyamine uptake. AZIN is ubiquitinated (**6**) and degraded (**7**) by the proteasome. The binding to AZ protects AZIN from its degradation by the proteasome. Polyamines (PAs) down-regulate the translation of ODC (**8**) and AZIN1 (**9**) mRNAs. (**10**) PAs stimulate the synthesis of AZs, by inducing the ribosomal frame-shifting at the stop codon of the ORF1 of AZ mRNA, allowing the translation of the ORF2.

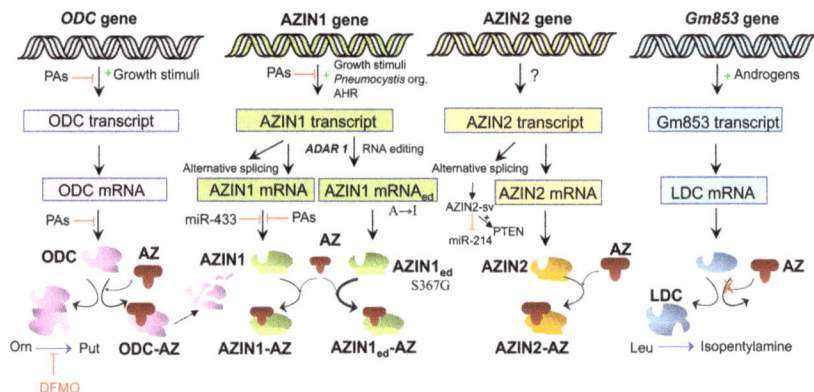

Figure 2. Scheme on the expression of ODC and its paralogues. The transcription of ODC and AZIN1 is up-regulated by different growth stimuli and down-regulated by polyamines (PAs). Translation of ODC mRNA is down-regulated by PAs and ODC activity and stability are decreased by AZ at the post-translational level. ODC activity is inhibited by α-difluoromethyl ornithine (DFMO). AZIN1 mRNA translation is inhibited by PAs and miR-433. PAs also affect AZIN1 splicing, and an edited form of AZIN1 with substitution of Ser367 by Gly has a higher affinity for AZ than AZIN1. AZIN2-sv is an lncRNA that interacts with miR-214 and activates phosphatase and tensin homolog (PTEN). *Gm853* is a paralogous gene of ODC and AZINs, that does not interact with AZs and that catalyses the decarboxylation of L-leucine to produce isopentylamine (See Section 3.5). AHR: aryl hydrocarbon receptor; AZ: antizyme; LDC: leucine decarboxylase.

2. Antizyme Inhibitor 1

The first antizyme inhibitor (here known as AZIN1) was originally characterized in rat liver extracts as a macromolecular inhibitor of the antizyme [36]. After its purification, it was demonstrated that it can bind to antizyme with higher affinity than ODC, releasing the enzyme from the ODC-antizyme complex [37,38]. The cloning of the rat and human genes contributed to deduce the protein sequence, showing that in spite of its high homology to ODC, AZIN1 is devoid of enzymatic activity [39,40]. This characteristic is shared by all AZIN1 orthologs studied, which have substitutions in several residues critical for ODC activity [41]. By negating the action of antizyme, AZIN1 can affect intracellular polyamine levels due to the concomitant increase of both ODC activity and polyamine uptake [42,43]. However, the possibility that AZIN1 could participate in the regulation of other processes by mechanisms unrelated to polyamines cannot be excluded.

2.1. Structural Aspects

Although initial studies suggested that AZIN1, like ODC, was able to form dimers, subsequent crystallographic and biochemical analyses revealed that under physiological conditions, AZIN1 exists as a monomer unable to bind pyridoxal 5-phosphate (a cofactor necessary for ODC activity), which could explain the lack of enzymatic activity and its high affinity to AZ [44]. More recently, it was described that the substitution of the residues Ser277, Ser331, Glu332 and Asp389 in AZIN1 for the corresponding residues of the putative dimer interface of ODC (Arg277, Tyr331, Asp332 and Tyr389, respectively) causes AZIN1 to behave as a dimer in solution [45]. Although both ODC and AZIN1 are proteins that can interact with AZ, AZIN1 has a higher AZ-binding affinity [42,46,47]. Mutational analyses demonstrated that the differences in certain residues in the AZ-binding element of ODC and AZIN1 are responsible for the differential AZ-binding affinities [48]. In fact, the substitution of residues N125 and M140 in ODC for lysines (corresponding residues in AZIN1) markedly increases the AZ-binding affinity to ODC. However, a more recent structural analysis of the AZIN1-AZ1 complex revealed that the residues A325 and S329, present in AZIN1 of all vertebrates, and that substitute N327 and Y331 in ODC may partially contribute to the higher affinity of AZIN1 for AZ1 [49]. Particularly interesting is the finding that the substitution of S367 by glycine leads to an AZIN1 variant with increased affinity for AZ1, likely by inducing a conformational change in its structure [50]. In addition, AZIN1 was able to interact not only with AZ1 but also with all members of the antizyme family, suggesting that AZIN1 may act as a general inhibitor of the function of antizymes [51]. On the other hand, AZIN1 variants unable to interact with AZs can still exert different cellular effects, suggesting that AZIN1 could also act by means of antizyme-independent mechanisms [52,53].

2.2. Tissue and Cellular Distribution

AZIN1, like ODC, is widely expressed as evidenced by the analysis of AZIN1 mRNA levels in different rat and mouse studies [39,54,55]. Although several types of AZIN1 mRNA have been found both in human and rodents, the ORF remains unaltered in most cases [39,40,56]. More recently, multiple forms of *Azin1* transcripts formed by alternative splicing and initiation of transcription from putative alternative start sites were reported in mice [57]. One of the novel splice variants encoded a truncated form of AZIN1 whose functional significance remains to be clarified. Remarkably, an edited transcript of AZIN1 was firstly detected in human hepatocellular carcinoma [50]. Although AZIN1 editing was also detected in healthy liver tissues, the level of editing increased with the pathological behaviour of the tumor. The AZIN1 mRNA A→I editing, which is mediated by a double stranded RNA specific adenosine deaminase (ADAR1), resulted in a Ser to Gly change at the residue 367 of AZIN1 protein that, as commented above, increased the affinity of AZIN1 for AZ1 [50]. At the cellular level AZIN1, like AZ1, has been found to be located in the centrosomes, where it can modulate centriole amplification. In fact, silencing of AZIN1 reduced centrosome abnormalities, whereas its overexpression produced centrosome overduplication [58]. In other cases, such as in HEK293T and COS7, cells transfected with AZIN1, AZIN1 protein were mainly located in the nuclei [59]. Changes in the subcellular location have been observed along the cell cycle or during development. Thus, AZIN1 was found to be present in the cytoplasm of hepatoma tissue culture (HTC) cells during interphase, and together with AZ1, at centrosomes during mitosis [60]. During the cell cycle AZIN1 was mainly accumulated at the early G1 period, likely to increase ODC activity, and in the G2/M phase, and its suppression increased the number of binucleated cells [60]. In addition, translocations between nucleus and cytoplasm were reported to take place during murine mammary gland development [61]. Together, these data suggest an important role of AZIN1 in cell division and differentiation.

2.3. Synthesis and Degradation

Early studies indicated that the amount of AZIN1 increased in rat liver in response to different nutritional stimuli [38]. More recent studies have shown that AZIN1 expression appears to be regulated at several levels by different factors related to cell growth. In mouse fibroblasts, *Azin1* mRNA content increased significantly following growth stimulation much earlier than the increase of ODC transcripts under the same conditions [56]. Furthermore, in breast cancer cells, AZIN1 was transiently up-regulated after induction of cell proliferation by diluting cells in fresh medium [62].

In alveolar macrophages, *Pneumocystis* organisms were found to induce AZIN1 expression, suggesting that this increase is related to the high polyamine content of these cells during pneumonia [63]. In fact, in mouse embryonic fibroblasts the inhibition of polyamine synthesis by α-difluoromethyl ornithine (DFMO), an irreversible ODC inhibitor, increased the *Azin1* mRNA levels, whereas the addition of polyamines resulted in opposite effects [57]. Remarkably, this effect of polyamines did not only affect *Azin1* transcription but also its splicing pattern. In addition, other studies have suggested that polyamines may mediate the repression of *Azin1* through its action at translational level on the short functional upstream ORF (uORF) existing in the *Azin1* mRNA [64]. Another element, a long-looped quadruplex detected in the 5′ untranslated region (5′UTR) in the *Azin1* mRNA, has been postulated as a regulator and sensor of polyamine levels [65]). Interestingly, it was reported that AZIN1 expression is also down-regulated by miR-433 [66], and therefore by sequestration of this miRNA, *Azin1* mRNA could indirectly mediate the expression of other genes targeted by miR-433. Recently it has been also reported that iron depletion up-regulates AZIN1 protein expression, although the mechanism is currently unknown [67].

Antizyme inhibitor 1, like its homologous ODC, is a short-lived protein, but in contrast to ODC, it is degraded by the proteasome by an ubiquitin-dependent mechanism [68]. Furthermore, although AZIN1 binds tightly to AZ1, this does not accelerate its degradation as in the case of ODC. Rather, antizyme binding stabilizes AZIN1 by preventing its ubiquitination [69].

2.4. Physiological Role

Antizyme inhibitor 1 is essential for survival, since transgenic mice with disruption of the gene died at birth showing abnormal liver morphology, slightly reduced body weight and decreased polyamine levels in several tissues [54]. This finding was in agreement with the notion, above commented, that AZIN1 exerts a positive effect on intracellular polyamine levels by repressing the inhibitory action of antizymes on the polyamine biosynthetic and transport pathways. Since there is ample evidence suggesting that increasing polyamine levels stimulates cell growth [2,3], it was postulated that AZIN1 could have a role in cell proliferation. This possibility was later confirmed by different experiments in both normal and transformed cells. Thus, stimulation of cell proliferation in both normal and cancer cells was associated to increased expression of AZIN1 [56,62,69]. In addition, induced overexpression of AZIN1 stimulates the growth, survival and oncogenic potential of tumor and non-tumor cells [42,52,62,69]. Conversely, silencing expression of AZIN1 by using RNA interference technology reduced intracellular polyamine levels and decreased the proliferation of cultured cancer cells [70] and prostate tumor growth in vivo [71]. Although most of the effects of AZIN1 appear to be related to its capacity to interfere the action of AZ1, it has been also reported that AZIN1 is able to interact directly with the cell cycle regulator cyclin D1, preventing the degradation of this cyclin [52]. It is then likely that AZIN1 may also affect cell proliferation by antizyme-independent mechanisms. In fact, the transcriptional profile of livers from *Azin1* knock-out mice at 19th day of gestation showed marked changes related to those of wild type mice, affecting genes related with cell cycle control and proliferation [72]. Finally, emerging evidence has suggested that AZIN1 may have also functions not directly related with cell proliferation. For instance, the similarity in the expression pattern between AZIN1 and certain reproductive related genes in the hypothalamus, ovary and uterus during the rat oestrous cycle [73] or in the avian ovarian follicles [74], suggested a possible role of AZIN1 in reproductive physiology. Transcriptomic studies have also revealed that *Azin1* is one of the genes that are consistently up-regulated by glucocorticoids in the brain [75]. Additionally, either the overexpression or the knock down of *Azin1*, in neurons of the paraventricular and supraoptic nuclei of the hypothalamus, revealed that AZIN1 could be important for the transcriptional regulation of arginine vasopressin [76].

2.5. Antizyme Inhibitor 1: Overexpression and RNA Editing in Cancer

Following the initial study showing that AZIN1 was highly expressed in human gastric tumor cells [77], current information available from the Oncomine (https://www.oncomine.org) or Gene Expression Profiling Interactive Analysis (GEPIA) (http://gepia.cancer-pku.cn) databases have revealed that AZIN1 is up-regulated in many different types of human malignancies. These findings are in agreement with experimental studies reporting that transformed NIH-3T3 fibroblast cells with over-expression of AZIN1 generated tumors after injection into nude mice [42], and that the overexpression of AZIN1 in rat prostate carcinoma cells enhanced their ability to grow in soft agar [52]. On the other hand, the knocking down of AZIN1 using shRNA in both human and rat prostate cancer cell lines decreased the ability of these cells to form tumors in vivo, after subcutaneous injection into nude mice [71].

Interestingly, a series of new findings have established that the RNA editing of AZIN1 may be a potential driver in the pathogenesis of human cancers. In the seminal paper [50], transcriptomic sequencing of several human hepatocellular carcinomas revealed that adenosine-to-inosine (A→I) RNA editing of AZIN1 was increased in tumors with respect to healthy liver tissue. This specific editing of the AZIN1 transcript resulted in the substitution of serine by glycine at residue 367 of human AZIN1. Remarkably, this change increased the affinity of AZIN1 toward antizyme, and induced the translocation of AZIN1 from the cytoplasm to the nucleus. When hepatocellular carcinoma cell lines were transduced with lentivirus carrying the edited version of AZIN1, they showed accelerated growth rates and higher frequency of colony formation. Edited AZIN1 cells also showed enhanced in vivo tumorigenic capacity [50]. This kind of epigenetic modification reported in liver tumors has also been described in other types of cancer. In oesophageal squamous cell carcinomas, overexpression of ADAR1 (the adenosine deaminase that converts adenosine into inosine acting on dsRNA) due to gene amplification was detected [78]. This resulted in hyper-editing of AZIN1, which conferred a gain-of-function phenotype associated with a more aggressive tumor behavior [78]. A close association between ADAR overexpression and AZIN1 editing was also observed in several non-small-cell lung cancer patient samples and lung cancer cell lines [79]. In these tumors, AZIN1 protein expression was higher in tumors with edited AZIN1 than in those with non-edited AZIN1. In lung cancer cell lines AZIN1 RNA editing induced proliferation, invasion and migration, both in vitro and in vivo [79]. More recently, AZIN1 RNA editing was analyzed in 392 colorectal tissues from multiple independent colorectal cancer patient cohorts. This study showed that AZIN1 RNA edited levels were higher in cancer tissue compared to normal mucosa, and that high levels of editing of AZIN1 may be considered as a prognostic factor for disease-free survival and overall survival, and also as an independent risk factor for lymph node and distant metastasis [80]. According to all these findings, it has been postulated that AZIN1 might be a critical target for cancer therapy [81]. However, since in these studies polyamine levels were not determined, it is difficult to ascertain whether all the changes observed could be explained by an increase in polyamine levels, or are rather linked to interactions with proteins not directly related with polyamine metabolism. It should be also necessary to know the influence of the hyper-editing of other genes in the changes observed in the cancer cells with overexpression of ADAR enzymes. The generation of conditional mouse models with wild type and edited *AZIN1* genes would provide valuable information on the relationship between AZIN1 and AZIN1 editing with tumor development. Interestingly, a recent report revealed that the aryl hydrocarbon receptor (AHR), acting as a transcriptional factor, activated the expression of both AZIN1 and ODC [82]. Moreover, in this work a new drug (clofazimine) was identified as a potent AHR antagonist that inhibited polyamine biosynthesis, decreased intracellular polyamine content and the growth of human multiple myeloma [82]. Consequently, targeting of AZIN1 through AHR inhibition appears as a promising strategy for cancer therapy.

2.6. Antizyme Inhibitor 1 and Fibrogenesis

Different studies have revealed the existence of a correlation between AZIN1 and fibrogenic processes in liver and kidney. Hepatic fibrosis is related with the conversion of hepatic stellate cells into myofibroblast-like cells. A single nucleotide polymorphism (SNP) in the AZIN1 gene seemed to be associated with slower rate of hepatic fibrosis in chronic viral hepatitis C [53]. This SNP increased the formation of a spliced variant of AZIN1 mRNA (AZIN1 SV2) that encoded a truncated version of the AZIN1 protein. In addition, the transfection of LX2 stellate hepatic cells with the variant AZIN1 SV2 inhibited fibrogenic gene expression through a polyamine-independent mechanism [53]. In another study, an association between a SNP of AZIN1 and liver cirrhosis risk in Chinese hepatitis B patients was also postulated [83].

Antizyme inhibitor 1 has also been implicated in the regulation of renal fibrosis, since AZIN1 overexpression suppressed transforming growth factor β (TGF-β)/Smad3 signalling pathway, a major player in tissue fibrosis [66]. In this process, micro RNA miR-433 was identified as an important component, since miR-433 overexpression suppressed *Azin1* expression and enhanced TGF-β1-induced fibrosis, whereas *Azin1* overexpression suppressed TGF-β signalling and the fibrotic response [66]. In this context, increased *Azin1* expression has been also correlated with the amelioration of renal fibrosis associated to diabetic nephropathy [84].

3. Antizyme Inhibitor 2

The second member of the AZIN family, named AZIN2, was identified in mouse and human [85,86]. This gene, earlier identified as *ODCp* or *ODC-like*, was mainly expressed in human brain and testes, as assessed by dot blot analysis [87]. Quantification of the expression levels of *Azin2* in mouse tissues by reverse transcription-polymerase chain reaction (RT-PCR) showed that its expression was about 23-fold higher than *Azin1* in the testes and 6-fold in the brain [88]. Due to this specific cellular distribution, it was initially suggested that AZIN2 could play a role in terminal differentiation rather than in cell proliferation.

3.1. Structural and Functional Aspects

Despite its homology with ODC, AZIN2 is not capable of decarboxylating ornithine [84–86], probably due to differences in some critical residues required for the enzymatic activity. On the other hand, AZIN2 binds efficiently to the three AZs, counteracting the negative regulation of these proteins on ODC and the polyamine uptake [35,85,86], although its binding to AZ1 and AZ3 appears to be less efficient than AZIN1 [88]. Regarding to this point, human AZIN2 has substitutions in the corresponding residues A325 and S329 of AZIN1 (N328 and F333 in AZIN2, respectively) that were shown to be relevant for the binding to AZs [49]. Sequence analogy with ODC and AZIN1 is high, mainly in the central part of the molecule and especially in the denominated antizyme-binding element (AZBE) [34]. Although the AZBE site is delimited between the residues 110 and 145, the specific residues involved in the direct interaction with AZs has not been fully identified. By multi-alignment sequence analysis of the AZBE region of AZIN2 orthologues and those corresponding paralogues provided by genome database, five conserved residues (K116, A124, E139, L140, and K142) were identified [89]. Whereas single mutations in these residues did not affect AZ binding, double or triple mutants markedly reduced the affinity of AZIN2 towards AZ1 [90]. AZIN2, in contrast to ODC, does not form homodimers, although its predicted monomeric tertiary structure was similar to that of ODC [90].

Antizyme inhibitor 2 is a short-lived protein. In human embryonic kidney (HEK) 293T-transfected cells the half-life of AZIN2 was much lower (≈90 min) than that ODC (>8 h) [90], but AZIN2 was less labile than AZIN1 [88,90]. The degradation of AZIN2 was reduced by the presence of any of the three AZs paralogues [90]. Interestingly, AZIN2 increased the stability of the three antizymes, as it was shown by co-transfections experiments [90]. Like AZIN1, it is degraded in an ubiquitin-dependent manner by a process that is inhibited by AZ1 [86,88], whereas in NIH-3T3 cells stably overexpressing *Azin2* the degradation of AZIN2 was inhibited by the proteasome inhibitor MG132 [88], and in transiently transfected HEK293T cells the effect of MG132 on AZIN2 was not so evident as in the cases of ODC and AZIN1 under the same experimental conditions [90]. In addition, the protective effect produced by inhibitors of the lysosomal degradative pathway suggested that AZIN2 may be also degraded by an alternative route to that of proteasome [90].

Regarding the subcellular localization of AZIN2 in both HEK293T and COS7 cells overexpressing AZIN2, the protein was mainly present in the ER-Golgi intermediate compartment (ERGIC) and in the *cis*-Golgi network [59]. In human neural Paju cells, immunostaining with rabbit antisera against AZIN2 revealed a vesicle-like expression pattern in the cytoplasm [91]. Co-localization studies with other subcellular markers in Paju cells transfected with AZIN2-FLAG indicated that AZIN2 localizes in the post-Golgi vesicular compartments of the secretory pathway [91]. Interestingly, co-expression of AZIN2 with any member of the AZs induced a shift of AZIN2 from the ERGIC to the cytosol [59]. Furthermore, whereas the deletion of the AZBE region did not alter AZIN2 location, the ablation of its N-terminal region abrogated the incorporation of the mutated AZIN2 to the ERGIC complex, revealing that this part of the protein plays a relevant role for the vesicular localization of AZIN2 [59]. Importantly, RNAi-mediated knockdown of AZIN2 produced a distorted morphology of the trans-Golgi network, although the functional impact of this change was not addressed [91]. Furthermore, AZIN2 has been detected by immunohistochemistry in granular or vesicle-like structures of the cytoplasm in different cell types, including neurons, human neural-crest-derived tumour cells, mast cells, ovarian hilus, corpus luteum and Leydig cells [92–94]. In mast cells, AZIN2 is specifically accumulated in serotonin-containing granules where its expression is rapidly induced after activation with phorbol 12-myristate 3-acetate or calcium ionophore A23187 [92]. This activation was associated with changes in the intracellular distribution of AZIN2, which relocated from the cytoplasm and nucleus to the peripheral areas of the cells, suggesting that AZIN2 might play a role in exocitosis [92].

Although transfection experiments clearly showed that AZIN2 stimulates ODC activity and polyamine uptake [35,85,86,88] little is known about its effects on polyamine levels in vivo. However, NIH-3T3 cells stably overexpressing AZIN2 grow more rapidly than control cells, but less than cells overexpressing AZIN1, indicating that AZIN2 provides cells some growth advantage [88]. Nowadays, the analysis of expression of AZIN2 in cancer database reveals that this gene is not up-regulated in most of the types of cancer examined (GEPIA (http://gepia.cancer-pku.cn). All these findings, together with the fact that AZIN2 is widely expressed in differentiated cells, support the contention that its major physiological role, in contrast to AZIN1, does not appear to be related with the stimulation of cell proliferation. In fact, transgenic mice with deleted *Azin2* gene are viable [95], in clear difference with *Azin1* knockout mutants [54].

3.2. Antizyme Inhibitor 2 in the Central Nervous System

Polyamines and their metabolic enzymes are present in the mammalian brain, showing a specific regional distribution [96,97], including the presence of antizyme in association with ODC [98,99]. Current evidence clearly indicates that AZIN2 is present in the brain showing a complex regional distribution, and that the expression pattern may be altered in some neurological diseases. Initial experiments detected high amounts of AZIN2/ODCp mRNA in different parts of the human adult brain, including cerebral cortex, cerebellum, hippocampus, substantia nigra, thalamus, corpus callosum and spinal cord [86]. A subsequent semiquantitative RT-PCR analysis using primers for AZIN2 mRNA detected the expression of AZIN2 in different regions of the rat brain (frontal cortex, hippocampus, hypothalamus, locus coeruleus, medulla and striatum) and the human brain (frontal cortex, hippocampus and nucleus accumbens), and in cultured rat neuronal cells (neurons, PC12, astrocytes and glioma cells) [100]. More recently, by using in situ hybridization and immunohistochemistry, a robust expression of AZIN2 was found in the soma and axon of human neurons from different areas of the central nervous system [94]. In this study, the subcellular localization of AZIN2 was dependent on the type of neuronal cell examined. In pyramidal neurons of the frontal cortex, AZIN2 staining was located in granular or vesicle-like structures, whereas in the Purkinje cells of the cerebellum the staining pattern was more diffuse. In the pyramidal neurons of the cortex, AZIN2 co-localized with the *N*-methyl-D-aspartate (NMDA) glutamate receptor. Interestingly, in some neurons of brains affected of Alzheimer disease, a robust expression of AZIN2 was observed [94]. We recently analysed the expression of AZIN2 in the brain of transgenic mice carrying an *Azin2-lacZ* construct under the control of the *Azin2* endogenous promoter. X-Gal staining of brain sections revealed a strong but heterogeneous AZIN2 expression pattern [101]. Labeled neurons showed various sized vesicles full of the LacZ reaction product, and some axons tracts also showed β-galactosidase staining in varying degrees. AZIN2 expression predominantly coincided with cholinergic and glutamatergic cells, and occasionally corresponded to GABAergic and glycinergic cells [101]. In spite of all these findings, the plausible physiological role of AZIN2 in the central nervous system remains to be elucidated.

A controversial issue on AZIN2/ODCp in the brain was related with the assertion that this gene encoded for an arginine decarboxylase (ADC) [102], the enzyme that in plants and bacteria catalyses the formation of agmatine from L-arginine. Although other studies from several independent laboratories clearly demonstrated that the product of this gene was devoid of ADC activity [85,86,103] the term ADC is still present in mammalian gene databanks, as a gene synonym of AZIN2. Even more, some of the effects ascribed to ADC in some expression or transfection experiments should be credited in all likelihood to AZIN2 [104–110].

3.3. Antizyme Inhibitor 2 in Reproductive Tissues

The specific role of polyamines in reproductive tissues is not fully understood [111]. As commented above, preliminary studies showed a high expression of AZIN2 in the testes of adult humans and mice [85,87]. Subsequent comparative analyses revealed that the testis is the tissue with the highest levels of *Azin2* mRNA among the murine tissues examined [55]. Importantly, *Azin2* mRNA was undetectable in the testes of newborn mice, but it markedly increased along the first wave of spermatogenesis to reach constant values after the 7th week of age [112]. Interestingly, in situ RNA hybridization and immunochemical analyses revealed that mouse *Azin2* in mainly expressed in the inner part of the seminiferous tubules, where spermatids at different stages of differentiation and spermatozoa are located [112]. The fact that this spatial and temporal expression pattern was similar to that of *Az3*, the testis-specific antizyme isoform [21,22], suggested that AZIN2 may have a role in spermiogenesis, likely by affecting polyamine homeostasis [112]. However, unlike *Az3* knockout mice, which were infertile despite showing unaffected testicular polyamine [113], *Azin2* knockout mice were fertile [94]. It is plausible that the functions of these two proteins, apparently antagonists, could be related with targeting other proteins not related to polyamine metabolism. In fact, AZ3 can interact with gametogenetin, a testicular protein [114], and with MYPT3, a regulatory protein of the protein phosphatases PP1β and PP1γ2 in the testis [115].

On the other hand, immunochemical analyses revealed the presence of AZIN2 protein in the steroidogenic cells of human gonads [93]. Thus, a robust expression of AZIN2 was found in the testicular Leydig cells, and in ovarian luteinized cells, suggesting a role of AZIN2 in steroidogenesis [93]. In this context, *Azin2* knockout mice showed decreased testosterone levels in plasma and testis, as well as decreased sperm motility [116]. Furthermore, in addition to mammalian gonads, AZIN2 expression has been also reported in ovarian follicles in the goose [117]. All these results support that AZIN2 may play a relevant role in the reproductive system.

3.4. Antizyme Inhibitor 2: Expression in Other Tissues

Although the initial studies on AZIN2 expression in mammalian organs appeared to indicate an exclusive distribution of the AZIN2 mRNA in brain and testis, in more recent analyses by real-time RT-PCR lower levels of AZIN2 messenger were detected in other mouse tissues, including epididymis, pancreas, adrenal gland, kidney, lung, heart, intestine and liver [55]. Interestingly, the comparison of the relative expression of *Azin2* with *Azin1* in each tissue revealed that *Azin2* mRNA was more abundant than *Azin1* mRNA in testis, adrenal gland, lung, brain and epididymis. In addition, the analysis of the gene expression in different renal zones revealed that *Azin2* mRNA levels were lower than those of *Azin1* in all zones, without differences between male and female kidneys [118]. More recently, a novel lncRNA that is up-regulated in human adult hearts was identified as a splice variant of the AZIN2 gene [119]. It has been postulated that this AZIN2-sv transcript decreases the activation of the PI3K/Akt signalling pathway by directly activating phosphatase and tensin homolog (PTEN), and also by negating the negative effect of the miR-214 on PTEN expression. Interestingly, knockdown of AZIN2-sv attenuated ventricular remodelling and improved cardiac function after myocardial infarction of adult rats [119]. Whether AZIN2 mRNA, which contains as AZIN2-sv the complementary sequence to the seed sequence of miR-214, is able to mimic some of these effects, remains to be tested.

By immunological characterization, the AZIN2 protein has been detected in the mast cells of sections of human skin samples from patients with cutaneous mastocytosis and in the cytoplasm and nuclei of different human and murine mast cell lines [92]. Remarkably, AZIN2 was selectively expressed in serotonin-containing mast cell granules, in which serotonin release was polyamine dependent [92]. Additional information has been provided by studies using bone marrow-derived mast cells from both wild type and transgenic *Azin2* knockout mice. Compared to wild-type controls, mast cells derived from *Azin2*-deficient mice showed reduced levels of spermidine and spermine, associated to decreased levels on intracellular and extracellular serotonin and increased histamine [120]. All these findings support a role for AZIN2 as regulator of biogenic amines such as serotonin and histamine in mast cells.

Later, in an extended analysis on the expression of AZIN2 in human tissues, the protein was also detected in several types of specific cells: Pulmonary type two pneumocytes, megakaryocytes, gastric parietal cells, acinar cells of sweat glands, in selected enteroendocrine cells, and different types of renal cells [121]. Remarkably, the transgenic *Azin2* knockout mice model provides an interesting tool to study the expression of AZIN2 protein in tissues or organs different to brain and testis in which moderated levels of *Azin2* mRNA had been detected. In fact, by using this strategy AZIN2 was shown to be also expressed in the endocrine pancreas and adrenal glands [95]. In these organs, AZIN2 was restricted to specific regions and cell types. In the adrenal gland, only the cells of the adrenal medulla displayed a positive X-Gal staining with a cytosolic and granular localization. Regarding to the pancreas, AZIN2 expression was located mainly in the islets of Langerhans, showing a heterogeneous pattern in the β-cells from unstained cells to cells showing strong granular and cytosolic staining. Interestingly, plasma insulin levels were significantly reduced in the *Azin2* knockout mice [95]. Altogether, these results support the idea that AZIN2 may have a role in secretory processes.

In spite of all these findings, the mechanism(s) by which AZIN2 might participate in the regulation of the functions of these specialized cells remains elusive. Its intracellular localization, associated to granular or vesicular structures, suggested that AZIN2 may act as a regulator of vesicular trafficking by locally activating polyamine biosynthesis [91]. Additionally, its proven effect antagonizing the effect of AZs on polyamine uptake would imply that AZIN2 could influence polyamine internalization and compartmentalization. In fact, elevated levels of polyamines are present in vesicular structures, like in mast cell secretory granules [122], although the mechanism for the vesicular uptake of polyamines is mostly unknown. A putative mechanism proposed the participation of a vesicular antiporter polyamine-proton for the vesicular sequestration of free cytosolic polyamines [123]. One can speculate that AZIN2 could have some direct or indirect effect on the regulation of both the plasma membrane polyamine transporter and the vesicular transporter, and hence the polyamine content of vesicular structures.

3.5. Gm853 as a New Paralogue of Odc/Azins with Leucine Decarboxylase Activity

Although AZIN1 and AZIN2 are two well-established ODC homologues with a widely expression in many different animal tissues, recent genomic studies predicted the existence of a new paralog of the family, that in mice received the name of *Gm853*. This murine gene is located in the same chromosome than *Azin2*, a characteristic shared by all their corresponding orthologs, suggesting that *Azin2* and *Gm853* presumably evolved from a common ancestor gene. However, overexpression of the gene in HEK293T revealed that the protein encoded by *Gm853* not only did not act as an antizyme inhibitor, but also presented enzymatic activity [124]. This is a not surprising fact since *Gm853* contains all the amino acid residues that are critical for the enzymatic activity of ODC [41]. Interestingly, the protein, which was mainly expressed in the mouse male kidney, was unable to decarboxylate ornithine or lysine, but instead was very active catalysing the decarboxylation of L-leucine to produce isopentylamine, an aliphatic monoamine with unknown biological function. The biological significance of this novel leucine decarboxylase remains to be elucidated.

4. Concluding Remarks, Controversies and Future Perspectives

As commented above, numerous in vitro experiments have clearly showed that both AZIN1 and AZIN2 share the capacity to interact with the three AZs, and as a result both proteins are able to increase polyamine biosynthesis and uptake. However, the physiological role of these two AZINs in mammalian cells appears to be quite different. Thus, whereas AZIN1 is required for normal embryonic development and clearly related with cell proliferation, AZIN2 presents a more restricted cellular distribution (differentiated cells) and is dispensable for embryonic development. Both abnormal AZIN1 expression and mRNA editing have been detected in different types of cancer, suggesting that AZIN1 may be considered as a potential carcinogenic molecule [81]. Besides, AZIN2, through the regulation of local intracellular levels of polyamines, might affect vesicle trafficking and secretory processes, although pathologies related to this protein have not been described so far.

More recent evidence has revealed that AZs can bind to different molecular targets that do not belong to the polyamine metabolic pathway. Accordingly, AZINs, by negating the action of AZs on these targets, might exert polyamine independent effects. Several proteins, different to ODC or AZINs, have been identified as AZ-binding proteins (Figure 3). In some cases, the binding of the protein to AZ1 promoted its proteasomal degradation without ubiquitination. Several of these new targets of AZ1 are growth-related proteins such as Smad1, a transcriptional regulator of genes responsive to bone morphogenetic proteins [125,126]; cyclin D1, the activator of the cell cycle kinase CDK4 required for the transition G1/S [127]; the protein kinase Aurora A [128]; the mitotic check point kinase Mps1 [129], and the antiapoptotic DNp73, an amino terminally truncated form of the proapoptotic p73 [130]. All these new findings suggested that AZ1 might affect cell proliferation by polyamine-independent mechanisms. Other studies, however, did not find any significant effect of AZ1 or AZ2 on the degradation of cyclin D1, Aurora A or DNp73 in comparison with that of ODC by using a co-degradation assay [131]. In addition, they did not detect any antiproliferative effect of antizyme, when polyamine levels were maintained constant, suggesting that AZIN1 affects cell proliferation exclusively by affecting polyamine metabolism [131]. Nevertheless, some of the discrepancies might be related to the differential affinity of AZ1 towards their target proteins. In fact, a mechanistic study showed that cyclin D1 has a 4-fold lower binding affinity for AZ1 than does ODC and about 40-fold lower than AZIN1 [47]. It is then likely that in presence of ODC or AZIN1 the effect of AZ1 on the degradation of cyclin D1 would be negligible. In addition to AZ1, both AZ2 and AZ3 are also able to interact with specific proteins, affecting their degradation or modulating their activities. Thus, AZ2 binds to the oncogenic protein c-Myc and accelerates its proteasome-mediated degradation without ubiquitination [132]. AZ2 also interacts with ATP citrate lyase, the enzyme that catalyses the production of acetyl-CoA, a metabolite used for lipid biosynthesis or acetylation of cellular components [133]. In the case of AZ3, it has been reported that this testis specific AZ isoform can bind to other testicular proteins such as gametogenetin [114] and MYPT3, a regulator of protein phosphatases [115]. In all these cases, AZINs could indirectly modulate again the functions of these proteins. Finally, an emerging aspect on the possible additional regulatory activities of AZINs is related to the interaction of their mRNAs with certain micro RNAs (i.e., miR-433, miR-214), which may affect different biological processes such as fibrogenesis [66,84] or cardiac regeneration [119].

In conclusion, new efforts should be addressed to make progress in our understanding on the regulatory mechanisms that control AZINs expression as well as in the knowledge of the physiopathological repercussions of these proteins. In particular, although more studies are required for a better understanding of the implication of AZIN1 on cancer development, specifically the role of AZIN1 mRNA editing, targeting of AZIN1 for cancer therapy appears as a promising strategy that requires a rigorous validation.

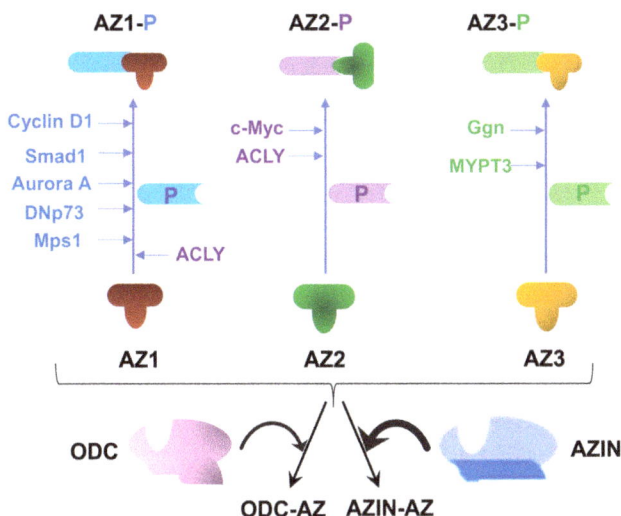

Figure 3. AZ-interacting proteins. The three AZs interact with ODC and AZINs, proteins implicated in polyamine metabolism. In addition, AZs also bind to other proteins not directly related with polyamines. In particular, AZ1 interacts with cell cycle proteins (cyclin D1, Smad, protein kinases such as Aurora A and Msp1), apoptosis related proteins (DNp73, an amino terminally truncated form of p73), or with the acetyl-CoA forming enzyme ATP citrate lyase (ACLY). AZ2 also interacts with ACLY and with the oncogenic protein c-Myc. AZ3 binds to the testicular protein gametogenetin (Ggn), and to MYPT3, a regulatory protein of protein phosphatases.

Author Contributions: Writing—original draft preparation, B.R.-M., A.L. and R.P.; writing—review and editing, B.R.-M., A.L. and R.P.; funding acquisition, R.P.

Funding: This work was supported by the Spanish Ministry of Economy and Competitiveness, SAF2011-29051 (with European Community FEDER support) and by Seneca Foundation (Autonomous Community of Murcia), 19875/GERM/15.

References

1. Igarashi, K.; Kashiwagi, K. Modulation of cellular function by polyamines. *Int. J. Biochem. Cell Biol.* **2010**, *42*, 39–51. [CrossRef] [PubMed]
2. Pegg, A.E. Functions of Polyamines in Mammals. *J. Biol. Chem.* **2016**, *291*, 14904–14912. [CrossRef] [PubMed]

3. Miller-Fleming, L.; Olin-Sandoval, V.; Campbell, K.; Ralser, M. Remaining Mysteries of Molecular Biology: The Role of Polyamines in the Cell. *J. Mol. Biol.* **2015**, *427*, 3389–3406. [CrossRef] [PubMed]

4. Pegg, A.E. Polyamine metabolism and its importance in neoplastic growth and a target for chemotherapy. *Cancer Res.* **1988**, *48*, 759–774. [PubMed]

5. Cason, A.L.; Ikeguchi, Y.; Skinner, C.; Wood, T.C.; Holden, K.R.; Lubs, H.A.; Martinez, F.; Simensen, R.J.; Stevenson, R.E.; Pegg, A.E.; et al. X-linked spermine synthase gene (SMS) defect: The first polyamine deficiency syndrome. *Eur. J. Hum. Genet.* **2003**, *11*, 937–944. [CrossRef] [PubMed]

6. Gerner, E.W.; Meyskens, F.L., Jr. Polyamines and cancer: Old molecules, new understanding. *Nat. Rev. Cancer* **2004**, *4*, 781–792. [CrossRef] [PubMed]

7. Babbar, N.; Murray-Stewart, T.; Casero, R.A., Jr. Inflammation and polyamine catabolism: The good, the bad and the ugly. *Biochem. Soc. Trans.* **2007**, *35 Pt 2*, 300–304. [CrossRef]

8. Lewandowski, N.M.; Ju, S.; Verbitsky, M.; Ross, B.; Geddie, M.L.; Rockenstein, E.; Adame, A.; Muhammad, A.; Vonsattel, J.P.; Ringe, D.; et al. Polyamine pathway contributes to the pathogenesis of Parkinson disease. *Proc. Natl. Acad. Sci. USA* **2010**, *107*, 16970–16975. [CrossRef] [PubMed]

9. Minois, N.; Carmona-Gutierrez, D.; Madeo, F. Polyamines in aging and disease. *Aging* **2011**, *3*, 716–732. [CrossRef] [PubMed]

10. Guerra, G.P.; Rubin, M.A.; Mello, C.F. Modulation of learning and memory by natural polyamines. *Pharmacol. Res.* **2016**, *112*, 99–118. [CrossRef] [PubMed]

11. Casero, R.A.; Jr Murray Stewart, T.; Pegg, A.E. Polyamine metabolism and cancer: Treatments, challenges and opportunities. *Nat. Rev. Cancer* 2018. [CrossRef] [PubMed]

12. Pegg, A.E. Mammalian polyamine metabolism and function. *IUBMB Life* **2009**, *61*, 880–894. [CrossRef] [PubMed]

13. Pegg, A.E. Regulation of ornithine decarboxylase. *J. Biol. Chem.* **2006**, *281*, 14529–14532. [CrossRef] [PubMed]

14. Coffino, P. Regulation of cellular polyamines by antizyme. *Nat. Rev. Mol. Cell Biol.* **2001**, *2*, 188–194. [CrossRef] [PubMed]

15. Shantz, L.M. Transcriptional and translational control of ornithine decarboxylase during Ras transformation. *Biochem. J.* **2004**, *377 Pt 1*, 257–264. [CrossRef]

16. Kahana, C. Protein degradation, the main hub in the regulation of cellular polyamines. *Biochem. J.* **2016**, *473*, 4551–4558. [CrossRef] [PubMed]

17. Kahana, C. Antizyme and antizyme inhibitor, a regulatory tango. *Cell. Mol. Life Sci.* **2009**, *66*, 2479–2488. [CrossRef] [PubMed]

18. Mangold, U. The antizyme family: Polyamines and beyond. *IUBMB Life* **2005**, *57*, 671–676. [CrossRef] [PubMed]

19. Ivanov, I.P.; Gesteland, R.F.; Atkins, J.F. A second mammalian antizyme: Conservation of programmed ribosomal frameshifting. *Genomics* **1998**, *52*, 119–129. [CrossRef] [PubMed]

20. Zhu, C.; Lang, D.W.; Coffino, P. Antizyme2 is a negative regulator of ornithine decarboxylase and polyamine transport. *J. Biol. Chem.* **1999**, *274*, 26425–26430. [CrossRef] [PubMed]

21. Tosaka, Y.; Tanaka, H.; Yano, Y.; Masai, K.; Nozaki, M.; Yomogida, K.; Otani, S.; Nojima, H.; Nishimune, Y. Identification and characterization of testis specific ornithine decarboxylase antizyme (OAZ-t) gene: Expression in haploid germ cells and polyamine-induced frameshifting. *Genes Cells* **2000**, *5*, 265–276. [CrossRef] [PubMed]

22. Ivanov, I.P.; Rohrwasser, A.; Terreros, D.A.; Gesteland, R.F.; Atkins, J.F. Discovery of a spermatogenesis stage-specific ornithine decarboxylase antizyme: Antizyme 3. *Proc. Natl. Acad. Sci. USA* **2000**, *97*, 4808–4813. [CrossRef] [PubMed]

23. Murai, N.; Murakami, Y.; Matsufuji, S. Identification of nuclear export signals in antizyme-1. *J. Biol. Chem.* **2003**, *278*, 44791–44798. [CrossRef] [PubMed]

24. Murai, N.; Shimizu, A.; Murakami, Y.; Matsufuji, S. Subcellular localization and phosphorylation of antizyme 2. *J. Cell. Biochem.* **2009**, *108*, 1012–1021. [CrossRef] [PubMed]

25. Snapir, Z.; Keren-Paz, A.; Bercovich, Z.; Kahana, C. Antizyme 3 inhibits polyamine uptake and ornithine decarboxylase (ODC) activity, but does not stimulate ODC degradation. *Biochem. J.* **2009**, *419*, 99–103. [CrossRef] [PubMed]

26. Rom, E.; Kahana, C. Polyamines regulate the expression of ornithine decarboxylase antizyme in vitro by inducing ribosomal frame-shifting. *Proc. Natl. Acad. Sci. USA* **1994**, *91*, 3959–3963. [CrossRef] [PubMed]

27. Matsufuji, S.; Matsufuji, T.; Miyazaki, Y.; Murakami, Y.; Atkins, J.F.; Gesteland, R.F.; Hayashi, S.I. Autoregulatory frameshifting in decoding mammalian ornithine decarboxylase antizyme. *Cell* **1995**, *80*, 51–60. [CrossRef]

28. Murakami, Y.; Matsufuji, S.; Kameji, T.; Hayashi, S.I.; Igarashi, K.; Tamura, T.; Tanaka, K.; Ichihara, A. Ornithine decarboxylase is degraded by the 26S proteasome without ubiquitination. *Nature* **1992**, *360*, 597–599. [CrossRef] [PubMed]

29. Mamroud-Kidron, E.; Omer-Itsicovich, M.; Bercovich, Z.; Tobias, K.E.; Rom, E.; Kahana, C. A unified pathway for the degradation of ornithine decarboxylase in reticulocyte lysate requires interaction with the polyamine-induced protein, ornithine decarboxylase antizyme. *Eur. J. Biochem.* **1994**, *226*, 547–554. [CrossRef] [PubMed]

30. Mitchell, J.L.; Judd, G.G.; Bareyal-Leyser, A.; Ling, S.Y. Feedback repression of polyamine transport is mediated by antizyme in mammalian tissue-culture cells. *Biochem. J.* **1994**, *299 Pt 1*, 19–22. [CrossRef]

31. Hoshino, K.; Momiyama, E.; Yoshida, K.; Nishimura, K.; Sakai, S.; Toida, T.; Kashiwagi, K.; Igarashi, K. Polyamine transport by mammalian cells and mitochondria: Role of antizyme and glycosaminoglycans. *J. Biol. Chem.* **2005**, *280*, 42801–42808. [CrossRef] [PubMed]

32. Ramos-Molina, B.; Lopez-Contreras, A.J.; Lambertos, A.; Dardonville, C.; Cremades, A.; Penafiel, R. Influence of ornithine decarboxylase antizymes and antizyme inhibitors on agmatine uptake by mammalian cells. *Amino Acids* **2015**, *47*, 1025–1034. [CrossRef] [PubMed]

33. Mangold, U. Antizyme inhibitor: Mysterious modulator of cell proliferation. *Cell. Mol. Life Sci.* **2006**, *63*, 2095–2101. [CrossRef] [PubMed]

34. Lopez-Contreras, A.J.; Ramos-Molina, B.; Cremades, A.; Penafiel, R. Antizyme inhibitor 2: Molecular, cellular and physiological aspects. *Amino Acids* **2010**, *38*, 603–611. [CrossRef] [PubMed]

35. Lopez-Contreras, A.J.; Ramos-Molina, B.; Cremades, A.; Penafiel, R. Antizyme inhibitor 2 (AZIN2/ODCp) stimulates polyamine uptake in mammalian cells. *J. Biol. Chem.* **2008**, *283*, 20761–20769. [CrossRef] [PubMed]

36. Fujita, K.; Murakami, Y.; Hayashi, S. A macromolecular inhibitor of the antizyme to ornithine decarboxylase. *Biochem. J.* **1982**, *204*, 647–652. [CrossRef] [PubMed]

37. Kitani, T.; Fujisawa, H. Purification and characterization of antizyme inhibitor of ornithine decarboxylase from rat liver. *Biochim. Biophys. Acta* **1989**, *991*, 44–49. [CrossRef]

38. Murakami, Y.; Matsufuji, S.; Nishiyama, M.; Hayashi, S. Properties and fluctuations in vivo of rat liver antizyme inhibitor. *Biochem. J.* **1989**, *259*, 839–845. [CrossRef] [PubMed]

39. Murakami, Y.; Ichiba, T.; Matsufuji, S.; Hayashi, S. Cloning of antizyme inhibitor, a highly homologous protein to ornithine decarboxylase. *J. Biol. Chem.* **1996**, *271*, 3340–3342. [CrossRef] [PubMed]

40. Koguchi, K.; Kobayashi, S.; Hayashi, T.; Matsufuji, S.; Murakami, Y.; Hayashi, S. Cloning and sequencing of a human cDNA encoding ornithine decarboxylase antizyme inhibitor. *Biochim. Biophys. Acta* **1997**, *1353*, 209–216. [CrossRef]

41. Ivanov, I.P.; Firth, A.E.; Atkins, J.F. Recurrent emergence of catalytically inactive ornithine decarboxylase homologous forms that likely have regulatory function. *J. Mol. Evol.* **2010**, *70*, 289–302. [CrossRef] [PubMed]

42. Keren-Paz, A.; Bercovich, Z.; Porat, Z.; Erez, O.; Brener, O.; Kahana, C. Overexpression of antizyme-inhibitor in NIH3T3 fibroblasts provides growth advantage through neutralization of antizyme functions. *Oncogene* **2006**, *25*, 5163–5172. [CrossRef] [PubMed]

43. Mitchell, J.L.; Thane, T.K.; Sequeira, J.M.; Marton, L.J.; Thokala, R. Antizyme and antizyme inhibitor activities influence cellular responses to polyamine analogs. *Amino Acids* **2007**, *33*, 291–297. [CrossRef] [PubMed]

44. Albeck, S.; Dym, O.; Unger, T.; Snapir, Z.; Bercovich, Z.; Kahana, C. Crystallographic and biochemical studies revealing the structural basis for antizyme inhibitor function. *Protein Sci.* **2008**, *17*, 793–802. [CrossRef] [PubMed]

45. Su, K.L.; Liao, Y.F.; Hung, H.C.; Liu, G.Y. Critical factors determining dimerization of human antizyme inhibitor. *J. Biol. Chem.* **2009**, *284*, 26768–26777. [CrossRef] [PubMed]

46. Keren-Paz, A.; Bercovich, Z.; Kahana, C. Antizyme inhibitor: A defective ornithine decarboxylase or a physiological regulator of polyamine biosynthesis and cellular proliferation. *Biochem. Soc. Trans.* **2007**, *35 Pt 2*, 311–313. [CrossRef]

47. Liu, Y.C.; Lee, C.Y.; Lin, C.L.; Chen, H.Y.; Liu, G.Y.; Hung, H.C. Multifaceted interactions and regulation between antizyme and its interacting proteins cyclin D1, ornithine decarboxylase and antizyme inhibitor. *Oncotarget* **2015**, *6*, 23917–23929. [CrossRef] [PubMed]

48. Liu, Y.C.; Liu, Y.L.; Su, J.Y.; Liu, G.Y.; Hung, H.C. Critical factors governing the difference in antizyme-binding affinities between human ornithine decarboxylase and antizyme inhibitor. *PLoS ONE* **2011**, *6*, e19253. [CrossRef] [PubMed]

49. Wu, H.Y.; Chen, S.F.; Hsieh, J.Y.; Chou, F.; Wang, Y.H.; Lin, W.T.; Lee, P.Y.; Yu, Y.J.; Lin, L.Y.; Lin, T.S.; et al. Structural basis of antizyme-mediated regulation of polyamine homeostasis. *Proc. Natl. Acad. Sci. USA* **2015**, *112*, 11229–11234. [CrossRef] [PubMed]

50. Chen, L.; Li, Y.; Lin, C.H.; Chan, T.H.; Chow, R.K.; Song, Y.; Liu, M.; Yuan, Y.F.; Fu, L.; Kong, K.L.; et al. Recoding RNA editing of AZIN1 predisposes to hepatocellular carcinoma. *Nat. Med.* **2013**, *19*, 209–216. [CrossRef] [PubMed]

51. Mangold, U.; Leberer, E. Regulation of all members of the antizyme family by antizyme inhibitor. *Biochem. J.* **2005**, *385 Pt 1*, 21–28. [CrossRef]

52. Kim, S.W.; Mangold, U.; Waghorne, C.; Mobascher, A.; Shantz, L.; Banyard, J.; Zetter, B.R. Regulation of cell proliferation by the antizyme inhibitor: Evidence for an antizyme-independent mechanism. *J. Cell Sci.* **2006**, *119 Pt 12*, 2583–2591. [CrossRef]

53. Paris, A.J.; Snapir, Z.; Christopherson, C.D.; Kwok, S.Y.; Lee, U.E.; Ghiassi-Nejad, Z.; Kocabayoglu, P.; Sninsky, J.J.; Llovet, J.M.; Kahana, C.; et al. A polymorphism that delays fibrosis in hepatitis C promotes alternative splicing of AZIN1, reducing fibrogenesis. *Hepatology* **2011**, *54*, 2198–2207. [CrossRef] [PubMed]

54. Tang, H.; Ariki, K.; Ohkido, M.; Murakami, Y.; Matsufuji, S.; Li, Z.; Yamamura, K.I. Role of ornithine decarboxylase antizyme inhibitor in vivo. *Genes Cells* **2009**, *14*, 79–87. [CrossRef] [PubMed]

55. Ramos-Molina, B.; Lopez-Contreras, A.J.; Cremades, A.; Penafiel, R. Differential expression of ornithine decarboxylase antizyme inhibitors and antizymes in rodent tissues and human cell lines. *Amino Acids* **2012**, *42*, 539–547. [CrossRef] [PubMed]

56. Nilsson, J.; Grahn, B.; Heby, O. Antizyme inhibitor is rapidly induced in growth-stimulated mouse fibroblasts and releases ornithine decarboxylase from antizyme suppression. *Biochem. J.* **2000**, *346 Pt 3*, 699–704. [CrossRef]

57. Murakami, Y.; Ohkido, M.; Takizawa, H.; Murai, N.; Matsufuji, S. Multiple forms of mouse antizyme inhibitor 1 mRNA differentially regulated by polyamines. *Amino Acids* **2014**, *46*, 575–583. [CrossRef] [PubMed]

58. Mangold, U.; Hayakawa, H.; Coughlin, M.; Munger, K.; Zetter, B.R. Antizyme, a mediator of ubiquitin-independent proteasomal degradation and its inhibitor localize to centrosomes and modulate centriole amplification. *Oncogene* **2008**, *27*, 604–613. [CrossRef] [PubMed]

59. Lopez-Contreras, A.J.; Sanchez-Laorden, B.L.; Ramos-Molina, B.; de la Morena, M.E.; Cremades, A.; Penafiel, R. Subcellular localization of antizyme inhibitor 2 in mammalian cells: Influence of intrinsic sequences and interaction with antizymes. *J. Cell. Biochem.* **2009**, *107*, 732–740. [CrossRef] [PubMed]

60. Murakami, Y.; Suzuki, J.I.; Samejima, K.; Kikuchi, K.; Hascilowicz, T.; Murai, N.; Matsufuji, S.; Oka, T. The change of antizyme inhibitor expression and its possible role during mammalian cell cycle. *Exp. Cell Res.* **2009**, *315*, 2301–2311. [CrossRef] [PubMed]

61. Murakami, Y.; Suzuki, J.; Samejima, K.; Oka, T. Developmental alterations in expression and subcellular localization of antizyme and antizyme inhibitor and their functional importance in the murine mammary gland. *Amino Acids* **2010**, *38*, 591–601. [CrossRef] [PubMed]

62. Silva, T.M.; Cirenajwis, H.; Wallace, H.M.; Oredsson, S.; Persson, L. A role for antizyme inhibitor in cell proliferation. *Amino Acids* **2015**, *47*, 1341–1352. [CrossRef] [PubMed]

63. Liao, C.P.; Lasbury, M.E.; Wang, S.H.; Zhang, C.; Durant, P.J.; Murakami, Y.; Matsufuji, S.; Lee, C.H. Pneumocystis mediates overexpression of antizyme inhibitor resulting in increased polyamine levels and apoptosis in alveolar macrophages. *J. Biol. Chem.* **2009**, *284*, 8174–8184. [CrossRef] [PubMed]

64. Ivanov, I.P.; Loughran, G.; Atkins, J.F. uORFs with unusual translational start codons autoregulate expression of eukaryotic ornithine decarboxylase homologs. *Proc. Natl. Acad. Sci. USA* **2008**, *105*, 10079–10084. [CrossRef] [PubMed]

65. Lightfoot, H.L.; Hagen, T.; Clery, A.; Allain, F.H.; Hall, J. Control of the polyamine biosynthesis pathway by G2-quadruplexes. *Elife* **2018**, *7*. [CrossRef] [PubMed]

66. Li, R.; Chung, A.C.; Dong, Y.; Yang, W.; Zhong, X.; Lan, H.Y. The microRNA miR-433 promotes renal fibrosis by amplifying the TGF-β/Smad3-Azin1 pathway. *Kidney Int.* **2013**, *84*, 1129–1144. [CrossRef] [PubMed]

67. Lane, D.J.; Bae, D.H.; Siafakas, A.R.; Rahmanto, Y.S.; Al-Akra, L.; Jansson, P.J.; Casero, R.A., Jr.; Richardson, D.R. Coupling of the polyamine and iron metabolism pathways in the regulation of proliferation: Mechanistic links to alterations in key polyamine biosynthetic and catabolic enzymes. *Biochim. Biophys. Acta* **2018**, *1864 Pt B*, 2793–2813. [CrossRef]

68. Bercovich, Z.; Kahana, C. Degradation of antizyme inhibitor, an ornithine decarboxylase homologous protein, is ubiquitin-dependent and is inhibited by antizyme. *J. Biol. Chem.* **2004**, *279*, 54097–54102. [CrossRef] [PubMed]

69. Olsen, R.R.; Zetter, B.R. Evidence of a role for antizyme and antizyme inhibitor as regulators of human cancer. *Mol. Cancer Res.* **2011**, *9*, 1285–1293. [CrossRef] [PubMed]

70. Choi, K.S.; Suh, Y.H.; Kim, W.H.; Lee, T.H.; Jung, M.H. Stable siRNA-mediated silencing of antizyme inhibitor: Regulation of ornithine decarboxylase activity. *Biochem. Biophys. Res. Commun.* **2005**, *328*, 206–212. [CrossRef] [PubMed]

71. Olsen, R.R.; Chung, I.; Zetter, B.R. Knockdown of antizyme inhibitor decreases prostate tumor growth in vivo. *Amino Acids* **2012**, *42*, 549–558. [CrossRef] [PubMed]

72. Wan, T.; Hu, Y.; Zhang, W.; Huang, A.; Yamamura, K.; Tang, H. Changes in liver gene expression of Azin1 knock-out mice. *Z. Naturforsch. C* **2010**, *65*, 519–527. [CrossRef] [PubMed]

73. Fernandes, J.R.D.; Jain, S.; Banerjee, A. Expression of ODC1, SPD, SPM and AZIN1 in the hypothalamus, ovary and uterus during rat estrous cycle. *Gen. Comp. Endocrinol.* **2017**, *246*, 9–22. [CrossRef] [PubMed]

74. Ma, R.; Jiang, D.; Kang, B.; Bai, L.; He, H.; Chen, Z.; Yi, Z. Molecular cloning and mRNA expression analysis of antizyme inhibitor 1 in the ovarian follicles of the Sichuan white goose. *Gene* **2015**, *568*, 55–60. [CrossRef] [PubMed]

75. Juszczak, G.R.; Stankiewicz, A.M. Glucocorticoids, genes and brain function. *Prog. Neuropsychopharmacol. Biol. Psychiatry* **2018**, *82*, 136–168. [CrossRef] [PubMed]

76. Greenwood, M.P.; Greenwood, M.; Paton, J.F.; Murphy, D. Control of Polyamine Biosynthesis by Antizyme Inhibitor 1 Is Important for Transcriptional Regulation of Arginine Vasopressin in the Male Rat Hypothalamus. *Endocrinology* **2015**, *156*, 2905–2917. [CrossRef] [PubMed]

77. Jung, M.H.; Kim, S.C.; Jeon, G.A.; Kim, S.H.; Kim, Y.; Choi, K.S.; Park, S.I.; Joe, M.K.; Kimm, K. Identification of differentially expressed genes in normal and tumor human gastric tissue. *Genomics* **2000**, *69*, 281–286. [CrossRef] [PubMed]

78. Qin, Y.R.; Qiao, J.J.; Chan, T.H.; Zhu, Y.H.; Li, F.F.; Liu, H.; Fei, J.; Li, Y.; Guan, X.Y.; Chen, L. Adenosine-to-inosine RNA editing mediated by ADARs in esophageal squamous cell carcinoma. *Cancer Res.* **2014**, *74*, 840–851. [CrossRef] [PubMed]

79. Hu, X.; Chen, J.; Shi, X.; Feng, F.; Lau, K.W.; Chen, Y.; Chen, Y.; Jiang, L.; Cui, F.; Zhang, Y.; et al. RNA editing of AZIN1 induces the malignant progression of non-small-cell lung cancers. *Tumour Biol.* **2017**, *39*. [CrossRef] [PubMed]

80. Shigeyasu, K.; Okugawa, Y.; Toden, S.; Miyoshi, J.; Toiyama, Y.; Nagasaka, T.; Takahashi, N.; Kusunoki, M.; Takayama, T.; Yamada, Y.; et al. AZIN1 RNA editing confers cancer stemness and enhances oncogenic potential in colorectal cancer. *JCI Insight* **2018**, *3*. [CrossRef] [PubMed]

81. Qiu, S.; Liu, J.; Xing, F. Antizyme inhibitor 1: A potential carcinogenic molecule. *Cancer Sci.* **2017**, *108*, 163–169. [CrossRef] [PubMed]

82. Bianchi-Smiraglia, A.; Bagati, A.; Fink, E.E.; Affronti, H.C.; Lipchick, B.C.; Moparthy, S.; Long, M.D.; Rosario, S.R.; Lightman, S.M.; Moparthy, K.; et al. Inhibition of the aryl hydrocarbon receptor/polyamine biosynthesis axis suppresses multiple myeloma. *J. Clin. Investig.* **2018**, in press. [CrossRef] [PubMed]

83. Peng, L.; Guo, J.; Zhang, Z.; Liu, L.; Cao, Y.; Shi, H.; Wang, J.; Wang, J.; Friedman, S.L.; Sninsky, J.J. A candidate gene study for the association of host single nucleotide polymorphisms with liver cirrhosis risk in chinese hepatitis B patients. *Genet. Test. Mol. Biomark.* **2013**, *17*, 681–686. [CrossRef] [PubMed]

84. Zhu, D.; Zhang, L.; Cheng, L.; Ren, L.; Tang, J.; Sun, D. Pancreatic Kininogenase Ameliorates Renal Fibrosis in Streptozotocin Induced-Diabetic Nephropathy Rat. *Kidney Blood Press Res.* **2016**, *41*, 9–17. [CrossRef] [PubMed]

85. Lopez-Contreras, A.J.; Lopez-Garcia, C.; Jimenez-Cervantes, C.; Cremades, A.; Penafiel, R. Mouse ornithine decarboxylase-like gene encodes an antizyme inhibitor devoid of ornithine and arginine decarboxylating activity. *J. Biol. Chem.* **2006**, *281*, 30896–30906. [CrossRef] [PubMed]

86. Kanerva, K.; Makitie, L.T.; Pelander, A.; Heiskala, M.; Andersson, L.C. Human ornithine decarboxylase paralogue (ODCp) is an antizyme inhibitor but not an arginine decarboxylase. *Biochem. J.* **2008**, *409*, 187–192. [CrossRef] [PubMed]
87. Pitkanen, L.T.; Heiskala, M.; Andersson, L.C. Expression of a novel human ornithine decarboxylase-like protein in the central nervous system and testes. *Biochem. Biophys. Res. Commun.* **2001**, *287*, 1051–1057. [CrossRef] [PubMed]
88. Snapir, Z.; Keren-Paz, A.; Bercovich, Z.; Kahana, C. ODCp, a brain- and testis-specific ornithine decarboxylase paralogue, functions as an antizyme inhibitor, although less efficiently than AzI1. *Biochem. J.* **2008**, *410*, 613–619. [CrossRef] [PubMed]
89. Ramos-Molina, B.; Lambertos, A.; Lopez-Contreras, A.J.; Penafiel, R. Mutational analysis of the antizyme-binding element reveals critical residues for the function of ornithine decarboxylase. *Biochim. Biophys. Acta* **2013**, *1830*, 5157–5165. [CrossRef] [PubMed]
90. Ramos-Molina, B.; Lambertos, A.; Lopez-Contreras, A.J.; Kasprzak, J.M.; Czerwoniec, A.; Bujnicki, J.M.; Cremades, A.; Peñafiel, R. Structural and degradative aspects of ornithine decarboxylase antizyme inhibitor 2. *FEBS Open Bio* **2014**, *4*, 510–521. [CrossRef] [PubMed]
91. Kanerva, K.; Makitie, L.T.; Back, N.; Andersson, L.C. Ornithine decarboxylase antizyme inhibitor 2 regulates intracellular vesicle trafficking. *Exp. Cell Res.* **2010**, *316*, 1896–1906. [CrossRef] [PubMed]
92. Kanerva, K.; Lappalainen, J.; Makitie, L.T.; Virolainen, S.; Kovanen, P.T.; Andersson, L.C. Expression of antizyme inhibitor 2 in mast cells and role of polyamines as selective regulators of serotonin secretion. *PLoS ONE* **2009**, *4*, e6858. [CrossRef] [PubMed]
93. Makitie, L.T.; Kanerva, K.; Sankila, A.; Andersson, L.C. High expression of antizyme inhibitor 2, an activator of ornithine decarboxylase in steroidogenic cells of human gonads. *Histochem. Cell Biol.* **2009**, *132*, 633–638. [CrossRef] [PubMed]
94. Makitie, L.T.; Kanerva, K.; Polvikoski, T.; Paetau, A.; Andersson, L.C. Brain neurons express ornithine decarboxylase-activating antizyme inhibitor 2 with accumulation in Alzheimer's disease. *Brain Pathol.* **2010**, *20*, 571–580. [CrossRef] [PubMed]
95. Lopez-Garcia, C.; Ramos-Molina, B.; Lambertos, A.; Lopez-Contreras, A.J.; Cremades, A.; Penafiel, R. Antizyme inhibitor 2 hypomorphic mice. New patterns of expression in pancreas and adrenal glands suggest a role in secretory processes. *PLoS ONE* **2013**, *8*, e69188. [CrossRef] [PubMed]
96. Seiler, N.; Schmidt-Glenewinkel, T. Regional distribution of putrescine, spermidine and spermine in relation to the distribution of RNA and DNA in the rat nervous system. *J. Neurochem.* **1975**, *24*, 791–795. [CrossRef] [PubMed]
97. Bernstein, H.G.; Muller, M. The cellular localization of the L-ornithine decarboxylase/polyamine system in normal and diseased central nervous systems. *Prog. Neurobiol.* **1999**, *57*, 485–505. [CrossRef]
98. Laitinen, P.H.; Hietala, O.A.; Pulkka, A.E.; Pajunen, A.E. Purification of mouse brain ornithine decarboxylase reveals its presence as an inactive complex with antizyme. *Biochem. J.* **1986**, *236*, 613–616. [CrossRef] [PubMed]
99. Kilpelainen, P.; Rybnikova, E.; Hietala, O.; Pelto-Huikko, M. Expression of ODC and its regulatory protein antizyme in the adult rat brain. *J. Neurosci. Res.* **2000**, *62*, 675–685. [CrossRef]
100. Iyo, A.H.; Zhu, M.Y.; Ordway, G.A.; Regunathan, S. Expression of arginine decarboxylase in brain regions and neuronal cells. *J. Neurochem.* **2006**, *96*, 1042–1050. [CrossRef] [PubMed]
101. Martinez-de-la-Torre, M.; Lambertos, A.; Penafiel, R.; Puelles, L. An exercise in brain genoarchitectonics: Analysis of AZIN2-Lacz expressing neuronal populations in the mouse hindbrain. *J. Neurosci. Res.* **2018**, *96*, 1490–1517. [CrossRef] [PubMed]
102. Zhu, M.Y.; Iyo, A.; Piletz, J.E.; Regunathan, S. Expression of human arginine decarboxylase, the biosynthetic enzyme for agmatine. *Biochim. Biophys. Acta* **2004**, *1670*, 156–164. [CrossRef] [PubMed]
103. Coleman, C.S.; Hu, G.; Pegg, A.E. Putrescine biosynthesis in mammalian tissues. *Biochem. J.* **2004**, *379 Pt 3*, 849–855. [CrossRef]
104. Wang, X.; Ying, W.; Dunlap, K.A.; Lin, G.; Satterfield, M.C.; Burghardt, R.C.; Wu, G.; Bazer, F.W. Arginine decarboxylase and agmatinase: An alternative pathway for de novo biosynthesis of polyamines for development of mammalian conceptuses. *Biol. Reprod.* **2014**, *90*, 84. [CrossRef] [PubMed]

105. Elmetwally, M.A.; Halawa, A.A.; Lenis, Y.Y.; Tang, W.; Wu, G.; Bazer, F.W. Effects of Bisphenol-A on proliferation and expression of genes related to synthesis of polyamines, interferon tau and insulin-like growth factor 2 by ovine trophectoderm cells. *Reprod. Toxicol.* **2018**, *78*, 90–96. [CrossRef] [PubMed]

106. Elmetwally, M.A.; Lenis, Y.; Tang, W.; Wu, G.; Bazer, F.W. Effects of catecholamines on secretion of interferon tau and expression of genes for synthesis of polyamines and apoptosis by ovine trophectoderm. *Biol. Reprod.* **2018**. [CrossRef] [PubMed]

107. Peters, D.; Berger, J.; Langnaese, K.; Derst, C.; Madai, V.I.; Krauss, M.; Fischer, K.D.; Veh, R.W.; Laube, G. Arginase and Arginine Decarboxylase—Where Do the Putative Gate Keepers of Polyamine Synthesis Reside in Rat Brain? *PLoS ONE* **2013**, *8*, e66735. [CrossRef] [PubMed]

108. Bokara, K.K.; Kwon, K.H.; Nho, Y.; Lee, W.T.; Park, K.A.; Lee, J.E. Retroviral expression of arginine decarboxylase attenuates oxidative burden in mouse cortical neural stem cells. *Stem Cells Dev.* **2011**, *20*, 527–537. [CrossRef] [PubMed]

109. Bokara, K.K.; Kim, J.H.; Kim, J.Y.; Lee, J.E. Transfection of arginine decarboxylase gene increases the neuronal differentiation of neural progenitor cells. *Stem Cell Res.* **2016**, *17*, 256–265. [CrossRef] [PubMed]

110. Moon, S.U.; Kwon, K.H.; Kim, J.H.; Bokara, K.K.; Park, K.A.; Lee, W.T.; Lee, J.E. Recombinant hexahistidine arginine decarboxylase (hisADC) induced endogenous agmatine synthesis during stress. *Mol. Cell. Biochem.* **2010**, *345*, 53–60. [CrossRef] [PubMed]

111. Lefevre, P.L.; Palin, M.F.; Murphy, B.D. Polyamines on the reproductive landscape. *Endocr. Rev.* **2011**, *32*, 694–712. [CrossRef] [PubMed]

112. López-Contreras, A.J.; Ramos-Molina, B.; Martínez-de-la-Torre, M.; Peñafiel-Verdú, C.; Puelles, L.; Cremades, A.; Peñafiel, R. Expression of antizyme inhibitor 2 in male haploid germinal cells suggests a role in spermiogenesis. *Int. J. Biochem. Cell Biol.* **2009**, *41*, 1070–1078. [CrossRef] [PubMed]

113. Tokuhiro, K.; Isotani, A.; Yokota, S.; Yano, Y.; Oshio, S.; Hirose, M.; Wada, M.; Fujita, K.; Ogawa, Y.; Okabe, M.; et al. OAZ-t/OAZ3 is essential for rigid connection of sperm tails to heads in mouse. *PLoS Genet.* **2009**, *5*, e1000712. [CrossRef] [PubMed]

114. Zhang, J.; Wang, Y.; Zhou, Y.; Cao, Z.; Huang, P.; Lu, B. Yeast two-hybrid screens imply that GGNBP1, GGNBP2 and OAZ3 are potential interaction partners of testicular germ cell-specific protein GGN1. *FEBS Lett.* **2005**, *579*, 559–566. [CrossRef] [PubMed]

115. Ruan, Y.; Cheng, M.; Ou, Y.; Oko, R.; van der Hoorn, F.A. Ornithine decarboxylase antizyme Oaz3 modulates protein phosphatase activity. *J. Biol. Chem.* **2011**, *286*, 29417–29427. [CrossRef] [PubMed]

116. Lambertos, A.; Ramos-Molina, B.; Lopez-Contreras, A.J.; Cremades, A.; Peñafiel, R. New insights of polyamine metabolism in testicular physiology: A role of ornithine decarboxylase antizyme inhibitor 2 (AZIN2) in the modulation of testosterone levels and sperm motility. *PLoS ONE* **2018**. under review.

117. Kang, B.; Deng, T.; Chen, Z.; Wang, X.; Yi, Z.; Jiang, D. Molecular Cloning of AZIN2 and its Expression Profiling in Goose Tissues and Follicles. *Folia Biol.* **2018**, *66*, 25–31. [CrossRef]

118. Levillain, O.; Ramos-Molina, B.; Forcheron, F.; Penafiel, R. Expression and distribution of genes encoding for polyamine-metabolizing enzymes in the different zones of male and female mouse kidneys. *Amino Acids* **2012**, *43*, 2153–2163. [CrossRef] [PubMed]

119. Li, X.; He, X.; Wang, H.; Li, M.; Huang, S.; Chen, G.; Jing, Y.; Wang, S.; Chen, Y.; Liao, W.; et al. Loss of AZIN2 splice variant facilitates endogenous cardiac regeneration. *Cardiovasc. Res.* 2018. [CrossRef] [PubMed]

120. Acosta-Andrade, C.; Lambertos, A.; Urdiales, J.L.; Sanchez-Jimenez, F.; Penafiel, R.; Fajardo, I. A novel role for antizyme inhibitor 2 as a regulator of serotonin and histamine biosynthesis and content in mouse mast cells. *Amino Acids* **2016**, *48*, 2411–2421. [CrossRef] [PubMed]

121. Rasila, T.; Lehtonen, A.; Kanerva, K.; Makitie, L.T.; Haglund, C.; Andersson, L.C. Expression of ODC Antizyme Inhibitor 2 (AZIN2) in Human Secretory Cells and Tissues. *PLoS ONE* **2016**, *11*, e0151175. [CrossRef] [PubMed]

122. García-Faroldi, G.; Rodríguez, C.E.; Urdiales, J.L.; Pérez-Pomares, J.M.; Dávila, J.C.; Pejler, G.; Sánchez-Jiménez, F.; Fajardo, I. Polyamines are present in mast cell secretory granules and are important for granule homeostasis. *PLoS ONE* **2010**, *5*, e15071. [CrossRef] [PubMed]

123. Poulin, R.; Casero, R.A.; Soulet, D. Recent advances in the molecular biology of metazoan polyamine transport. *Amino Acids* **2012**, *42*, 711–723. [CrossRef] [PubMed]

124. Lambertos, A.; Ramos-Molina, B.; Cerezo, D.; Lopez-Contreras, A.J.; Penafiel, R. The mouse *Gm853* gene encodes a novel enzyme: Leucine decarboxylase. *Biochim. Biophys. Acta* **2018**, *1862*, 365–376. [CrossRef] [PubMed]

125. Gruendler, C.; Lin, Y.; Farley, J.; Wang, T. Proteasomal degradation of Smad1 induced by bone morphogenetic proteins. *J. Biol. Chem.* **2001**, *276*, 46533–46543. [CrossRef] [PubMed]

126. Lin, Y.; Martin, J.; Gruendler, C.; Farley, J.; Meng, X.; Li, B.Y.; Lechleider, R.; Huff, C.; Kim, R.H.; Grasser, W.; et al. A novel link between the proteasome pathway and the signal transduction pathway of the bone morphogenetic proteins (BMPs). *BMC Cell Biol.* **2002**, *3*, 15. [CrossRef]

127. Newman, R.M.; Mobascher, A.; Mangold, U.; Koike, C.; Diah, S.; Schmidt, M.; Finley, D.; Zetter, B.R. Antizyme targets cyclin D1 for degradation. A novel mechanism for cell growth repression. *J. Biol. Chem.* **2004**, *279*, 41504–41511. [CrossRef] [PubMed]

128. Lim, S.K.; Gopalan, G. Antizyme1 mediates AURKAIP1-dependent degradation of Aurora-A. *Oncogene* **2007**, *26*, 6593–6603. [CrossRef] [PubMed]

129. Kasbek, C.; Yang, C.H.; Fisk, H.A. Antizyme restrains centrosome amplification by regulating the accumulation of Mps1 at centrosomes. *Mol. Biol. Cell* **2010**, *21*, 3878–3889. [CrossRef] [PubMed]

130. Dulloo, I.; Gopalan, G.; Melino, G.; Sabapathy, K. The antiapoptotic DeltaNp73 is degraded in a c-Jun-dependent manner upon genotoxic stress through the antizyme-mediated pathway. *Proc. Natl. Acad. Sci. USA* **2010**, *107*, 4902–4907. [CrossRef] [PubMed]

131. Bercovich, Z.; Snapir, Z.; Keren-Paz, A.; Kahana, C. Antizyme affects cell proliferation and viability solely through regulating cellular polyamines. *J. Biol. Chem.* **2011**, *286*, 33778–33783. [CrossRef] [PubMed]

132. Murai, N.; Murakami, Y.; Tajima, A.; Matsufuji, S. Novel ubiquitin-independent nucleolar c-Myc degradation pathway mediated by antizyme 2. *Sci. Rep.* **2018**, *8*, 3005. [CrossRef] [PubMed]

133. Tajima, A.; Murai, N.; Murakami, Y.; Iwamoto, T.; Migita, T.; Matsufuji, S. Polyamine regulating protein antizyme binds to ATP citrate lyase to accelerate acetyl-CoA production in cancer cells. *Biochem. Biophys. Res. Commun.* **2016**, *471*, 646–651. [CrossRef] [PubMed]

![medical sciences logo] *medical sciences*

MDPI

Review

Myc, Oncogenic Protein Translation, and the Role of Polyamines

Andrea T. Flynn [1],* and Michael D. Hogarty [1,2]

[1]   Children's Hospital of Philadelphia, Philadelphia, PA 19104, USA
[2]   Perelman School of Medicine, University of Pennsylvania, Philadelphia, PA 19104, USA;
      hogartym@email.chop.edu
*   Correspondence: flynna@email.chop.edu; Tel.: +1-267-425-1920

Received: 9 April 2018; Accepted: 22 May 2018; Published: 25 May 2018

Abstract: Deregulated protein synthesis is a common feature of cancer cells, with many oncogenic signaling pathways directly augmenting protein translation to support the biomass needs of proliferating tissues. MYC's ability to drive oncogenesis is a consequence of its essential role as a governor linking cell cycle entry with the requisite increase in protein synthetic capacity, among other biomass needs. To date, direct pharmacologic inhibition of MYC has proven difficult, but targeting oncogenic signaling modules downstream of MYC, such as the protein synthetic machinery, may provide a viable therapeutic strategy. Polyamines are essential cations found in nearly all living organisms that have both direct and indirect roles in the control of protein synthesis. Polyamine metabolism is coordinately regulated by MYC to increase polyamines in proliferative tissues, and this is further augmented in the many cancer cells harboring hyperactivated MYC. In this review, we discuss MYC-driven regulation of polyamines and protein synthetic capacity as a key function of its oncogenic output, and how this dependency may be perturbed through direct pharmacologic targeting of components of the protein synthetic machinery, such as the polyamines themselves, the eukaryotic translation initiation factor 4F (eIF4F) complex, and the eukaryotic translation initiation factor 5A (eIF5A).

Keywords: polyamines; *MYC*; protein synthesis in cancer; neuroblastoma

---

1. Introduction

The polyamines (putrescine, spermidine, and spermine) are small organic cations required in nearly all organisms, from bacteria to mammals, to support cell growth and proliferation [1]. Polyamine abundance is increased in many human cancers, as the polyamine synthetic pathway is a direct downstream target of several oncogenes, including the *MYC* family [2–5]. The *MYC* proto-oncogenes (*MYC, MYCN, MYCL*) have been extensively studied since their discovery in the early 1980's [6] and they continue to be of great interest as the most commonly deregulated genes in human cancer [7]. *MYC* genes encode highly homologous helix-loop-helix leucine zipper transcription factors, and MYC overexpression correlates with aggressive tumor behavior and poor prognosis in a wide array of cancers [8–12]. MYC plays a central role in creating the biomass necessary to drive cell progression, including significant increases in protein translation. Deregulated protein synthesis is a common feature of human cancers, and recent work has led to a more complete understanding of how qualitative and quantitative alterations in translation control are sentinel to cancer development, maintenance, and progression [13,14].

While MYC oncoproteins provide an attractive target for cancer therapy because of their frequent somatic activation and the addiction that MYC-driven tumors have to these oncoproteins (reviewed in Gabay et al. [15]), they have been difficult to directly inhibit pharmacologically [7,16]. Attempts to target MYC using microRNAs or antisense RNAs, or small molecules that interfere with MYC–MAX dimerization, DNA binding or stability are under investigation. Alternatively, therapeutically targeting the principal oncogenic outputs of hyperactivated MYC, such as those driving protein synthesis, may provide an alternative anti-cancer approach. The polyamine synthetic pathway is one such pathway [17]. In this review, we focus on the intersection of MYC-driven translation, the effects of polyamine depletion on protein translation, and the cellular dependencies that exist at this juncture as potential avenues of therapeutic intervention.

2. MYC-Driven Oncogenesis

MYC family of proto-oncogenes (*MYC, MYCN, MYCL*) encode highly homologous transcription factors whose activities are tightly regulated in normal cells but frequently deregulated through translocations, amplifications, or alterations of upstream signaling pathways in approximately half of all human cancers [7,18]. MYC genes function at a central node of cellular signaling to link the commitment to enter the cell cycle (stimulated by diverse inputs to MYC) with the requisite biomass production and energetics necessary to do so [19–21]. Indeed, MYC overexpression alone is sufficient to push quiescent cells through the cell cycle [22], while depleting cells of MYC greatly impedes protein translation and markedly reduces proliferation [23]. This non-redundant role in the control of cell proliferation also imbues MYC with significant oncogenic potential when it is deregulated.

Capitalizing on this, many mouse models of cancer have been developed using tissue-specific transgenic overexpression of *MYC* genes, providing insight into the role of MYC in cancer. Such models include the *Eμ-MYC* model of B cell leukemia/lymphoma, the *TH-MYCN* model of neuroblastoma, the Lo-MYC and Hi-MYC models of prostate cancer, and the involucrin-c-mycER skin cancer model [24–27]. These and other models provide important tools to dissect the molecular mechanisms by which MYC overexpression drives oncogenesis within a tissue and a platform for identifying cooperating lesions and testing therapeutic agents for the prevention or treatment of these malignancies.

MYC and Protein Synthesis

MYC is a promiscuous transcription factor regulating thousands of target genes through both canonical target gene promoter binding and accumulation at promoters of actively transcribed genes, leading to transcriptional amplification [28–31]. Yet, much of MYC's output supports biomass production. Since 55–75% of the dry biomass of a cell is protein or involved in protein processing [32], much of MYC's output involves the regulation of genes that support translation: ribosomal proteins, rRNAs, tRNAs, and initiation and elongation factors [21]. Transcriptome analyses in cells in which MYC expression was modulated from null through supraphysiological [33] and in diverse cell types [34] confirm MYC's primordial function in regulating ribosome biogenesis and protein synthesis. This enrichment in the regulation of ribosomal proteins and rRNA, along with genes involved in protein synthesis and turnover, has been similarly shown for *MYC*-driven cancers such as neuroblastoma [35,36]. Genome-wide, MYC binding at the promoters of genes involved in ribosome biogenesis or protein synthesis accounts for roughly the same number of binding events observed for genes involved in cell cycle progression, highlighting the importance of proteogenesis in MYC-driven oncogenesis [37].

Protein processing not only features prominently downstream of MYC signaling, but also leads to a MYC dependency in MYC cancers, as supported by multiple lines of evidence. In switchable models of MYC-induced lymphoma and osteosarcoma, the ability of MYC to induce ribosomal gene products and enhance protein translation correlates with its ability to sustain tumorigenesis [38]. This has been studied in the *Eμ-MYC* lymphoma model, where MYC dramatically increases global protein synthesis, cell size, cell cycle entry, and lymphomatous transformation [39]. Intercrossing *Eμ-MYC* mice into an L24 ribosomal protein hemizygous background (*RPL24+/−*) blocked MYC's effects on protein translation and cell size, induced cell cycle entry, and initiated transformation (and increased cell death among MYC-activated lymphoid cells) [39]. Specifically, cap-dependent protein translation was enhanced by deregulated MYC, which persisted through mitosis rather than switching to internal ribosome entry site (IRES)-dependent translation [40]. This coordinated switch normally allows a subset of IRES-dependent mitosis-specific proteins to be translated, and their loss downstream of activated MYC may contribute to genomic instability. This switch was restored in the *RPL24+/−* background or through treatment of *Eμ-MYC* cells with mTOR inhibitors that impair cap-dependent translation. This same *RPL24* haploinsuffient restriction of protein synthesis did not impact oncogenesis in *TP53* null mice, supporting it as an attribute of *MYC* oncogenic signaling [39].

Cap-dependent translation initiation is rate-limiting for most translated proteins and is stimulated by the eukaryotic translation initiation 4F (eIF4F) complex that contains eIF4E that binds the mRNA cap structure, eIF4A, an RNA helicase that prepares the template for ribosome loading, and eIF4G, which provides a scaffold for bridging the mRNA and ribosome pre-initiation complex (reviewed in Lin et al. [41]). MYC stimulates cap-dependent translation through regulation of eIF4E, eIF4A, and eIF4G, while mTORC1 activity also regulates eIF4E. mTORC1-dependent phosphorylation of 4E binding protein-1 (4EBP1), an eIF4E-binding protein, induces the release of eIF4E for translation initiation. Just as backcrossing *Eμ-MYC* mice into a ribosomal protein-insufficient background (*RPL34+/−* or *RPL38+/−*) or treating with an mTORC1 inhibitor like rapamycin reduce MYC-initiated tumorigenesis, so too does backcrossing into a non-phosphorylatable 4EBP1 background in which eIF4E remains inhibited [42]. Thus, MYC exerts both a quantitative and a qualitative control over protein translation to drive oncogenesis.

3. Polyamines

Polyamines are small organic cations found in nearly all living organisms that are essential for cell growth and survival. The three principal polyamines synthesized by mammals are putrescine, spermidine, and spermine, and these polycations support cell growth and proliferation through cationic, chaperone-like interactions with anionic macromolecules such as DNA, RNA, proteins, and phospholipids. Polyamines also have sequence-dependent DNA interactions and play roles in specific cellular processes such as DNA methylation, chromatin structure, transcription, ion channel function, and scavenging of reactive oxygen species. While their chaperone functions are important for maximizing protein translation efficiency, polyamines also play a direct role in translation by the absolute requirement for spermidine to activate the eukaryotic translation initiation factor 5A (eIF5A) elongation factor and by indirect regulation of mTORC1 and eIF4E activity. The maintenance of normal intracellular polyamine levels is essential to support cell proliferation [43], and polyamine homeostasis is maintained by a multi-level regulation of their transport, biosynthesis, and catabolism.

3.1. Polyamine Biosynthesis and Metabolism

The first, and in most conditions, rate-limiting step in polyamine biosynthesis is the conversion of ornithine (a product of the urea cycle or formed from the catabolism of glutamine) to putrescine catalyzed by ornithine decarboxylase (ODC, encoded by *ODC1*). Putrescine is then converted to spermidine via spermidine synthase (SRM), and spermidine to spermine by spermine synthase (SMS). The activity of the second major regulatory enzyme in polyamine synthesis, adenosylmethionine decarboxylase (AMD, encoded by *AMD1*), provides the aminopropyl groups transferred by SRM and SMS. Both ODC and AMD have the shortest half-lives (10–30 min) of any mammalian enzyme, allowing fine-tuning of polyamine synthesis. Ornithine decarboxylase itself is tightly regulated by transcription, post-transcriptional processing, modified translation efficiency, and altered protein stability, with both antizymes (OAZ1 and OAZ2) and an antizyme inhibitor (AZIN1) contributing [44–48]. Polyamine abundance also negatively feeds back on ODC activity by inducing a +1 ribosomal frameshift that is required to translate functional OAZ1 by reading over a stop codon [49].

Catabolic flux is regulated by the inducible spermidine/spermine-N-acetyltransferase (SAT1) that acetylates higher-order polyamines for export through specific transmembrane solute carriers or for oxidation to lower-order polyamines via polyamine oxidase (PAOX), or by spermine oxidase (SMOX) activities, enhancing homeotic control over the repertoire of cellular polyamines [50]. Finally, an as of yet incompletely characterized energy-dependent polyamine transporter imports polyamines from the microenvironment and can function to rescue polyamine synthesis deficits. These pathways are more completely reviewed in Miller-Fleming et. al. [51]. Indeed, the remarkable investment made to regulate polyamine homeostasis underscores the importance of this pathway in mammalian biology.

3.2. MYC, Polyamines, and Cancer

Deregulated polyamine metabolism has been implicated in several pathologies, notably cancer. Polyamines are abundant in proliferative tissues, including cancer tissues, and are present at much lower levels in senescent and non-proliferative tissues [52–54]. Neuroblastoma is a highly lethal pediatric malignancy of neural crest cells and a prototypical MYC-driven cancer. *MYCN* is deregulated through genomic amplification in ~40% of high-risk neuroblastomas [9,55], while MYC is frequently deregulated in a high proportion of the remainder [56,57]. Additionally, *ODC1* itself has been shown to be amplified in ~6% of high-risk tumors along with *MYCN* and associates with exceptionally poor tumor survival ([3,58,59] and unpublished data). *ODC1* is a direct MYC target and bona fide oncogene [60–62], providing the first demonstration of dual amplification of an oncogenic transcription factor and its oncogenic target gene. Whether this confers an enhanced dependency on polyamines or a resistance mechanism against efforts to deplete polyamines, is under further investigation, since this distinction will inform enrichment strategies for polyamine-depletion cancer trials.

Beyond this somatic activation of polyamine synthesis, transcriptome analyses identified polyamine metabolism as coordinately deregulated by MYC in neuroblastomas. Aggressive tumor behavior and poor outcome is correlated with high expression of synthetic genes (*ODC1, SRM, SMS, AMD1, AZIN1*) and low expression of catabolic genes (*OAZ1, OAZ2, SAT1, PAOX*), and these were identified as direct MYC effects by promoter activity and MYCN binding studies ([3]). Many of these correlations retained independent significance in multivariate analyses controlled for key prognostic variables. These data validate polyamine homeostasis as a key oncogenic output of hyperactivated MYC signaling.

3.3. Polyamines and Protein Translation

Polyamines are important at both the initiation and elongation stages of mRNA translation, though all of their direct and indirect effects on protein synthesis have yet to be fully elucidated [63]. Studies in fractionated mammalian cell-free translation systems demonstrated that the addition of polyamines stimulated protein translation of diverse mRNA species up to eightfold [64,65]. Spermidine and spermine were more potent in this role than putrescine, and it was noted that some mRNAs appeared particularly dependent on the presence of polyamines for their translation, forming a polyamine modulon [66] or polyamine-dependent translatome.

Polyamines regulate protein translation in cells via effects on translation initiation, by modulation of eIF4F complex activity, and on a specific translation factor, namely, eIF5A. Deficits in these processes likely account for the essential roles of polyamines in cell proliferation [63]. While polyamines function to chaperone other macromolecules, the link between their homeostatic control and protein translation was underscored by studies in which *SAT1* was overexpressed via adenoviral transduction in HeLa cells. Acute hyperactivation of SAT1 cells led to a rapid depletion of cellular spermidine and spermine that coincided with abolished global protein synthesis, without a change in DNA or RNA synthesis [67]. Treatment with difluoromethylornithine (DFMO, an inhibitor of ornithine decarboxylase that depletes putrescine and spermidine from cells) also caused a reduction in global protein synthesis in vitro, further underscoring the connection between polyamines and translation [63]. In the *Eu-MYC* model, treating lymphoma-prone mice with DFMO extended survival [4] to a similar extent as genetic crosses into the *RPL24+/−* background [39] or a dominant-negative 4EBP1 mutant background [42], consistent with the notion that the major mechanism downstream of DFMO is disruption of protein translation. Newer tools now being employed to define the effects of polyamine perturbation on ribosomes and their activities are leading to more refined insights into these pathways.

4. Targeting Polyamine Homeostasis and Protein Translation as a Therapy for *MYC*-Driven Cancers

MYC-driven malignancies have a significant dependence on the creation of biomass to support cellular proliferation, specifically on the upregulation of genes that support protein synthesis: ribosome biogenesis, tRNAs, and elongation factors (Figure 1). Indeed, in an analysis across >1000 cancer cell lines from the Cancer Cell Line Encyclopedia, a bioribogenesis and protein translation gene set score was found to be highly correlated with *MYC* expression, with a correlation coefficient of 0.48 at $p < 0.0001$ [68]. This dependence on enhanced protein synthesis for oncogenicity and the concomitant requirement for polyamines to support protein translation provide therapeutic opportunities. Discussed below are strategies to target polyamine metabolism directly as well as key protein translation factors that are influenced by polyamines, as potential points of therapeutic synergy.

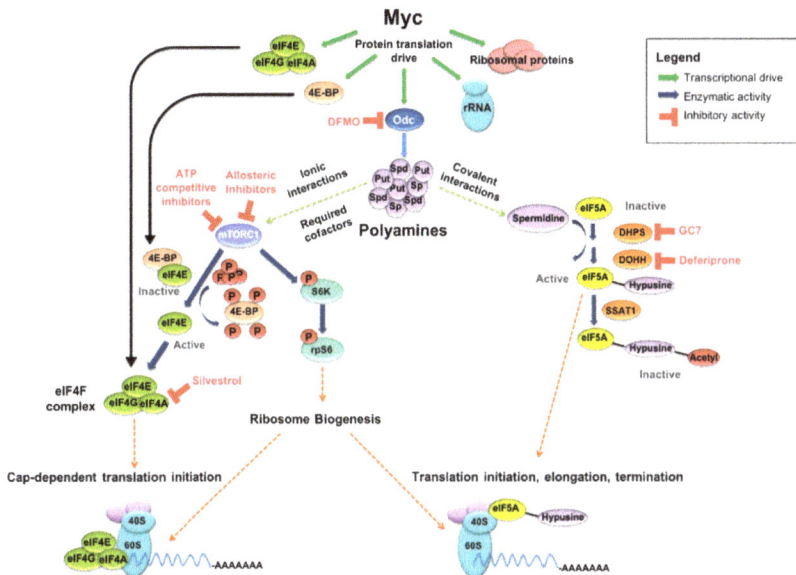

Figure 1. MYC drives protein synthetic output and polyamine metabolism. MYC-driven output includes ribosomal proteins, rRNA, and translation factors, among others. MYC also transcriptionally upregulates ornithine decarboxylase (ODC), a rate-limiting enzyme in polyamine synthesis, as well as spermidine synthase (SRM) and other key polyamine enzymes. Polyamines support protein synthesis in several ways, including spermidine-dependent activation of eukaryotic initiation factor 5A (eIF5A) and as cofactors in the mTORC1-driven release of eIF4E from eIF4E-binding protein 1 (4EBP1) and phosphorylation of S6K. Inhibitors exist for several key enzymes in these polyamine-supported pathways, allowing for the possibility of indirect MYC inhibition through a multi-faceted pharmacologic approach.

4.1. Targeting Polyamine Homeostasis

Polyamine metabolism is upregulated in proliferating tissues, including cancers. Increased polyamine production can be inhibited through impairing synthesis, decreasing import, or increasing export. Not surprisingly, compounds affecting each of these facets of polyamine homeostasis, alone or in combination, provide attractive anticancer approaches currently under development.

DFMO, or eflornithine, is an irreversible inhibitor of ornithine decarboxylase. As a modified ornithine analog, DFMO works through enzyme-activated covalent binding to its target enzyme, ODC. Intravenous DFMO is Food and Drug Administration (FDA)- and European Medicines Agency

(EMA)-approved for the treatment of Trypanosomiasis (African sleeping sickness), as ODC activity is essential for the survival of that protozoan, and has also shown anti-cancer activity in a variety of preclinical models [3,4,69,70]. DFMO is orally bioavailable and has acceptable tolerability even at high doses, which may be required to inactivate ODC in cells with *MYC* and/or *ODC1* amplification. The mechanisms of DFMO's anti-cancer activity include tumor cell polyamine depletion and impaired protein translation, although this remains an area of ongoing investigation. DFMO alone significantly delayed tumor initiation in neuroblastoma-prone *TH-MYCN* mice (homozygous for the *MYCN* transgene), although all mice eventually succumbed to the tumor. More surprisingly, initiating DFMO treatment after tumor onset in this rapidly lethal tumor model also extended survival [69]. In that setting, progressing tumors had reduced putrescine, as evidence of ODC inhibition, but preserved spermidine and spermine because of compensatory mechanisms, like activation of AMD and enhanced polyamine uptake to restore polyamine levels, demonstrating that tumor progression downstream of MYC requires sufficient polyamines.

Adenosylmethionine decarboxylase, a second major regulatory enzyme in the polyamine synthetic pathway, has oncogenic activities in prostate cancer where it is stabilized and activated by mTORC1, showing increased levels in mTORC1-activated tumors and decreased levels in tumors from patients treated with the mTORC1 inhibitor everolimus [71]. Pharmacologic inhibition of AMD is achieved with SAM486, and combining this agent with DFMO to inhibit both rate-limiting polyamine synthetic enzymes in *TH-MYCN* homozygous mice was able to prevent tumor initiation in 40% of mice when used pre-emptively, confirming the requirement for polyamine sufficiency downstream of MYC in tumor initiation [69]. However, the same synergistic activity was not seen when DFMO and SAM486 were used to treat established and progressing tumors in this model.

4.2. Targeting Polyamine Uptake and Export

Additional efforts to augment polyamine depletion in concert with DFMO therapy include approaches that enhance polyamine export or block uptake mechanisms. Sat1 is the key enzyme regulating the catabolism of the polyamines spermine and spermidine, as this acetyltransferase acetylates them for export from the cell. SAT1 activity can be induced by non-steroidal anti-inflammatory drugs (NSAIDs) such as celecoxib or sulindac, and synergy with DFMO has been shown in preclinical neuroblastoma models where the combination enhanced tumor polyamine depletion and extended survival. More importantly, a randomized, placebo-controlled trial of DFMO in combination with sulindac was shown to markedly reduce the recurrence of adenomas, particularly advanced adenomas, in at-risk adults [2]. Several clinical trials are currently open using DFMO in the treatment of a variety of pre-cancer and cancer states, both alone and in combination with chemotherapeutics and NSAIDs, including in patients with neuroblastoma.

Various polyamine transport inhibitors are being developed for anti-tumor activity in combination with DFMO. AMXT1501 is one such inhibitor that in combination with DFMO has been shown to have anti-tumor and immune stimulatory activities in a mouse model of squamous cell carcinoma [72,73] and also to inhibit neuroblastoma proliferation in vitro [69] and extend survival of *TH-MYCN* mice with neuroblastoma in vivo. Most recently, the polyamine transport inhibitor Trimer44NMe was used in combination with DFMO in an orthotopic mouse model of pancreatic ductal adenocarcinoma, demonstrating tumor growth inhibition as well as decreased MYC expression [74]. Transport inhibitors in combination with polyamine synthesis inhibition may be a promising pharmacologic strategy for the treatment of polyamine-dependent tumors, as this approach blocks the primary rescue pathway utilized by cancer cells undergoing polyamine deprivation stress.

It is possible that inhibiting more than one aspect of polyamine metabolism may provide an effective strategy for targeting the dependency of malignant tissue having high levels of polyamines. Work is ongoing for the transition of several of these polyamine-inhibitory compounds, alone and in various combinations, from a preclinical to clinical setting.

4.3. The eIF4F Complex and Cap-Dependent Translation

In addition to synergistically targeting polyamine homeostatic pathways to augment the extent of polyamine depletion, another approach is to look for synergy opportunities in the principal oncogenic pathways disrupted by polyamine depletion. Given the relationship between polyamines and protein translation, this is an area of active investigation.

The eIF4F cap-binding translation initiation complex is responsible for cap-dependent mRNA translation. eIF4F is composed of eF4E (cap-binding protein), eIF4A (ATP-dependent RNA helicase), and eIF4G (scaffolding protein). The formation of the eIF4F complex is highly regulated and dependent on several signaling pathways that are also involved in oncogenesis. eIF4E is the least abundant of the eIF4F complex proteins and therefore is the rate-limiting element in eIF4F formation. 4E-binding proteins sequester eIF4E, and phosphorylation of 4EBP1 by mTORC1 is required for the release of eIF4E. The central role of eIF4E as a highly regulated component of the eIF4F complex and its overexpression in several cancers make eIF4E an attractive target for the development of therapeutic agents [75]. Overexpression of eIF4E transforms immortalized mouse NIH-3T3 cell [76] and cooperates with the viral homolog of MYC to transform rat embryo fibroblasts [77]. Subsequently, eIF4E was found to cooperate with MYC in the *Eμ-MYC* transgenic mouse model to enhance lymphomagenesis and drug resistance in vivo through antagonism of MYC-dependent apoptosis [78,79]. Additionally, induced overexpression of eIF4E leads to the recruitment of ribosomes to a subset of mRNAs that promote oncogenesis, specifically, mRNAs with highly structured 5' untranslated regions (5'UTRs) [80–82]. Of note, mRNAs with highly structured 5'UTRs, that are translated more efficiently in the setting of eIF4E overexpression, include two relevant oncogenes, i.e., *ODC1* and *MYC* [83,84].

Increased MYC expression leads directly to increased transcription of three core components of the eIF4F complex (eIF4E, eIF4AI, and eIF4GI), mediated through canonical E-boxes in their promoters. Enhanced expression of these core components by MYC results in a feed-forward loop with increased eIF4F expression, leading to increased translation of MYC [85]. The polyamine putrescine was found to increase the activation of mTORC1 and the subsequent phosphorylation of 4EBP1, and increasing concentrations of putrescine led to increased protein synthesis and cell proliferation in a dose-dependent fashion. These changes in protein production and proliferation were reversed in the presence of the mTOR inhibitor rapamycin [86]. Further linking mTORC1 with polyamine signaling is its role as an amino acid sensor. Oncogenic polyamine synthesis maintains methionine and s-adenosyl-methionine (SAM) at levels sufficient to inhibit SAMTOR, a nutrient sensor in the mTORC1 pathway, and supports mTORC1 signaling [87]. In the setting of DFMO, there is enhanced AMD activity that leads to reduced pools of methionine and SAM, de-repression of SAMTOR, and inhibition of mTORC1 activation. Additionally, in ODC over-producing cells, there was an increase in eIF4E phosphorylation and subsequent helicase activity of the eIF4F complex [88]. This, in combination with increased efficiency of translation of *ODC1* in the setting of eIF4E overexpression, suggests an additional feed-forward loop of the eIF4F components and an oncogene, here *ODC1*, possibly of relevance to oncogenic transformation.

4.4. Targeting The eIF4F Complex

Several agents that inhibit the components of the eIF4F complex have shown preclinical antitumor activity and may have enhanced activity in combination with polyamine depletion agents like DFMO. A 4E-antisense oligonucleotide to reduce the expression of eIF4E was shown to inhibit tumor growth in a prostate cancer xenograft model [89] and was safe in humans in a Phase I clinical trial, though it did not demonstrate anti-tumor activity as a single agent [90]. Pateamine A is a macrolide originally isolated from marine sponges in the 1990s, and an analog of pateamine led to tumor regression in two melanoma mouse xenograft models [91]. Inhibitors of eIF4A helicase activity include hippuristanol, pateamine A, and other flavaglines and rocaglates derived from natural products.

Hippuristanol inhibits the binding of mRNA to eIF4A and has been shown to reverse the resistance to chemotherapy and induce apoptosis in synergy with ABT-737 (BCL-2 inhibitor) in the *Eμ-MYC* lymphoma model [92]. The most extensively studied of the agents targeting the eIF4A helicase is silvestrol, which belongs to the flavagline class and enhances eIF4A binding to mRNA, thus preventing its participation in the eIF4F complex. This agent extended survival when combined with chemotherapy in both the *Eμ-MYC/eIF4E* and *Eμ-MYC/PTEN+/−* lymphoma models, in which MYC and mTOR are deregulated to drive lymphoma onset [93]. In the *NOTCH1*-driven murine T-ALL model of acute leukemia, in which *MYC* is also deregulated [94], silvestrol was found to preferentially inhibit the translation of mRNAs with G-quadruplex structures, which include several transcription factors and epigenetic regulators [95].

4.5. Eukaryotic Translation Initiation Factor 5A and Its Spermidine-Dependent Activation

Eukaryotic translation initiation factor 5A is a unique RNA-binding protein that is evolutionarily conserved from fungi through plants, insects, and mammals and is essential to eukaryotic cell viability. There are two paralogous genes encoding isoforms of eIF5A: eIF5A1 at chromosome band 17p13 and eIF5A2 at 3q26 [96]. eIF5A1 is ubiquitously expressed, while eIF5A2 expression is largely restricted to the brain and testis. Either isoform can rescue yeast with genetic eIF5A deletion, consistent with their functional redundancy. eIF5A was originally implicated in having stimulatory effects in the formation of the first peptide bond between Met-tRNA and puromycin in translation initiation [97,98]. However, its effects are likely far broader on the basis of polysome analyses demonstrating ribosomal stalling in eIF5A-depleted conditions in yeast [99,100]. The bacterial orthologue, elongation factor P (EF-P), is structurally and functionally similar to eIF5A. EF-P was found to be important in resolving ribosomal stalling at sites of consecutive proline residues [101,102], and eIF5A's effect on the resolution of polyproline stretches was subsequently established in yeast [103]. Most recently, eIF5A has been shown to have a role in accelerating peptidyl transfer more generally at ribosome stalling sites, though tripeptide motifs have differential dependence on the activity of this factor (with prolines featuring prominently in disruptive motifs), and to promote peptide release from the ribosome at termination [104]. eIF5A appears to effect multiple facets of protein synthesis, and the biologic and therapeutic implications of its functions remain areas of active investigation. The extent to which eIF5A contributes to global protein translation may also be a function of genome complexity, as proline-rich motifs occur at higher frequency in the proteomes of higher organisms [105].

In addition to roles as cationic chaperones, the polyamine spermidine is absolutely required for the activation of eIF5A [106]. eIF5A undergoes a two-step post-translational modification termed hypusination, in which a butyl amine moiety of spermidine is covalently bound to a lysine residue (K50) of eIF5A to yield its hypusinated form (Figure 1). Hypusination of eIF5A is essential for most, if not all, of its activities, therefore making spermidine an essential substrate for eIF5A activation. The two enzymes necessary for hypusination of eIF5A, i.e., deoxyhypusine synthase (DHS) and deoxyhypusine hydroxylase (DOHH), are dedicated exclusively to this pathway and, like eIF5A, are essential genes [107]. No other eukaryotic protein is known to require hypusination, further underscoring the unique importance of this polyamine-dependent modification [108]. Hypusination of eIF5A occurs co-translationally and appears to be irreversible, suggesting a non-dynamic role in regulating eIF5A activity under normal conditions. Inactivation through acetylation by the SAT1 acetyltransferase may provide additional regulation, though this remains incompletely understood [109]. Indeed, while SAT1 can acetylate the hypusine moiety of eIF5A, rendering the factor inactive, prolonged overexpression of SAT1 in HeLa cells abolished protein translation but inactivated less than 10% of the total hypusinated-eIF5A pool [67].

Both eIF5A isoforms have been linked to cancer functions, though, given the weak or undetectable expression of eIF5A2 in most tissues, their overexpression in various cancers is more readily identified. Moreover, high expression of eIF5A2 has been found in numerous cancer in association with *EIF5A2* gene amplification and correlates with advanced cancer stage and worse prognosis for many patients, leading to its identification as an oncogene (reviewed in Nakanishi et al. [110]). However, a targeted shRNA library screen identified *EIF5A, AMD1, SRM,* and *DHS* as tumor suppressors in the *Eμ-MYC* model [111]. This data has yet to be reconciled with extensive data linking polyamine sufficiency with malignant progression, though it is possible that, in B cells, an eIF5A-dependent translatome includes proteins with tumor suppressor functions disproportionately over those with putative oncogenic roles. Alternatively, there may be functional distinctions between the role of eIF5A in tumor initiation versus maintenance or progression. Indeed, as mentioned, AMD inhibition cooperates with ODC inhibition to block tumor initiation in the neural crest of *TH-MYCN* neuroblastoma-prone mice, yet it antagonizes DFMO activity in the setting of therapy for established and progressing tumors [69].

4.6. Inhibition of eIF5A Hypusination

The key enzymes in activating hypusination of eIF5A at lysine K-50 are DHS and DOHH, both of which have direct inhibitors. N1-Guanyl-1,7-diaminoheptane (GC7) inhibits DHS and has been studied in vitro in a variety of cancer cell lines. Indeed, GC7 treatment augments the activity of a variety of chemotherapeutic agents and has synergy with targeted agents such as imatinib in *BCR-ABL*-driven leukemia [112–114]. However, the clinical utility of GC7 beyond in vitro studies is uncertain, as it is a spermidine analog and can therefore affect other aspects of polyamine homeostasis not specific to hypusination of eIF5A. DOHH requires iron for the hydroxylation of hypusine, making DOHH susceptible to inhibition by iron chelators such as ciclopirox and deferiprone, both of which are clinically available agents. Ciclopirox is an antifungal agent and has been shown to inhibit growth and stimulate apoptosis in a variety of cancer types in vitro [115,116] and in vivo [117]. Deferiprone is an oral chelator used to treat iron overload secondary to chronic blood transfusion in thalassemia and was found to inhibit proliferation of cervical cancer cells in vitro [118]. A Phase I clinical trial in patients with advance hematologic malignancies to assess dose and side effect profile of ciclopirox demonstrated clinical tolerability of escalated doses and even evidence of disease stabilization in a few patients [119]. It is possible that a combination therapy with one of these agents blocking a key step in the hypusination of eIF5A and a polyamine antagonist may show additive or synergistic anti-tumor effects.

5. Conclusions: Current Gaps and Future Perspective

MYC-driven bioribogenesis and protein synthesis are central to its oncogenicity. While directly targeting MYC has proven difficult, downstream effectors driving increases in global and selective protein synthesis provide numerous targetable pathways that have potent oncogenic roles and an exploitable therapeutic index. Here, we have highlighted some of these pathways that are essential to the increased protein synthetic drive induced by MYC. Though polyamine homeostasis is essential for normal cell growth and proliferation, an increased drive for polyamine uptake and synthesis noted in human cancers may provide a targetable MYC-driven dependency. Gene expression signatures of MYC-driven cancers often demonstrate increased expression of enzymes involved in polyamine synthesis and decreased expression of polyamine catabolic enzymes, highlighting this coordinated signaling downstream of MYC. Polyamines play an important role in protein synthesis, both globally and selectively, by activating key translation factors such as eIF5A. However, the exact mechanisms by which polyamine depletion leads to protein translation stress and the mechanisms of anti-tumor activity when polyamines are depleted from tumor cells remain the subject of ongoing work.

Current studies in our laboratory and others are aimed at using a combination strategy to induce protein translation stress through direct polyamine depletion in addition to inhibition of the translation machinery in neuroblastoma and other MYC-driven tumor models. Polyamine depletion, alone or in combination with inhibitors of protein translation, may provide a potent stress on a MYC-driven cancer, which may result in clinically significant anti-tumor activity. Such preclinical work is required for the informed design of clinical protocols to test these approaches, with the hope of understanding which additional compounds, if any, provide the highest likelihood of clinically relevant anti-tumor effects.

Author Contributions: A.T.F. conceptualized, researched, wrote, and edited the paper. M.D.H. researched, wrote, reviewed, and edited the paper.

Acknowledgments: Support for this review was provided by Alex's Lemonade Stand Foundation (ALSF) Young Investigator Award (ATF), the ALSF Center of Excellence Award (ATF), and the ALSF REACH award (MDH), Hyundai Hope on Wheels Award (MDH), Wipe Out Kids Cancer Award (MDH), and Team Connor Award (MDH).

Conflicts of Interest: The authors declare no conflict of interest. The funding sponsors had no role in the writing of or the decision to publish the manuscript.

References

1. Igarashi, K.; Kashiwagi, K. Modulation of cellular function by polyamines. *Int. J. Biochem. Cell Biol.* **2010**, *42*, 39–51. [CrossRef] [PubMed]
2. Gerner, E.W.; Meyskens, F.L., Jr.; Goldschmid, S.; Lance, P.; Pelot, D. Rationale for, and design of, a clinical trial targeting polyamine metabolism for colon cancer chemoprevention. *Amino Acids* **2007**, *33*, 189–195. [CrossRef] [PubMed]
3. Hogarty, M.D.; Norris, M.D.; Davis, K.; Liu, X.; Evageliou, N.F.; Hayes, C.S.; Pawel, B.; Guo, R.; Zhao, H.; Sekyere, E.; et al. ODC1 is a critical determinant of MYCN oncogenesis and a therapeutic target in neuroblastoma. *Cancer Res.* **2008**, *68*, 9735–9745. [CrossRef] [PubMed]
4. Nilsson, J.A.; Keller, U.B.; Baudino, T.A.; Yang, C.; Norton, S.; Old, J.A.; Nilsson, L.M.; Neale, G.; Kramer, D.L.; Porter, C.W.; et al. Targeting ornithine decarboxylase in Myc-induced lymphomagenesis prevents tumor formation. *Cancer Cell* **2005**, *7*, 433–444. [CrossRef] [PubMed]
5. Thomas, T.; Thomas, T.J. Estradiol control of ornithine decarboxylase mRNA, enzyme activity, and polyamine levels in MCF-7 breast cancer cells: Therapeutic implications. *Breast Cancer Res. Treat.* **1994**, *29*, 189–201. [CrossRef] [PubMed]
6. Sheiness, D.K.; Hughes, S.H.; Varmus, H.E.; Stubblefield, E.; Bishop, J.M. The vertebrate homolog of the putative transforming gene of avian myelocytomatosis virus: Characteristics of the DNA locus and its RNA transcript. *Virology* **1980**, *105*, 415–424. [CrossRef]
7. Vita, M.; Henriksson, M. The Myc oncoprotein as a therapeutic target for human cancer. *Semin. Cancer Biol.* **2006**, *16*, 318–330. [CrossRef] [PubMed]
8. Barrans, S.; Crouch, S.; Smith, A.; Turner, K.; Owen, R.; Patmore, R.; Roman, E.; Jack, A. Rearrangement of MYC is associated with poor prognosis in patients with diffuse large B-cell lymphoma treated in the era of rituximab. *J. Clin. Oncol. Off. J. Am. Soc. Clin. Oncol.* **2010**, *28*, 3360–3365. [CrossRef] [PubMed]
9. Brodeur, G.; Seeger, R.; Schwab, M.; Varmus, H.; Bishop, J. Amplification of N-myc in untreated human neuroblastomas correlates with advanced disease stage. *Science* **1984**, *224*, 1121–1124. [CrossRef] [PubMed]
10. Chng, W.J.; Huang, G.F.; Chung, T.H.; Ng, S.B.; Gonzalez-Paz, N.; Troska-Price, T.; Mulligan, G.; Chesi, M.; Bergsagel, P.L.; Fonseca, R. Clinical and biological implications of MYC activation: A common difference between MGUS and newly diagnosed multiple myeloma. *Leukemia* **2011**, *25*, 1026–1035. [CrossRef] [PubMed]
11. Kanungo, A.; Medeiros, L.J.; Abruzzo, L.V.; Lin, P. Lymphoid neoplasms associated with concurrent t(14;18) and 8q24/c-MYC translocation generally have a poor prognosis. *Mod. Pathol.* **2006**, *19*, 25–33. [CrossRef] [PubMed]
12. Wolfer, A.; Wittner, B.S.; Irimia, D.; Flavin, R.J.; Lupien, M.; Gunawardane, R.N.; Meyer, C.A.; Lightcap, E.S.; Tamayo, P.; Mesirov, J.P.; et al. MYC regulation of a "poor-prognosis" metastatic cancer cell state. *Proc. Natl. Acad. Sci. USA* **2010**, *107*, 3698–3703. [CrossRef] [PubMed]
13. Bhat, M.; Robichaud, N.; Hulea, L.; Sonenberg, N.; Pelletier, J.; Topisirovic, I. Targeting the translation machinery in cancer. *Nat. Rev. Drug Discov.* **2015**, *14*, 261–278. [CrossRef] [PubMed]
14. Truitt, M.L.; Ruggero, D. New frontiers in translational control of the cancer genome. *Nat. Rev. Cancer* **2016**, *16*, 288–304. [CrossRef] [PubMed]
15. Gabay, M.; Li, Y.; Felsher, D.W. MYC activation is a hallmark of cancer initiation and maintenance. *Cold Spring Harb. Perspect. Med.* **2014**, *4*. [CrossRef] [PubMed]
16. Prochownik, E.V.; Vogt, P.K. Therapeutic targeting of Myc. *Genes Cancer* **2010**, *1*, 650–659. [CrossRef] [PubMed]
17. Casero, R.A., Jr.; Marton, L.J. Targeting polyamine metabolism and function in cancer and other hyperproliferative diseases. *Nat. Rev. Drug Discov.* **2007**, *6*, 373–390. [CrossRef] [PubMed]
18. Nesbit, C.E.; Tersak, J.M.; Prochownik, E.V. MYC oncogenes and human neoplastic disease. *Oncogene* **1999**, *18*, 3004–3016. [CrossRef] [PubMed]

19. Eilers, M.; Eisenman, R.N. Myc's broad reach. *Genes Dev* **2008**, *22*, 2755–2766. [CrossRef] [PubMed]
20. Fernandez, P.C.; Frank, S.R.; Wang, L.; Schroeder, M.; Liu, S.; Greene, J.; Cocito, A.; Amati, B. Genomic targets of the human c-Myc protein. *Genes Dev* **2003**, *17*, 1115–1129. [CrossRef] [PubMed]
21. van Riggelen, J.; Yetil, A.; Felsher, D.W. MYC as a regulator of ribosome biogenesis and protein synthesis. *Nat. Rev. Cancer* **2010**, *10*, 301–309. [CrossRef] [PubMed]
22. Bouchard, C.; Staller, P.; Eilers, M. Control of cell proliferation by Myc. *Trends Cell Biol.* **1998**, *8*, 202–206. [CrossRef]
23. Mateyak, M.K.; Obaya, A.J.; Adachi, S.; Sedivy, J.M. Phenotypes of c-Myc-deficient rat fibroblasts isolated by targeted homologous recombination. *Cell Growth Differ. Mol. Biol. J. Am. Assoc. Cancer Res.* **1997**, *8*, 1039–1048.
24. Adams, J.M.; Harris, A.W.; Pinkert, C.A.; Corcoran, L.M.; Alexander, W.S.; Cory, S.; Palmiter, R.D.; Brinster, R.L. The c-Myc oncogene driven by immunoglobulin enhancers induces lymphoid malignancy in transgenic mice. *Nature* **1985**, *318*, 533–538. [CrossRef] [PubMed]
25. Ellwood-Yen, K.; Graeber, T.G.; Wongvipat, J.; Iruela-Arispe, M.L.; Zhang, J.; Matusik, R.; Thomas, G.V.; Sawyers, C.L. Myc-driven murine prostate cancer shares molecular features with human prostate tumors. *Cancer Cell* **2003**, *4*, 223–238. [CrossRef]
26. Pelengaris, S.; Littlewood, T.; Khan, M.; Elia, G.; Evan, G. Reversible activation of c-Myc in skin: Induction of a complex neoplastic phenotype by a single oncogenic lesion. *Mol. Cell* **1999**, *3*, 565–577. [CrossRef]
27. Weiss, W.A.; Aldape, K.; Mohapatra, G.; Feuerstein, B.G.; Bishop, J.M. Targeted expression of MYCN causes neuroblastoma in transgenic mice. *EMBO J.* **1997**, *16*, 2985–2995. [CrossRef] [PubMed]
28. Lin, C.Y.; Loven, J.; Rahl, P.B.; Paranal, R.M.; Burge, C.B.; Bradner, J.E.; Lee, T.I.; Young, R.A. Transcriptional amplification in tumor cells with elevated c-Myc. *Cell* **2012**, *151*, 56–67. [CrossRef] [PubMed]
29. Nie, Z.; Hu, G.; Wei, G.; Cui, K.; Yamane, A.; Resch, W.; Wang, R.; Green, D.R.; Tessarollo, L.; Casellas, R.; et al. c-Myc is a universal amplifier of expressed genes in lymphocytes and embryonic stem cells. *Cell* **2012**, *151*, 68–79. [CrossRef] [PubMed]
30. Sabo, A.; Kress, T.R.; Pelizzola, M.; de Pretis, S.; Gorski, M.M.; Tesi, A.; Morelli, M.J.; Bora, P.; Doni, M.; Verrecchia, A.; et al. Selective transcriptional regulation by Myc in cellular growth control and lymphomagenesis. *Nature* **2014**, *511*, 488–492. [CrossRef] [PubMed]
31. Walz, S.; Lorenzin, F.; Morton, J.; Wiese, K.E.; von Eyss, B.; Herold, S.; Rycak, L.; Dumay-Odelot, H.; Karim, S.; Bartkuhn, M.; et al. Activation and repression by oncogenic MYC shape tumour-specific gene expression profiles. *Nature* **2014**, *511*, 483–487. [CrossRef] [PubMed]
32. Mourant, J.R.; Short, K.W.; Carpenter, S.; Kunapareddy, N.; Coburn, L.; Powers, T.M.; Freyer, J.P. Biochemical differences in tumorigenic and nontumorigenic cells measured by Raman and infrared spectroscopy. *J. Biomed. Opt.* **2005**, *10*, 031106. [CrossRef] [PubMed]
33. O'Connell, B.C.; Cheung, A.F.; Simkevich, C.P.; Tam, W.; Ren, X.; Mateyak, M.K.; Sedivy, J.M. A large scale genetic analysis of c-Myc-regulated gene expression patterns. *J. Biol. Chem.* **2003**, *278*, 12563–12573. [CrossRef] [PubMed]
34. Ji, H.; Wu, G.; Zhan, X.; Nolan, A.; Koh, C.; De Marzo, A.; Doan, H.M.; Fan, J.; Cheadle, C.; Fallahi, M.; et al. Cell-type independent MYC target genes reveal a primordial signature involved in biomass accumulation. *PLoS ONE* **2011**, *6*, e26057. [CrossRef] [PubMed]
35. Boon, K.; Caron, H.N.; van Asperen, R.; Valentijn, L.; Hermus, M.C.; van Sluis, P.; Roobeek, I.; Weis, I.; Voute, P.A.; Schwab, M.; et al. N-myc enhances the expression of a large set of genes functioning in ribosome biogenesis and protein synthesis. *EMBO J.* **2001**, *20*, 1383–1393. [CrossRef] [PubMed]
36. Chen, Q.R.; Song, Y.K.; Yu, L.R.; Wei, J.S.; Chung, J.Y.; Hewitt, S.M.; Veenstra, T.D.; Khan, J. Global genomic and proteomic analysis identifies biological pathways related to high-risk neuroblastoma. *J. Proteome Res.* **2010**, *9*, 373–382. [CrossRef] [PubMed]
37. Ruggero, D. The role of Myc-induced protein synthesis in cancer. *Cancer Res.* **2009**, *69*, 8839–8843. [CrossRef] [PubMed]
38. Wu, C.H.; Sahoo, D.; Arvanitis, C.; Bradon, N.; Dill, D.L.; Felsher, D.W. Combined analysis of murine and human microarrays and ChIP analysis reveals genes associated with the ability of MYC to maintain tumorigenesis. *PLoS Genet.* **2008**, *4*, e1000090. [CrossRef] [PubMed]

39. Barna, M.; Pusic, A.; Zollo, O.; Costa, M.; Kondrashov, N.; Rego, E.; Rao, P.H.; Ruggero, D. Suppression of Myc oncogenic activity by ribosomal protein haploinsufficiency. *Nature* **2008**, *456*, 971–975. [CrossRef] [PubMed]

40. Pyronnet, S.; Sonenberg, N. Cell-cycle-dependent translational control. *Curr. Opin. Genet. Dev.* **2001**, *11*, 13–18. [CrossRef]

41. Lin, C.J.; Malina, A.; Pelletier, J. c-Myc and eIF4F constitute a feedforward loop that regulates cell growth: Implications for anticancer therapy. *Cancer Res.* **2009**, *69*, 7491–7494. [CrossRef] [PubMed]

42. Pourdehnad, M.; Truitt, M.L.; Siddiqi, I.N.; Ducker, G.S.; Shokat, K.M.; Ruggero, D. Myc and mTOR converge on a common node in protein synthesis control that confers synthetic lethality in myc-driven cancers. *Proc. Natl. Acad. Sci. USA* **2013**, *110*, 11988–11993. [CrossRef] [PubMed]

43. Pegg, A.E. Functions of polyamines in mammals. *J. Biol. Chem.* **2016**, *291*, 14904–14912. [CrossRef] [PubMed]

44. Albeck, S.; Dym, O.; Unger, T.; Snapir, Z.; Bercovich, Z.; Kahana, C. Crystallographic and biochemical studies revealing the structural basis for antizyme inhibitor function. *Protein Sci. Publ. Protein Soc.* **2008**, *17*, 793–802. [CrossRef] [PubMed]

45. Kurian, L.; Palanimurugan, R.; Godderz, D.; Dohmen, R.J. Polyamine sensing by nascent ornithine decarboxylase antizyme stimulates decoding of its mRNA. *Nature* **2011**, *477*, 490–494. [CrossRef] [PubMed]

46. Nowotarski, S.L.; Shantz, L.M. Cytoplasmic accumulation of the RNA-binding protein HuR stabilizes the ornithine decarboxylase transcript in a murine nonmelanoma skin cancer model. *J. Biol. Chem.* **2010**, *285*, 31885–31894. [CrossRef] [PubMed]

47. Pegg, A.E. Regulation of ornithine decarboxylase. *J. Biol. Chem.* **2006**, *281*, 14529–14532. [CrossRef] [PubMed]

48. Shantz, L.M. Transcriptional and translational control of ornithine decarboxylase during ras transformation. *Biochem. J.* **2004**, *377*, 257–264. [CrossRef] [PubMed]

49. Palanimurugan, R.; Scheel, H.; Hofmann, K.; Dohmen, R.J. Polyamines regulate their synthesis by inducing expression and blocking degradation of ODC antizyme. *EMBO J.* **2004**, *23*, 4857–4867. [CrossRef] [PubMed]

50. Kramer, D.L.; Diegelman, P.; Jell, J.; Vujcic, S.; Merali, S.; Porter, C.W. Polyamine acetylation modulates polyamine metabolic flux, a prelude to broader metabolic consequences. *J. Biol. Chem.* **2008**, *283*, 4241–4251. [CrossRef] [PubMed]

51. Miller-Fleming, L.; Olin-Sandoval, V.; Campbell, K.; Ralser, M. Remaining mysteries of molecular biology: The role of polyamines in the cell. *J. Mol. Biol.* **2015**, *427*, 3389–3406. [CrossRef] [PubMed]

52. Evageliou, N.F.; Hogarty, M.D. Disrupting polyamine homeostasis as a therapeutic strategy for neuroblastoma. *Clin. Cancer Res. Off. J. Am. Assoc. Cancer Res.* **2009**, *15*, 5956–5961. [CrossRef] [PubMed]

53. Gamble, L.D.; Hogarty, M.D.; Liu, X.; Ziegler, D.S.; Marshall, G.; Norris, M.D.; Haber, M. Polyamine pathway inhibition as a novel therapeutic approach to treating neuroblastoma. *Front. Oncol.* **2012**, *2*, 62. [CrossRef] [PubMed]

54. Thomas, T.; Thomas, T.J. Polyamines in cell growth and cell death: Molecular mechanisms and therapeutic applications. *Cell. Mol. Life Sci.* **2001**, *58*, 244–258. [CrossRef] [PubMed]

55. Seeger, R.C.; Brodeur, G.M.; Sather, H.; Dalton, A.; Siegel, S.E.; Wong, K.Y.; Hammond, D. Association of multiple copies of the N-myc oncogene with rapid progression of neuroblastomas. *New Engl. J. Med.* **1985**, *313*, 1111–1116. [CrossRef] [PubMed]

56. Fredlund, E.; Ringner, M.; Maris, J.M.; Pahlman, S. High Myc pathway activity and low stage of neuronal differentiation associate with poor outcome in neuroblastoma. *Proc. Natl. Acad. Sci. USA* **2008**, *105*, 14094–14099. [CrossRef] [PubMed]

57. Westermann, F.; Muth, D.; Benner, A.; Bauer, T.; Henrich, K.O.; Oberthuer, A.; Brors, B.; Beissbarth, T.; Vandesompele, J.; Pattyn, F.; et al. Distinct transcriptional MYCN/c-MYC activities are associated with spontaneous regression or malignant progression in neuroblastomas. *Genome Biol.* **2008**, *9*, R150. [CrossRef] [PubMed]

58. Depuydt, P.; Boeva, V.; Hocking, T.D.; Cannoodt, R.; Ambros, I.M.; Ambros, P.F.; Asgharzadeh, S.; Attiyeh, E.F.; Combaret, V.; Defferrari, R.; et al. Genomic amplifications and distal 6q loss: Novel markers for poor survival in high-risk neuroblastoma patients. *J. Natl. Cancer Inst.* **2018**, *110*, 1–10. [CrossRef] [PubMed]

59. George, R.E.; Kenyon, R.; McGuckin, A.G.; Kohl, N.; Kogner, P.; Christiansen, H.; Pearson, A.D.; Lunec, J. Analysis of candidate gene co-amplification with MYCN in neuroblastoma. *Eur. J. Cancer* **1997**, *33*, 2037–2042. [CrossRef]

60. Auvinen, M.; Paasinen, A.; Andersson, L.C.; Holtta, E. Ornithine decarboxylase activity is critical for cell transformation. *Nature* **1992**, *360*, 355–358. [CrossRef] [PubMed]

61. Bello-Fernandez, C.; Packham, G.; Cleveland, J.L. The ornithine decarboxylase gene is a transcriptional target of c-Myc. *Proc. Natl. Acad. Sci. USA* **1993**, *90*, 7804–7808. [CrossRef] [PubMed]

62. Moshier, J.A.; Dosescu, J.; Skunca, M.; Luk, G.D. Transformation of NIH/3T3 cells by ornithine decarboxylase overexpression. *Cancer Res.* **1993**, *53*, 2618–2622. [PubMed]

63. Landau, G.; Bercovich, Z.; Park, M.H.; Kahana, C. The role of polyamines in supporting growth of mammalian cells is mediated through their requirement for translation initiation and elongation. *J. Biol. Chem.* **2010**, *285*, 12474–12481. [CrossRef] [PubMed]

64. Atkins, J.F.; Lewis, J.B.; Anderson, C.W.; Gesteland, R.F. Enhanced differential synthesis of proteins in a mammalian cell-free system by addition of polyamines. *J. Biol. Chem.* **1975**, *250*, 5688–5695. [PubMed]

65. Igarashi, K.; Hikami, K.; Sugawara, K.; Hirose, S. Effect of polyamines on polypeptide synthesis in rat liver cell-free system. *Biochim. Biophys. Acta* **1973**, *299*, 325–330. [CrossRef]

66. Igarashi, K.; Kashiwagi, K. Modulation of protein synthesis by polyamines. *IUBMB Life* **2015**, *67*, 160–169. [CrossRef] [PubMed]

67. Mandal, S.; Mandal, A.; Johansson, H.E.; Orjalo, A.V.; Park, M.H. Depletion of cellular polyamines, spermidine and spermine, causes a total arrest in translation and growth in mammalian cells. *Proc. Natl. Acad. Sci. USA* **2013**, *110*, 2169–2174. [CrossRef] [PubMed]

68. Manier, S.; Huynh, D.; Shen, Y.J.; Zhou, J.; Yusufzai, T.; Salem, K.Z.; Ebright, R.Y.; Shi, J.; Park, J.; Glavey, S.V.; et al. Inhibiting the oncogenic translation program is an effective therapeutic strategy in multiple myeloma. *Sci. Transl. Med.* **2017**, *9*. [CrossRef] [PubMed]

69. Evageliou, N.F.; Haber, M.; Vu, A.; Laetsch, T.W.; Murray, J.; Gamble, L.D.; Cheng, N.C.; Liu, K.; Reese, M.; Corrigan, K.A.; et al. Polyamine antagonist therapies inhibit neuroblastoma initiation and progression. *Clin. Cancer Res. Off. J. Am. Assoc. Cancer Res.* **2016**, *22*, 4391–4404. [CrossRef] [PubMed]

70. Rounbehler, R.J.; Li, W.; Hall, M.A.; Yang, C.; Fallahi, M.; Cleveland, J.L. Targeting ornithine decarboxylase impairs development of MYCN-amplified neuroblastoma. *Cancer Res.* **2009**, *69*, 547–553. [CrossRef] [PubMed]

71. Zabala-Letona, A.; Arruabarrena-Aristorena, A.; Martin-Martin, N.; Fernandez-Ruiz, S.; Sutherland, J.D.; Clasquin, M.; Tomas-Cortazar, J.; Jimenez, J.; Torres, I.; Quang, P.; et al. mTORC1-dependent AMD1 regulation sustains polyamine metabolism in prostate cancer. *Nature* **2017**, *547*, 109–113. [CrossRef] [PubMed]

72. Samal, K.; Zhao, P.; Kendzicky, A.; Yco, L.P.; McClung, H.; Gerner, E.; Burns, M.; Bachmann, A.S.; Sholler, G. AMXT-1501, a novel polyamine transport inhibitor, synergizes with DFMO in inhibiting neuroblastoma cell proliferation by targeting both ornithine decarboxylase and polyamine transport. *Int. J. Cancer* **2013**, *133*, 1323–1333. [CrossRef] [PubMed]

73. Hayes, C.S.; Shicora, A.C.; Keough, M.P.; Snook, A.E.; Burns, M.R.; Gilmour, S.K. Polyamine-blocking therapy reverses immunosuppression in the tumor microenvironment. *Cancer Immunol. Res.* **2014**, *2*, 274–285. [CrossRef] [PubMed]

74. Gitto, S.B.; Pandey, V.; Oyer, J.L.; Copik, A.J.; Hogan, F.C.; Phanstiel, O.; Altomare, D.A. Difluoromethylornithine combined with a polyamine transport inhibitor is effective against gemcitabine resistant pancreatic cancer. *Mol. Pharm.* **2018**, *15*, 369–376. [CrossRef] [PubMed]

75. Malka-Mahieu, H.; Newman, M.; Desaubry, L.; Robert, C.; Vagner, S. Molecular pathways: The eIF4F translation initiation complex-new opportunities for cancer treatment. *Clin. Cancer Res. Off. J. Am. Assoc. Cancer Res.* **2017**, *23*, 21–25. [CrossRef] [PubMed]

76. Lazaris-Karatzas, A.; Montine, K.S.; Sonenberg, N. Malignant transformation by a eukaryotic initiation factor subunit that binds to mRNA 5′ cap. *Nature* **1990**, *345*, 544–547. [CrossRef] [PubMed]

77. Lazaris-Karatzas, A.; Sonenberg, N. The mRNA 5′ cap-binding protein, eIF-4E, cooperates with v-myc or E1A in the transformation of primary rodent fibroblasts. *Mol. Cell. Biol.* **1992**, *12*, 1234–1238. [CrossRef] [PubMed]

78. Ruggero, D.; Montanaro, L.; Ma, L.; Xu, W.; Londei, P.; Cordon-Cardo, C.; Pandolfi, P.P. The translation factor eIF-4E promotes tumor formation and cooperates with c-Myc in lymphomagenesis. *Nat. Med.* **2004**, *10*, 484–486. [CrossRef] [PubMed]

79. Wendel, H.G.; De Stanchina, E.; Fridman, J.S.; Malina, A.; Ray, S.; Kogan, S.; Cordon-Cardo, C.; Pelletier, J.; Lowe, S.W. Survival signalling by Akt and eIF4E in oncogenesis and cancer therapy. *Nature* **2004**, *428*, 332–337. [CrossRef] [PubMed]

80. Koromilas, A.E.; Lazaris-Karatzas, A.; Sonenberg, N. Mrnas containing extensive secondary structure in their 5′ non-coding region translate efficiently in cells overexpressing initiation factor eIF-4E. *EMBO J.* **1992**, *11*, 4153–4158. [PubMed]

81. Larsson, O.; Li, S.; Issaenko, O.A.; Avdulov, S.; Peterson, M.; Smith, K.; Bitterman, P.B.; Polunovsky, V.A. Eukaryotic translation initiation factor 4E induced progression of primary human mammary epithelial cells along the cancer pathway is associated with targeted translational deregulation of oncogenic drivers and inhibitors. *Cancer Res.* **2007**, *67*, 6814–6824. [CrossRef] [PubMed]

82. Smith, M.R.; Jaramillo, M.; Liu, Y.L.; Dever, T.E.; Merrick, W.C.; Kung, H.F.; Sonenberg, N. Translation initiation factors induce DNA synthesis and transform NIH 3T3 cells. *New Biol.* **1990**, *2*, 648–654. [PubMed]

83. De Benedetti, A.; Graff, J.R. eIF-4E expression and its role in malignancies and metastases. *Oncogene* **2004**, *23*, 3189–3199. [CrossRef] [PubMed]

84. Shantz, L.M.; Coleman, C.S.; Pegg, A.E. Expression of an ornithine decarboxylase dominant-negative mutant reverses eukaryotic initiation factor 4E-induced cell transformation. *Cancer Res.* **1996**, *56*, 5136–5140. [PubMed]

85. Lin, C.J.; Cencic, R.; Mills, J.R.; Robert, F.; Pelletier, J. c-Myc and eIF4F are components of a feedforward loop that links transcription and translation. *Cancer Res.* **2008**, *68*, 5326–5334. [CrossRef] [PubMed]

86. Kong, X.; Wang, X.; Yin, Y.; Li, X.; Gao, H.; Bazer, F.W.; Wu, G. Putrescine stimulates the mTOR signaling pathway and protein synthesis in porcine trophectoderm cells. *Biol. Reprod.* **2014**, *91*, 106. [CrossRef] [PubMed]

87. Gu, X.; Orozco, J.M.; Saxton, R.A.; Condon, K.J.; Liu, G.Y.; Krawczyk, P.A.; Scaria, S.M.; Harper, J.W.; Gygi, S.P.; Sabatini, D.M. SAMTOR is an S-adenosylmethionine sensor for the mTORC1 pathway. *Science* **2017**, *358*, 813–818. [CrossRef] [PubMed]

88. Shimogori, T.; Suzuki, T.; Kashiwagi, K.; Kakinuma, Y.; Igarashi, K. Enhancement of helicase activity and increase of eIF-4E phosphorylation in ornithine decarboxylase-overproducing cells. *Biochem. Biophys. Res. Commun.* **1996**, *222*, 748–752. [CrossRef] [PubMed]

89. Graff, J.R.; Konicek, B.W.; Vincent, T.M.; Lynch, R.L.; Monteith, D.; Weir, S.N.; Schwier, P.; Capen, A.; Goode, R.L.; Dowless, M.S.; et al. Therapeutic suppression of translation initiation factor eIF4E expression reduces tumor growth without toxicity. *J. Clin. Investig.* **2007**, *117*, 2638–2648. [CrossRef] [PubMed]

90. Hong, D.S.; Kurzrock, R.; Oh, Y.; Wheler, J.; Naing, A.; Brail, L.; Callies, S.; Andre, V.; Kadam, S.K.; Nasir, A.; et al. A phase 1 dose escalation, pharmacokinetic, and pharmacodynamic evaluation of eIF-4E antisense oligonucleotide LY2275796 in patients with advanced cancer. *Clin. Cancer Res. Off. J. Am. Assoc. Cancer Res.* **2011**, *17*, 6582–6591. [CrossRef] [PubMed]

91. Kuznetsov, G.; Xu, Q.; Rudolph-Owen, L.; Tendyke, K.; Liu, J.; Towle, M.; Zhao, N.; Marsh, J.; Agoulnik, S.; Twine, N.; et al. Potent in vitro and in vivo anticancer activities of des-methyl, des-amino pateamine A, a synthetic analogue of marine natural product pateamine A. *Mol. Cancer Ther.* **2009**, *8*, 1250–1260. [CrossRef] [PubMed]

92. Cencic, R.; Robert, F.; Galicia-Vazquez, G.; Malina, A.; Ravindar, K.; Somaiah, R.; Pierre, P.; Tanaka, J.; Deslongchamps, P.; Pelletier, J. Modifying chemotherapy response by targeted inhibition of eukaryotic initiation factor 4A. *Blood Cancer J.* **2013**, *3*, e128. [CrossRef] [PubMed]

93. Bordeleau, M.E.; Robert, F.; Gerard, B.; Lindqvist, L.; Chen, S.M.; Wendel, H.G.; Brem, B.; Greger, H.; Lowe, S.W.; Porco, J.A.; et al. Therapeutic suppression of translation initiation modulates chemosensitivity in a mouse lymphoma model. *J. Clin. Investig.* **2008**, *118*, 2651–2660. [CrossRef] [PubMed]

94. Palomero, T.; Lim, W.K.; Odom, D.T.; Sulis, M.L.; Real, P.J.; Margolin, A.; Barnes, K.C.; O'Neil, J.; Neuberg, D.; Weng, A.P.; et al. NOTCH1 directly regulates c-MYC and activates a feed-forward-loop transcriptional network promoting leukemic cell growth. *Proc. Natl. Acad. Sci. USA* **2006**, *103*, 18261–18266. [CrossRef] [PubMed]

95. Wolfe, A.L.; Singh, K.; Zhong, Y.; Drewe, P.; Rajasekhar, V.K.; Sanghvi, V.R.; Mavrakis, K.J.; Jiang, M.; Roderick, J.E.; Van der Meulen, J.; et al. RNA G-quadruplexes cause eIF4A-dependent oncogene translation in cancer. *Nature* **2014**, *513*, 65–70. [CrossRef] [PubMed]

96. Jenkins, Z.A.; Haag, P.G.; Johansson, H.E. Human eIF5A2 on chromosome 3q25-q27 is a phylogenetically conserved vertebrate variant of eukaryotic translation initiation factor 5A with tissue-specific expression. *Genomics* **2001**, *71*, 101–109. [CrossRef] [PubMed]

97. Benne, R.; Hershey, J.W. The mechanism of action of protein synthesis initiation factors from rabbit reticulocytes. *J. Biol. Chem.* **1978**, *253*, 3078–3087. [PubMed]

98. Schreier, M.H.; Erni, B.; Staehelin, T. Initiation of mammalian protein synthesis. I. Purification and characterization of seven initiation factors. *J. Mol. Biol.* **1977**, *116*, 727–753. [CrossRef]

99. Henderson, A.; Hershey, J.W. Eukaryotic translation initiation factor (eIF) 5A stimulates protein synthesis in saccharomyces cerevisiae. *Proc. Natl. Acad. Sci. USA* **2011**, *108*, 6415–6419. [CrossRef] [PubMed]

100. Saini, P.; Eyler, D.E.; Green, R.; Dever, T.E. Hypusine-containing protein eIF5A promotes translation elongation. *Nature* **2009**, *459*, 118–121. [CrossRef] [PubMed]

101. Doerfel, L.K.; Wohlgemuth, I.; Kothe, C.; Peske, F.; Urlaub, H.; Rodnina, M.V. Ef-p is essential for rapid synthesis of proteins containing consecutive proline residues. *Science* **2013**, *339*, 85–88. [CrossRef] [PubMed]

102. Ude, S.; Lassak, J.; Starosta, A.L.; Kraxenberger, T.; Wilson, D.N.; Jung, K. Translation elongation factor EF-P alleviates ribosome stalling at polyproline stretches. *Science* **2013**, *339*, 82–85. [CrossRef] [PubMed]

103. Gutierrez, E.; Shin, B.S.; Woolstenhulme, C.J.; Kim, J.R.; Saini, P.; Buskirk, A.R.; Dever, T.E. eIF5A promotes translation of polyproline motifs. *Mol. Cell* **2013**, *51*, 35–45. [CrossRef] [PubMed]

104. Schuller, A.P.; Wu, C.C.; Dever, T.E.; Buskirk, A.R.; Green, R. eIF5A functions globally in translation elongation and termination. *Mol. Cell* **2017**, *66*, 194–205. [CrossRef] [PubMed]

105. Mandal, A.; Mandal, S.; Park, M.H. Genome-wide analyses and functional classification of proline repeat-rich proteins: Potential role of eIF5A in eukaryotic evolution. *PLoS ONE* **2014**, *9*, e111800. [CrossRef] [PubMed]

106. Wolff, E.C.; Kang, K.R.; Kim, Y.S.; Park, M.H. Posttranslational synthesis of hypusine: Evolutionary progression and specificity of the hypusine modification. *Amino Acids* **2007**, *33*, 341–350. [CrossRef] [PubMed]

107. Pallmann, N.; Braig, M.; Sievert, H.; Preukschas, M.; Hermans-Borgmeyer, I.; Schweizer, M.; Nagel, C.H.; Neumann, M.; Wild, P.; Haralambieva, E.; et al. Biological relevance and therapeutic potential of the hypusine modification system. *J. Biol. Chem.* **2015**, *290*, 18343–18360. [CrossRef] [PubMed]

108. Cooper, H.L.; Park, M.H.; Folk, J.E.; Safer, B.; Braverman, R. Identification of the hypusine-containing protein hy+ as translation initiation factor eIF-4D. *Proc. Natl. Acad. Sci. USA* **1983**, *80*, 1854–1857. [CrossRef] [PubMed]

109. Lee, S.B.; Park, J.H.; Folk, J.E.; Deck, J.A.; Pegg, A.E.; Sokabe, M.; Fraser, C.S.; Park, M.H. Inactivation of eukaryotic initiation factor 5A (eIF5A) by specific acetylation of its hypusine residue by spermidine/spermine acetyltransferase 1 (SSAT1). *Biochem. J.* **2011**, *433*, 205–213. [CrossRef] [PubMed]

110. Nakanishi, S.; Cleveland, J.L. Targeting the polyamine-hypusine circuit for the prevention and treatment of cancer. *Amino Acids* **2016**, *48*, 2353–2362. [CrossRef] [PubMed]

111. Scuoppo, C.; Miething, C.; Lindqvist, L.; Reyes, J.; Ruse, C.; Appelmann, I.; Yoon, S.; Krasnitz, A.; Teruya-Feldstein, J.; Pappin, D.; et al. A tumour suppressor network relying on the polyamine-hypusine axis. *Nature* **2012**, *487*, 244–248. [CrossRef] [PubMed]

112. Balabanov, S.; Gontarewicz, A.; Ziegler, P.; Hartmann, U.; Kammer, W.; Copland, M.; Brassat, U.; Priemer, M.; Hauber, I.; Wilhelm, T.; et al. Hypusination of eukaryotic initiation factor 5A (eIF5A): A novel therapeutic target in BCR-ABL-positive leukemias identified by a proteomics approach. *Blood* **2007**, *109*, 1701–1711. [CrossRef] [PubMed]

113. Preukschas, M.; Hagel, C.; Schulte, A.; Weber, K.; Lamszus, K.; Sievert, H.; Pallmann, N.; Bokemeyer, C.; Hauber, J.; Braig, M.; et al. Expression of eukaryotic initiation factor 5A and hypusine forming enzymes in glioblastoma patient samples: Implications for new targeted therapies. *PLoS ONE* **2012**, *7*, e43468. [CrossRef] [PubMed]

114. Xu, G.; Yu, H.; Shi, X.; Sun, L.; Zhou, Q.; Zheng, D.; Shi, H.; Li, N.; Zhang, X.; Shao, G. Cisplatin sensitivity is enhanced in non-small cell lung cancer cells by regulating epithelial-mesenchymal transition through inhibition of eukaryotic translation initiation factor 5A2. *BMC Pulm. Med.* **2014**, *14*, 174. [CrossRef] [PubMed]

115. Zhou, H.; Shen, T.; Luo, Y.; Liu, L.; Chen, W.; Xu, B.; Han, X.; Pang, J.; Rivera, C.A.; Huang, S. The antitumor activity of the fungicide ciclopirox. *Int. J. Cancer* **2010**, *127*, 2467–2477. [CrossRef] [PubMed]

116. Eberhard, Y.; McDermott, S.P.; Wang, X.; Gronda, M.; Venugopal, A.; Wood, T.E.; Hurren, R.; Datti, A.; Batey, R.A.; Wrana, J.; et al. Chelation of intracellular iron with the antifungal agent ciclopirox olamine induces cell death in leukemia and myeloma cells. *Blood* **2009**, *114*, 3064–3073. [CrossRef] [PubMed]

117. Kim, Y.; Schmidt, M.; Endo, T.; Lu, D.; Carson, D.; Schmidt-Wolf, I.G. Targeting the wnt/beta-catenin pathway with the antifungal agent ciclopirox olamine in a murine myeloma model. *In Vivo* **2011**, *25*, 887–893. [PubMed]
118. Memin, E.; Hoque, M.; Jain, M.R.; Heller, D.S.; Li, H.; Cracchiolo, B.; Hanauske-Abel, H.M.; Pe'ery, T.; Mathews, M.B. Blocking eIF5A modification in cervical cancer cells alters the expression of cancer-related genes and suppresses cell proliferation. *Cancer Res.* **2014**, *74*, 552–562. [CrossRef] [PubMed]
119. Minden, M.D.; Hogge, D.E.; Weir, S.J.; Kasper, J.; Webster, D.A.; Patton, L.; Jitkova, Y.; Hurren, R.; Gronda, M.; Goard, C.A.; et al. Oral ciclopirox olamine displays biological activity in a phase i study in patients with advanced hematologic malignancies. *Am. J. Hematol.* **2014**, *89*, 363–368. [CrossRef] [PubMed]

medical
sciences

MDPI

Review

Cellular and Animal Model Studies on the Growth Inhibitory Effects of Polyamine Analogues on Breast Cancer

T. J. Thomas [1,*] and Thresia Thomas [2,†]

[1] Department of Medicine, Rutgers Robert Wood Johnson Medical School and Rutgers Cancer Institute of New Jersey, Rutgers, The State University of New Jersey, 675 Hoes Lane West, KTL Room N102, Piscataway, NJ 08854, USA

[2] Retired from Department of Environmental and Occupational Medicine, Rutgers Robert Wood Johnson Medical School and Rutgers Cancer Institute of New Jersey, Rutgers, The State University of New Jersey, 675 Hoes Lane West, Piscataway, NJ 08854, USA; thomasthresia@gmail.com

* Correspondence: thomastj@rwjms.rutgers.edu; Tel.: +1-732-235-5852
† Present address: 40 Caldwell Drive, Princeton, NJ 08540, USA.

Received: 28 January 2018; Accepted: 6 March 2018; Published: 13 March 2018

Abstract: Polyamine levels are elevated in breast tumors compared to those of adjacent normal tissues. The female sex hormone, estrogen is implicated in the origin and progression of breast cancer. Estrogens stimulate and antiestrogens suppress the expression of polyamine biosynthetic enzyme, ornithine decarboxylate (ODC). Using several bis(ethyl)spermine analogues, we found that these analogues inhibited the proliferation of estrogen receptor-positive and estrogen receptor negative breast cancer cells in culture. There was structure-activity relationship in the efficacy of these compounds in suppressing cell growth. The activity of ODC was inhibited by these compounds, whereas the activity of the catabolizing enzyme, spermidine/spermine N^1-acetyl transferase (SSAT) was increased by 6-fold by bis(ethyl)norspermine in MCF-7 cells. In a transgenic mouse model of breast cancer, bis(ethyl)norspermine reduced the formation and growth of spontaneous mammary tumor. Recent studies indicate that induction of polyamine catabolic enzymes SSAT and spermine oxidase (SMO) play key roles in the anti-proliferative and apoptotic effects of polyamine analogues and their combinations with chemotherapeutic agents such as 5-fluorouracil (5-FU) and paclitaxel. Thus, polyamine catabolic enzymes might be important therapeutic targets and markers of sensitivity in utilizing polyamine analogues in combination with other therapeutic agents.

Keywords: polyamines; ornithine decarboxylase; polyamine analogs; spermidine/spermine N^1-acetyl transferase; spermine oxidase; bis(ethyl)polyamine analogs; breast cancer; MCF-7 cells; transgenic mice

1. Introduction

Breast cancer is a major public health problem and is the most common cancer in women worldwide, with nearly 1.7 million new cases diagnosed in 2012 (second most common cancer overall) [1]. For women in the United States of America in 2017, the estimation is 252,710 new cases of invasive breast cancer, 63,410 new cases of breast carcinoma in situ and 40,610 breast cancer deaths per year [2]. Although breast cancer incidence rates are highest in non-Hispanic white women, breast cancer death rates are highest in African American women. Breast cancer is a heterogeneous disease and harbors different receptors and in some cases no recognized receptor driving the disease [3]. The female hormone, estrogen is implicated in the origin and progression of breast cancer and approximately 70% of breast tumors harbor the receptor protein, estrogen receptor

(ER) [4–6]. There are two forms of ER, ERα and ERβ, with several isoforms of each of these subtypes [6,7]. Binding of estrogen with ERα provokes conformational changes in ERα, enabling it to recruit coactivator proteins and recognize the estrogen response element (ERE) in responsive genes to facilitate breast cancer growth. ERα-positive tumors are responsive to antiestrogens, such as tamoxifen [8]. Aromatase inhibitors are also available to treat these tumors by blocking the synthesis of estrogens [9]. Approximately 15% of breast tumors harbor the human epidermal receptor-2 (HER-2) and these tumors respond to a monoclonal antibody, Herceptin, targeted to HER-2 [10]. A subset of 15–20% of breast tumors lacks ER, HER-2 and progesterone receptor (PR) and this subtype is classified as triple negative breast cancer (TNBC). There is no targeted therapy for TNBCs [11]. In the case of ERα- and HER-2 positive breast tumors, drug resistance develops in the course of therapy and hence new drugs are needed for treating all forms of breast cancer [12]. Research on polyamine analogues have been undertaken in this context [13–17].

2. Polyamine Metabolism and Breast Cancer

Natural polyamines (putrescine, spermidine and spermine) are ubiquitous cellular cations and play important roles in cell growth and differentiation [18,19]. Polyamine levels are elevated in cancer cells compared to that in adjacent tissues [14,20,21]. Rise in cellular polyamine levels, associated with up-regulation of polyamine biosynthetic enzymes, is characteristic of increased proliferation of cancer cells. Polyamines are positively charged at physiological pH and ionic conditions and hence they can interact with a variety of biological targets, including membrane phospholipids, proteins and nucleic acids through electrostatic interactions with negatively charged groups on these macromolecules [22–25]. Polyamines are capable of facilitating the interaction of transcription factors—such as ERα and nuclear factor kappa B (NF-κB)—with their response elements in breast cancer [26–28]. In addition, polyamines enhanced the bending of estrogen response element in the presence of ERα [29].

Cellular polyamine levels are exquisitely regulated by biosynthetic and catabolic enzymes (Figure 1) [17–19]. The biosynthetic enzymes are ornithine decarboxylase (ODC), adenosylmethionine decarboxylase (AdoMetDC) and aminopropyl transferase (spermidine synthase and spermine synthase). Increased ODC activity in human breast cancer tissues was found to be an independent adverse prognostic factor for recurrence and death. The catabolic enzymes are spermidine/spermine N^1-acetyltransferase (SSAT), N^1-acetylpolyamine oxidase (APAO) and spermine oxidase (SMO) [30–32]. Induction of SMO and the concomitant production of reactive oxygen species (ROS) have been linked to inflammation-associated cancers [32]. Multiple amine oxidases are also implicated in the degradation of diamines and polyamines, resulting in the production of H_2O_2 and aldehydes [33]. In addition, polyamine transport systems control the cellular import and export of polyamines to maintain polyamine homeostasis [34]. Antizyme proteins further control polyamine levels by binding to ODC and facilitating its degradation and thereby regulating polyamine pools in the cell [35]. The polyamine metabolic and transport pathways have been attractive targets for therapeutic intervention in cancers, including breast cancer [14,18,20,21,36,37].

Estradiol stimulates ODC at the mRNA, enzyme activity and polyamine biosynthesis levels and ODC knockout diminishes the mRNA and protein expression of ERα in ERα-positive MCF-7 and T-47D human breast cancer cells [38,39]. Inhibition of ODC by α-difluoromethylornithine (DFMO), an irreversible inhibitor of the enzyme, blocked the growth promoting activity of estradiol in *N*-methyl-*N*-nitrosourea-induced Sprague-Dawley rat mammary tumor grown in the soft agar clonogenic assay as well as xenograft growth in animals [40]. This drug was once considered as a promising anticancer drug, interfering with polyamine pathway and it was evaluated in Phase I/II clinical trials as a single agent as well as in combination treatment protocols [14]. However, this enzyme inhibitor exerted only limited therapeutic effects in breast and other cancers, probably due to the high concentrations required to inhibit ODC, poor cellular uptake and the versatility of polyamine pathway to replenish cellular polyamine pools. Interestingly, the use of DFMO in combination with other chemopreventive agents showed promising results in colorectal cancer [41]. AdoMetDC inhibitors also suppressed breast cancer cell growth in a structure-dependent manner [42].

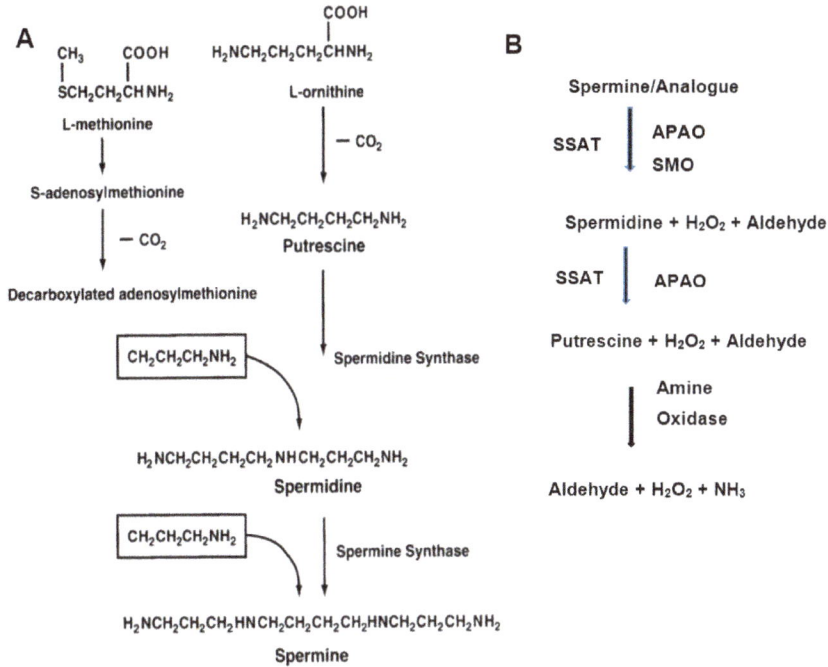

Figure 1. Schematic representation of the polyamine biosynthetic (**A**) and catabolic (**B**) pathways. (**A**) Putrescine is formed by the decarboxylation of ornithine by ornithine decarboxylate (ODC). Spermidine is formed by the action of spermidine synthase that links putrescine to an aminopropyl group derived from decarboxylated S-adenosylmethionine, a reaction product of AdoMetDC. Spermine is synthesized from spermidine by a similar process by spermine synthase; (**B**) Spermine and spermidine are first acetylated spermidine/spermine N^1-acetyltransferase (SSAT) and then oxidized by N^1-acetylpolyamine oxidase (APAO). Spermine oxidase (SMO) degrades unmodified spermine/internalized analogue. H_2O_2 and 3-aceto-aminopropanal are among the degradation products [17]. Multiple amine oxidases are also involved in the degradation of diamines and polyamines, producing H_2O_2 and aldehydes [33].

Recent studies showed that endoxifen, an active metabolite of tamoxifen, suppressed the activity of ODC and AdoMetDC and induced SMO and APAO in MCF-7 cells [37]. Cellular putrescine and spermidine levels were reduced in response to endoxifen treatment. Results of this study indicated that in addition to the reduction of polyamine biosynthetic enzymes, induction of catabolic enzymes could play a role in antiproliferative effects of tamoxifen and endoxifen. Induction of SMO and APAO lead to degradation of polyamines and the production of H_2O_2, contributing of cell death by apoptosis [36,43,44]. Thus, the polyamine metabolic pathway plays a critical role in the mechanism of action of antiestrogens. Although most polyamine-related breast cancer studies were conducted using ERα-positive cells, polyamine biosynthetic inhibitors suppressed the proliferation of ERα-negative breast cancer cells also [45,46].

3. Polyamine Analogues and Breast Cancer Therapeutics

Despite initial success in cellular and animal models of several cancers, including breast cancer, clinical advance of DFMO and related polyamine biosynthetic inhibitors was hampered by lack of efficacy and adverse side effects [47,48]. Polyamine analogues were synthesized for cancer therapeutics on the premise that these molecules could utilize the polyamine transport pathway for cellular internalization and disrupt cellular functions of natural polyamines [13–17,20,49–51]. Early developments in this area involved the synthesis of polyamine analogues with structural alterations in the number of methylene groups between the amino and imino groups of natural polyamines. Porter and Bergeron found that homologues of putrescine and spermidine were taken up by L1210 leukemia cells and that analogues with small changes in carbon chain length could reverse DFMO-mediated cell growth inhibition, whereas large changes in chain length deprived the analogue's ability to support cell growth [49]. A differential effect of putrescine analogues in preventing DFMO-mediated cell growth inhibition and normal immune response was also reported [52]. DFMO inhibited cell growth and suppressed putrescine and spermidine levels in MCF-7 breast cancer cells, whereas putrescine and its close homologues, diaminopropane and diaminopentane partially reversed the growth inhibitory effects of DFMO [38]. However, diaminoethane was not able to reverse DFMO's growth inhibitory effect on MCF-7 cells. Similar results were obtained with a series of spermidine homologues of the structure, $H_2N(CH_2)_nNH(CH_2)_3NH_2$ (where $n = 2$ to 8; abbreviated as AP_n with $n = 4$ for spermidine). Spermidine was most effective in reversing the effects of DFMO, whereas compounds with shorter or longer methylene bridging regions were less effective. The homologue abbreviated as AP8 ($n = 8$) was ineffective in reversing the growth inhibitory effects of DFMO. In addition, AP8 inhibited DNA synthesis by 66% as a single agent, as measured by [^{3}H]-thymidine incorporation assay [38]. These data suggested that certain polyamine analogues could disrupt breast cancer growth in cell culture conditions.

Polyamines with terminal amino groups are good substrates for amine oxidases, including SMO and APAO [53,54]. Porter and Bergeron designed and synthesized bis(ethyl) spermine analogues (Figure 2) as cancer therapeutic agents, with the goal of exploiting the polyamine transport system to accumulate these analogues within the cancer cell, thereby downregulating polyamine biosynthesis and depleting natural polyamines [55–58]. Consequently bis(ethyl) substituted spermine and spermidine analogues were studied for their effectiveness in suppressing the growth of different cancer cell lines in culture and xenograft models [59–61]. Davidson et al. [62] found that bis(ethyl)spermine (BE-3-4-3) could inhibit the growth of six breast cancer cell lines, with half maximal inhibitory concentration (IC_{50}) values in the micromolar range. In addition to progressive depletion of intracellular polyamines over a period of 6 days, this compound induced polyamine catabolic enzyme, SSAT by 8-12-fold in selected cell lines.

Our laboratory conducted a detailed study of the effects of 6 bis(ethyl)spermine analogues (BE-3-4-3, BE-4-4-4, BE-3-3-3, BE-3-7-3, BE-3-3-3-3 and BE-4-4-4-4) on cell growth, activities of polyamine metabolic enzymes, intracellular polyamine levels and analogue uptake in breast cancer cells [63]. The IC_{50} values for cell growth inhibition of BE-3-4-3, BE-3-3-3 and BE-4-4-4 were in the range of 1–2 μM, whereas BE-3-7-3, BE-3-3-3-3 and BE-4-4-4-4 had IC_{50} values of ~5 μM. These values were comparable for three cell lines: ERα-positive MCF-7, HER-2-positive SK-BR-3 and triple negative MDA-MB-231 cells. Colony formation of MCF-7 cells in soft agar showed a concentration-dependent decrease in the number of colonies per well after 14 days of treatment. All compounds induced apoptosis of MCF-7 cells at 4–6 days of treatment with 10 μM drug concentration, although structural effects were evident. There was a facile transport of analogues within MCF-7 cells, although BR-3-4-3 had the highest level of transport because of its structural similarity to that of natural spermine (3-4-3).

All six bis(ethyl)spermine analogues selected in this study inhibited ODC activity and suppressed intracellular levels of putrescine and spermidine in MCF-7 cells [63]. Spermine levels were significantly reduced by BE-3-4-3, BE-3-3-3, BE-3-3-3-3 and BE-4-4-4-4, whereas BE-4-4-4 and BE-3-7-3 had no significant effect. SSAT activity was increased by 3- to 6-fold by BE-3-4-3, BE-3-3-3 and

BE-3-3-3-3, although other analogues exerted no significant effect. These results indicated that the polyamine metabolic pathway was affected by bis(ethyl)polyamine analogues in breast cancer cells. Molecular modeling studies further suggested a correlation between anti-proliferative activity of the analogues and their ability to dock into DNA major or minor grooves [63].

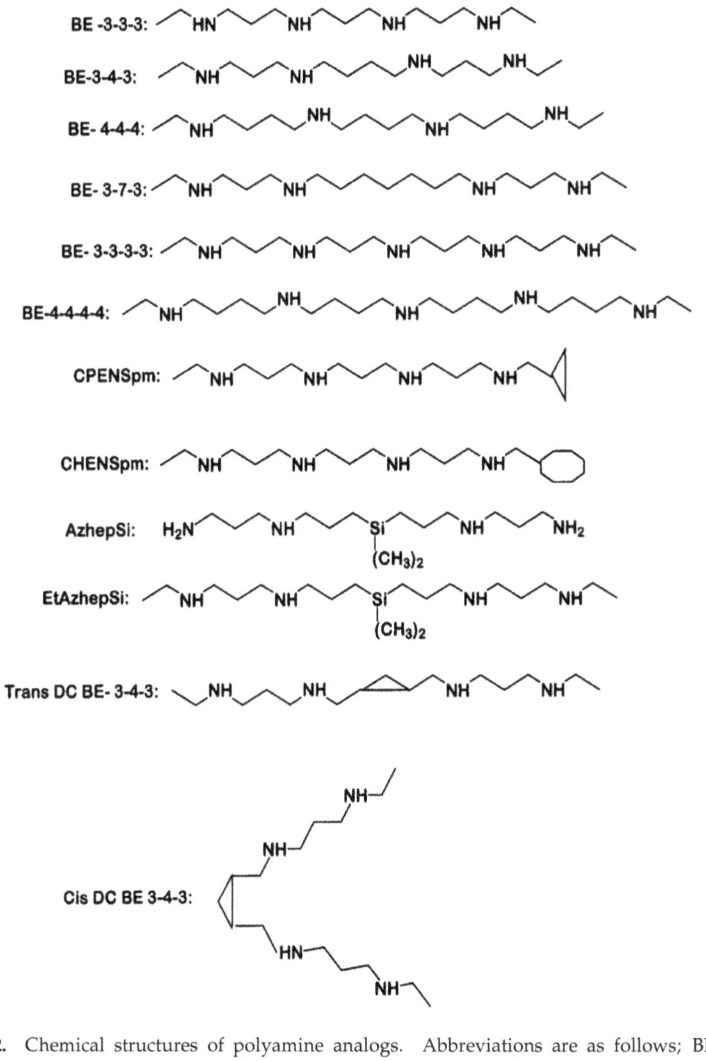

Figure 2. Chemical structures of polyamine analogs. Abbreviations are as follows; BE-3-3-3, N^1,N^{11}-bis(ethyl)norspermine (BENSpm or DENSpm); BE-3-4-3, N^1,N^{12}-bis(ethyl)spermine; BE-4-4-4, N^1,N^{14}-bis(ethyl)homospermine; BE-3-7-3, N^1,N^{15}-bis-[3-(ethylamino)propyl]-1-17-heptane diamine; BE-3-3-3-3, 1,15-bis(ethylamino)-4,8,12-triazapentadecane; BE-4-4-4-4, 1,19-bis(ethylamino)-5,10,15-triazanonadecane; CPENSpm, N^1-ethyl-N^{11}-(cyclopropyl)methyl-4,8-diazaundecane; CHENSpm, N^1-ethyl-N^{11}-(cycloheptyl)methyl)-4,8-diazaundecane; AzhepS1, bis(7-amino-4-azaheptyl) dimethyl-silane; EtAzhepSi, bis(7-ethylamino-4-azaheptyl)dimethylsilane; Trans DCBE-3-4-3, trans isomer of BE-3-4-3 with central 1,2, dimethylcyclopropyl residue; Cis DCBE-3-4-3, cis isomer of BE-3-4-3 with central 1,2-dimethylcyclopropyl residue.

We evaluated the anti-tumor effects of BE-3-3-3 and BE-3-3-3-3 using a HER-2-positive transgenic mouse model of breast cancer. Prior studies showed that increased polyamine biosynthetic activity critically interacted with HER2/neu in promoting human mammary cell transformation in culture [64]. FVB/NTgN (MMTVneu) transgenic mice developed mammary tumors at about four months of age, with a median incidence of 6.8 months [65]. Treatment of FVB/NTgN mice with BE-3-3-3 or BE-3-3-3-3 resulted in a 3- to 4-fold reduction in tumor volume compared to that of control mice (Figure 3) [66]. The activity of SSAT was determined from tumors and kidneys of control group and treatment groups. SSAT activity was significantly higher in tumors and kidneys of treatment groups than to that of controls (Figure 4). BE-3-3-3-3 was more effective than BE-3-3-3 in reducing tumor volume and inducing SSAT.

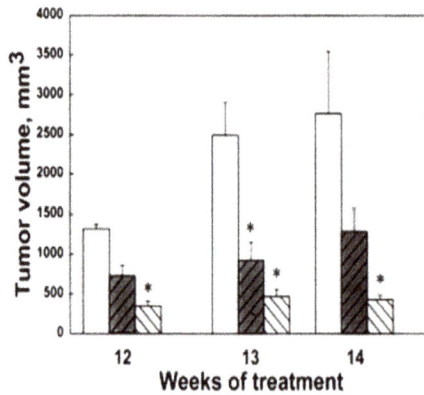

Figure 3. Effect of bis(ethyl)polyamines on tumor volume. The data presented are the average values determined on weeks 12, 13 and 14 of treatment of FVB/NTgN (MMTVneu) transgenic mice. Bars indicate control (unfilled bar), BE-3-3-3 treatment (darkened bar) and BE-3-3-3-3 treatment (striped bar) groups. * Statistically significant ($p < 0.05$) compared with controls, as determined by ANOVA followed by Dunnett's test. Reproduced with permission from [66].

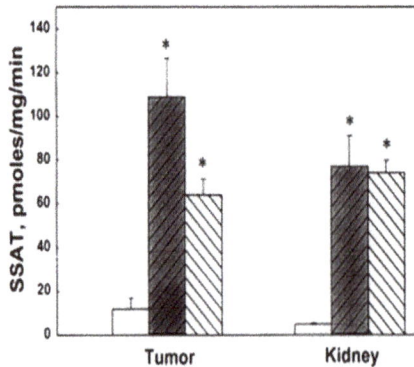

Figure 4. Effect of bis(ethyl)polyamines on SSAT activity in tumors and kidneys of FVB/NTgN (MMTVneu) mice. SSAT levels of control (open bar) and treatment groups, BE-3-3-3 (darkened bar) and BE-3-3-3-3 (striped bar). * Statistically significant ($p < 0.01$) compared with controls by ANOVA followed by Dunnett's test. Reproduced with permission from [66].

Among the group of bis(ethyl) polyamine analogues, BE-3-3-3, BE-3-4-3, BE-3-7-3 and BE-4-4-4-4, BE-3-3-3 was the most promising antitumor drug by in vitro studies. In Phase II trials, no evidence of clinical activity was detected, although this compound was reasonably tolerable [67]. Anti-proliferative action in the pre-clinical models suggested potential combination therapy approaches for this compound. Balabhadrapathruni et al. [68] showed that a combination of BE-3-3-3 and the pure antiestrogen, ICI 182780 caused down-regulation of the anti-apoptotic Bcl-2 and Bcl-XL proteins and increased the level of the pro-apoptotic Bax protein in MCF-7 and T-47D breast cancer cells. The efficacy of polyamine analogues on breast cancer cells might be governed in part by their effects on the expression of proapoptotic and antiapoptotic proteins in these cells [69]. In the case of ERα-positive tumors, the involvement of genomic and non-genomic pathways has also to be considered in mechanistic studies [70].

The activity of SMO is also affected by polyamine analogues. As shown in our recent study with endoxifen and by other studies, SMO induction is a remarkable chemotherapeutic target [36,71–73]. Purvalanol, a specific CDK inhibitor with apoptosis inducing activity in breast cancer cells, also induced SSAT, APAO and SMO in MCF-7 and MDA-MB-231 breast cancer cells [74]. Cervelli et al. [43] analyzed SMO mRNA and enzyme activity in breast cancer tissues and non-tumor samples. Lower levels of this enzyme were present in tumor samples than that in non-tumor tissues. Analogues BE-3-3-3 and CPENSpm were also found to be SMO inhibitors, a likely reason for the poor positive outcomes of these compounds in Phase I and Phase II clinical trials.

Combination treatment of BE-3-3-3 with 5-FU or paclitaxel resulted in the induction of SSAT mRNA and activity in MCF-7 and MDA-MB-231 cells compared to the effect of either drug alone [75]. Spermine oxidase mRNA and activity were increased by polyamine analogues in MDA-MB-231 cells. The in vivo therapeutic efficacy of B-3-3-3 alone and in combination with paclitaxel on tumor regression was reported from studies on xenograft mice models generated with MDA-MB-231 cells [71]. Intraperitoneal exposure to BE-3-3-3 or paclitaxel singly and in combination for 4 weeks resulted in significant inhibition in tumor growth. These findings suggested that synergistic drug response was realized with combinations of polyamine analogues and chemotherapeutic agents. Nair et al. [76] found a synergistic growth inhibitory effect of 2-methoxyestradiol and BE-3-3-3 on MCF-7 cells, as determined by Chou analysis for synergism. Synergistic growth inhibitory effect was also found with BE-3-3-3 and 5-FU [71]. These studies suggest that BE-3-3-3 might be a useful drug in combination therapeutic approaches for breast cancer treatment.

Palladination of polyamines analogues was also used as an effective strategy to inhibit breast cancer cell growth [77]. In contrast to the platination of CPENSpm, which reduced cytotoxicity, palladination of BENSpm resulted in enhanced cytotoxicity, which might be due to differences in the cellular uptake of Pd-BENSpm and Pt-CPENSpm. Palladinated bisethylnorspermine (Pd-BENSpm) was the most efficient compound in the induction of DNA damage and decrease in colony formation in soft agar. Our group has also studied the effects of a bis(benzyl)spermine analogue on MCF-7 cells growing in culture and nude mice xenografts [78]. Growth inhibitory effects were found in both cell culture and animal models.

A second generation of polyamine analogues are unsymmetrically substituted compounds that display structure-dependent and cell type specific effects on polyamine metabolism [17,79,80]. Another series of polyamine analogues are designated as conformationally restricted, cyclic oligoamines [81]. Some of these agents have limitation on the free rotation of single bonds in otherwise flexible molecules such as spermine or its linear analogues. Oligoamines consist of synthetic octa-, deca-, dodeca- and tetradecamines with longer chains than those of natural polyamines and some of them have conformational restriction. These novel polyamine analogues have shown significant activity against multiple human tumors both in vitro and in vivo [82–85]. Oligoamines do not highly induce polyamine catabolic enzymes but can still inhibit tumor cell growth and induce apoptosis. Multiple apoptotic mechanisms have been proposed for oligoamine-induced cytotoxic effects, indicating that cell growth inhibition and cell death might be governed by analogue structural specificity effects [82].

It is interesting to note that MCF-7 cells overexpressing Bcl-2 were resistant to paclitaxel but this resistance was overcome by co-treatment of paclitaxel with BE-3-3-3 [86]. Activation of the polyamine catabolic pathway appeared to play a role in inducing cell death by combination therapeutic approach. Polyamine analogues are known to induce reactive oxygen species (ROS) by the activation of polyamine catabolic pathways. Polyamine analogues are known to produce H_2O_2 and ROS by the activation of polyamine catabolic pathways and these species play an important role in analogue-induced apoptosis [87]. However, sub-lethal levels of H_2O_2 produced by SSAT activation increases susceptibility to skin carcinogenesis [88]. Increased SMO expression is also found in prostate cancer [89]. Polyamine analogues can also interfere with epigenetic modification and expression/re-expression of silenced genes [90]. Taken together, these reports suggest that the polyamine catabolic pathway is a "double-edged sword", that can participate in carcinogenesis or lead to apoptosis depending on the concentration of H_2O_2 and other ROS produced by SMO/SSAT induction [91,92].

4. Conclusions

The polyamine metabolic pathway has been an interesting area of research from a molecular biological and drug discovery perspective for half a century. Biosynthetic inhibitors were synthesized and investigated as cancer drug candidates; however, clinical effectiveness was not realized. Polyamine analogues received much attention as a new generation of drug candidates for different forms of cancer, including breast cancer. Several analogues showed excellent anti-cancer efficacy in cell culture and animal models of breast cancer. However, limited clinical studies showed no therapeutic efficacy when bis(ethyl)norspermine was used as a single agent. Combination therapeutic approaches provide new leads to the use of these molecules in breast cancer therapeutics, although clinical studies are yet to be pursued. Several analogues are finding use in nanoparticle strategies [31,93,94]. The metabolism and function of analogues are also being explored using selective deuteration of *N*-alkyl polyamine analogues [95]. These studies provide new insights into the mechanism of action of polyamines and their analogues in cell growth and cell death and point to the importance of additional research to realize the clinical potential of polyamine analogues in breast cancer.

Author Contributions: T.J.T. planned this review, conducted literature search and wrote the first draft (75% contribution); T.T. provided additional input to the draft (25% contribution). Both authors corrected the draft and prepared the final version.

Conflicts of Interest: The authors declare no conflict of interest.

References

1. World Cancer Research Fund International, Breast Cancer Statistics. 2017. Available online: http://www.wcrf. org/int/cancer-facts-figures/data-specific-cancers/breast-cancer-statistics (accessed on 15 January 2018).
2. American Cancer Society. *Cancer Facts and Figures 2017*; American Cancer Society: Atlanta, GA, USA, 2017.
3. Polyak, K. Heterogeneity in breast cancer. *J. Clin. Investig.* **2011**, *121*, 3786–3788. [CrossRef] [PubMed]
4. Lim, E.; Tarulli, G.; Portman, N.; Hickey, T.E.; Tilley, W.D.; Palmieri, C. Pushing estrogen receptor around in breast cancer. *Endocr. Relat. Cancer* **2016**, *23*, T227–T241. [CrossRef] [PubMed]
5. Nagini, S. Breast cancer: Current molecular therapeutic targets and new players. *Anticancer Agents Med. Chem.* **2017**, *17*, 152–163. [CrossRef] [PubMed]
6. Thomas, T.; Gallo, M.A.; Thomas, T.J. Estrogen receptors as targets for drug development for breast cancer, osteoporosis and cardiovascular diseases. *Curr. Cancer Drug Targets* **2004**, *4*, 483–499. [CrossRef] [PubMed]
7. Pearce, S.T.; Jordan, V.C. The biological role of estrogen receptors alpha and beta in cancer. *Crit. Rev. Oncol. Hematol.* **2004**, *50*, 3–22. [CrossRef] [PubMed]
8. Maximov, P.Y.; Lee, T.M.; Jordan, V.C. The discovery and development of selective estrogen receptor modulators (SERMs) for clinical practice. *Curr. Clin. Pharmacol.* **2013**, *8*, 135–155. [CrossRef] [PubMed]

9. Bhattacharjee, D.; Kumari, K.M.; Avin, S.; Babu, V.A.M. The evolutionary tale and future directions of aromatase inhibitors in breast carcinoma. *Anticancer Agents Med. Chem.* **2017**, *17*, 1487–1499. [CrossRef] [PubMed]
10. Baselga, J.; Coleman, R.E.; Cortés, J.; Janni, W. Advances in the management of HER2-positive early breast cancer. *Crit. Rev. Oncol. Hematol.* **2017**, *119*, 113–122. [CrossRef] [PubMed]
11. Yao, H.; He, G.; Yan, S.; Chen, C.; Song, L.; Rosol, T.J.; Deng, X. Triple-negative breast cancer: Is there a treatment on the horizon? *Oncotarget* **2017**, *8*, 1913–1924. [CrossRef] [PubMed]
12. Mancuso, M.R.; Massarweh, S.A. Endocrine therapy and strategies to overcome therapeutic resistance in breast cancer. *Curr. Probl. Cancer* **2016**, *40*, 95–105. [CrossRef] [PubMed]
13. Thomas, T.; Balabhadrapathruni, S.; Gallo, M.A.; Thomas, T.J. Development of polyamine analogs as cancer therapeutic agents. *Oncol. Res.* **2002**, *13*, 123–135. [PubMed]
14. Cervelli, M.; Pietropaoli, S.; Signore, F.; Amendola, R.; Mariottini, P. Polyamines metabolism and breast cancer: State of the art and perspectives. *Breast Cancer Res. Treat.* **2014**, *148*, 233–248. [CrossRef] [PubMed]
15. Vijayanathan, V.; Agostinelli, E.; Thomas, T.; Thomas, T.J. Innovative approaches to the use of polyamines for DNA nanoparticle preparation for gene therapy. *Amino Acids* **2014**, *46*, 499–509. [CrossRef] [PubMed]
16. Davidson, N.E.; Hahm, H.A.; McCloskey, D.E.; Woster, P.M.; Casero, R.A., Jr. Clinical aspects of cell death in breast cancer: The polyamine pathway as a new target for treatment. *Endocr. Relat. Cancer* **1999**, *6*, 69–73. [CrossRef] [PubMed]
17. Murray-Stewart, T.R.; Woster, P.M.; Casero, R.A., Jr. Targeting polyamine metabolism for cancer therapy and prevention. *Biochem. J.* **2016**, *473*, 2937–2953. [CrossRef] [PubMed]
18. Thomas, T.; Thomas, T.J. Polyamines in cell growth and cell death: Molecular mechanisms and therapeutic applications. *Cell. Mol. Life Sci.* **2001**, *58*, 244–258. [CrossRef] [PubMed]
19. Miller-Fleming, L.; Olin-Sandoval, V.; Campbell, K.; Ralser, M. Remaining mysteries of molecular biology: The role of polyamines in the cell. *J. Mol. Biol.* **2015**, *427*, 3389–3406. [CrossRef] [PubMed]
20. Thomas, T.; Thomas, T.J. Polyamine metabolism and cancer. *J. Cell. Mol. Med.* **2003**, *7*, 113–126. [CrossRef] [PubMed]
21. Nowotarski, S.L.; Woster, P.M.; Casero, R.A., Jr. Polyamines and cancer: Implications for chemotherapy and chemoprevention. *Expert Rev. Mol. Med.* **2013**, *15*, e3. [CrossRef] [PubMed]
22. Thomas, T.J.; Tajmir-Riahi, H.A.; Thomas, T. Polyamine-DNA interactions and development of gene delivery vehicles. *Amino Acids* **2016**, *48*, 2423–2431. [CrossRef] [PubMed]
23. Thomas, T.J.; Thomas, T. Collapse of DNA in packaging and cellular transport. *Int. J. Biol. Macromol.* **2018**, *109*, 36–48. [CrossRef] [PubMed]
24. Chanphai, P.; Thomas, T.J.; Tajmir-Riahi, H.A. Conjugation of biogenic and synthetic polyamines with serum proteins: A comprehensive review. *Int. J. Biol. Macromol.* **2016**, *92*, 515–522. [CrossRef] [PubMed]
25. Bignon, E.; Chan, C.H.; Morell, C.; Monari, A.; Ravanat, J.L.; Dumont, E. Molecular dynamics insights into polyamine-DNA binding modes: Implications for cross-link selectivity. *Chemistry* **2017**, *23*, 12845–12852. [CrossRef] [PubMed]
26. Shah, N.; Thomas, T.J.; Lewis, J.S.; Klinge, C.M.; Shirahata, A.; Gelinas, C.; Thomas, T. Regulation of estrogenic and nuclear factor κB functions by polyamines and their role in polyamine analog-induced apoptosis of breast cancer cells. *Oncogene* **2001**, *20*, 1715–1729. [CrossRef] [PubMed]
27. Shah, N.; Thomas, T.; Shirahata, A.; Sigal, L.H.; Thomas, T.J. Activation of nuclear factor κB by polyamines in breast cancer cells. *Biochemistry* **1999**, *38*, 14763–14774. [CrossRef] [PubMed]
28. Zhu, Q.; Jin, L.; Casero, R.A.; Davidson, N.E.; Huang, Y. Role of ornithine decarboxylase in regulation of estrogen receptor alpha expression and growth in human breast cancer cells. *Breast Cancer Res. Treat.* **2012**, *136*, 57–66. [CrossRef] [PubMed]
29. Vijayanathan, V.; Thomas, T.J.; Nair, S.K.; Shirahata, A.; Gallo, M.A.; Thomas, T. Bending of the estrogen response element by polyamines and estrogen receptors α and β: A fluorescence resonance energy transfer study. *Int. J. Biochem. Cell Biol.* **2006**, *38*, 1191–1206. [CrossRef] [PubMed]
30. Casero, R.A.; Pegg, A.E. Polyamine catabolism and disease. *Biochem. J.* **2009**, *421*, 323–338. [CrossRef] [PubMed]
31. Agostinelli, E.; Vianello, F.; Magliulo, G.; Thomas, T.; Thomas, T.J. Nanoparticle strategies for cancer therapeutics: Nucleic acids, polyamines, bovine serum amine oxidase and iron oxide nanoparticles. *Int. J. Oncol.* **2015**, *46*, 5–16. [CrossRef] [PubMed]

32. Hong, S.K.; Chaturvedi, R.; Piazuelo, M.J.; Coburn, L.A.; Williams, C.S.; Delgado, A.G.; Casero, R.A.; Pegg, A.G.; Schwartz, D.A.; Wilson, K.T. Increased expression and cellular localization of spermine oxidase in ulcerative colitis and relationship to disease activity. *Inflamm. Bowel Dis.* **2010**, *16*, 1557–1566. [CrossRef] [PubMed]

33. Agostinelli, E.; Arancia, G.; Vedova, L.D.; Belli, F.; Marra, M.; Salvi, M.; Toninello, A. The biological functions of polyamine oxidation products by amine oxidases: Perspectives of clinical applications. *Amino Acids* **2004**, *27*, 347–358. [CrossRef] [PubMed]

34. Palmer, A.J.; Wallace, H.M. The polyamine transport system as a target for anticancer drug development. *Amino Acids* **2010**, *38*, 415–422. [CrossRef] [PubMed]

35. Kahana, C. Antizyme and antizyme inhibitor, a regulatory tango. *Cell. Mol. Life Sci.* **2009**, *66*, 2479–2488. [CrossRef] [PubMed]

36. Murray-Stewart, T.; Casero, R.A. Regulation of polyamine metabolism by curcumin for cancer prevention and therapy. *Med. Sci. (Basel)* **2017**, *5*, 38. [CrossRef] [PubMed]

37. Thomas, T.J.; Thomas, T.; John, S.; Hsu, H.C.; Yang, P.; Keinänen, T.A.; Hyvönen, M.T. Tamoxifen metabolite endoxifen interferes with the polyamine pathway in breast cancer. *Amino Acids* **2016**, *48*, 2293–2302. [CrossRef] [PubMed]

38. Thomas, T.; Thomas, T.J. Estradiol control of ornithine decarboxylase mRNA, enzyme activity, and polyamine levels in MCF-7 breast cancer cells: Therapeutic implications. *Breast Cancer Res. Treat.* **1994**, *29*, 189–201. [CrossRef] [PubMed]

39. Thomas, T.; Thomas, T.J. Regulation of cyclin B1 by estradiol and polyamines in MCF-7 breast cancer cells. *Cancer Res.* **1994**, *54*, 1077–1084. [PubMed]

40. Manni, A.; Badger, B.; Glikman, P.; Bartholomew, M.; Santner, S.; Demers, L. Individual and combined effects of α-difluoromethylornithine and ovariectomy on the growth and polyamine milieu of experimental breast cancer in rats. *Cancer Res.* **1989**, *49*, 3529–3534. [PubMed]

41. Alexiou, G.A.; Lianos, G.D.; Ragos, V.; Galani, V.; Kyritsis, A.P. Difluoromethylornithine in cancer: New advances. *Future Oncol.* **2017**, *13*, 809–819. [CrossRef] [PubMed]

42. Thomas, T.; Faaland, C.A.; Adhikarakunnathu, S.; Thomas, T.J. Structure-activity relations of S-adenosylmethionine decarboxylase inhibitors on the growth of MCF-7 breast cancer cells. *Breast Cancer Res. Treat.* **1996**, *39*, 293–306. [CrossRef] [PubMed]

43. Cervelli, M.; Bellavia, G.; Fratini, E.; Amendola, R.; Polticelli, F.; Barba, M.; Federico, R.; Signore, F.; Gucciardo, G.; Grillo, R.; et al. Spermine oxidase (SMO) activity in breast tumor tissues and biochemical analysis of the anticancer spermine analogues BENSpm and CPENSpm. *BMC Cancer* **2010**, *10*, 555. [CrossRef] [PubMed]

44. Wallace, H.M. The physiological role of the polyamines. *Eur. J. Clin. Investig.* **2000**, *30*, 1–3. [CrossRef]

45. Hu, X.; Washington, S.; Verderame, M.F.; Demers, L.M.; Mauger, D.; Manni, A. Biological activity of the S-adenosylmethionine decarboxylase inhibitor SAM486A in human breast cancer cells in vitro and in vivo. *Int. J. Oncol.* **2004**, *25*, 1831–1838. [CrossRef] [PubMed]

46. Richert, M.M.; Phadke, P.A.; Matters, G.; DiGirolamo, D.J.; Washington, S.; Demers, L.M.; Bond, J.S.; Manni, A.; Welch, D.R. Metastasis of hormone-independent breast cancer to lung and bone is decreased by α-difluoromethylornithine treatment. *Breast Cancer Res.* **2005**, *7*, R819. [CrossRef] [PubMed]

47. Fabian, C.J.; Kimler, B.F.; Brady, D.A.; Mayo, M.S.; Chang, C.H.; Ferraro, J.A.; Zalles, C.M.; Stanton, A.L.; Masood, S.; Grizzle, W.E.; et al. A phase II breast cancer chemoprevention trial of oral α-difluoromethylornithine: Breast tissue, imaging, and serum and urine biomarkers. *Clin. Cancer Res.* **2002**, *8*, 3105–3117. [PubMed]

48. Lao, C.D.; Backoff, P.; Shotland, L.I.; McCarty, D.; Eaton, T.; Ondrey, F.G.; Viner, J.L.; Spechler, S.J.; Hawk, E.T.; Brenner, D.E. Irreversible ototoxicity associated with difluoromethylornithine. *Cancer Epidemiol. Biomarkers Prev.* **2004**, *13*, 1250–1252. [PubMed]

49. Porter, C.W.; Bergeron, R.J. Spermidine requirement for cell proliferation in eukaryotic cells: Structural specificity and quantitation. *Science* **1983**, *219*, 1083–1085. [CrossRef] [PubMed]

50. Israel, M.; Zol, E.C.; Muhammad, N.; Modest, E.J. Synthesis and antitumor evaluation of the presumed cytotoxic metabolites of spermine and N,N'-bis(3-aminopropyl)nonane-1,9-diamine. *J. Med. Chem.* **1973**, *16*, 1–5. [CrossRef] [PubMed]

51.   Keinänen, T.A.; Hyvönen, M.T.; Alhonen, L.; Vepsäläinen, J.; Khomutov, A.R. Selective regulation of polyamine metabolism with methylated polyamine analogues. *Amino Acids* **2014**, *46*, 605–620. [CrossRef] [PubMed]

52.   Singh, A.B.; Thomas, T.J.; Thomas, T.; Singh, M.; Mann, R.A. Differential effects of polyamine homologues on the prevention of DL-α-difluoromethylornithine-mediated inhibition of malignant cell growth and normal immune response. *Cancer Res.* **1992**, *52*, 1840–1847. [PubMed]

53.   Wang, Y.; Murray-Stewart, T.; Devereux, W.; Hacker, A.; Frydman, B.; Woster, P.M.; Casero, R.A., Jr. Properties of purified recombinant human polyamine oxidase, PAOh1/SMO. *Biochem. Biophys. Res. Commun.* **2003**, *304*, 605–611. [CrossRef]

54.   Sjögren, T.; Wassvik, C.M.; Snijder, A.; Aagaard, A.; Kumanomidou, T.; Barlind, L.; Kaminski, T.P.; Kashima, A.; Yokota, T.; Fjellström, O. The structure of murine N^1-acetylspermine oxidase reveals molecular details of vertebrate polyamine catabolism. *Biochemistry* **2017**, *56*, 458–467. [CrossRef] [PubMed]

55.   Porter, C.W.; Ganis, B.; Vinson, T.; Marton, L.J.; Kramer, D.L.; Bergeron, R.J. Comparison and characterization of growth inhibition in L1210 cells by α-difluoromethylornithine, an inhibitor of ornithine decarboxylase, and N^1,N^8-bis(ethyl)spermidine, an apparent regulator of the enzyme. *Cancer Res.* **1986**, *46*, 6279–6285. [PubMed]

56.   Porter, CW.; Bergeron, R.J. Regulation of polyamine biosynthetic activity by spermidine and spermine analogs—A novel antiproliferative strategy. *Adv. Exp. Med. Biol.* **1988**, *250*, 677–690. [PubMed]

57.   Bernacki, R.J.; Bergeron, R.J.; Porter, C.W. Antitumor activity of N,N′-bis(ethyl)spermine homologues against human MALME-3 melanoma xenografts. *Cancer Res.* **1992**, *52*, 2424–2430. [PubMed]

58.   Bergeron, R.J.; Neims, A.H.; McManis, J.S.; Hawthorne, T.R.; Vinson, J.R.; Bortell, R.; Ingeno, M.J. Synthetic polyamine analogues as antineoplastics. *J. Med. Chem.* **1988**, *31*, 1183–1190. [CrossRef] [PubMed]

59.   Fraser, A.V.; Goodwin, A.C.; Hacker-Prietz, A.; Sugar, E.; Woster, P.M.; Casero, R.A., Jr. Knockdown of ornithine decarboxylase antizyme 1 causes loss of uptake regulation leading to increased N^1,N^{11}-bis(ethyl)norspermine (BENSpm) accumulation and toxicity in NCI H157 lung cancer cells. *Amino Acids* **2012**, *42*, 529–538. [CrossRef] [PubMed]

60.   Huang, Y.; Pledgie, A.; Casero, R.A., Jr.; Davidson, N.E. Molecular mechanisms of polyamine analogs in cancer cells. *Anticancer Drugs* **2005**, *16*, 229–241. [CrossRef] [PubMed]

61.   Seiler, N. Thirty years of polyamine-related approaches to cancer therapy. Retrospect and prospect. Part 2. Structural analogues and derivatives. *Curr. Drug Targets* **2003**, *4*, 565–585. [CrossRef] [PubMed]

62.   Davidson, N.E.; Mank, A.R.; Prestigiacomo, L.J.; Bergeron, R.J.; Casero, R.A., Jr. Growth inhibition of hormone-responsive and -resistant human breast cancer cells in culture by N^1,N^{12}-bis(ethyl)spermine. *Cancer Res.* **1993**, *53*, 2071–2075. [PubMed]

63.   Faaland, C.A.; Thomas, T.J.; Balabhadrapathruni, S.; Langer, T.; Mian, S.; Shirahata, A.; Gallo, M.A.; Thomas, T. Molecular correlates of the action of bis(ethyl)polyamines in breast cancer cell growth inhibition and apoptosis. *Biochem. Cell Biol.* **2000**, *78*, 415–426. [CrossRef] [PubMed]

64.   Manni, A.; Wechter, R.; Verderame, M.F.; Mauger, D. Cooperativity between the polyamine pathway and HER2neu in transformation of human mammary epithelial cells in culture: Role of the MAPK pathway. *Int. J. Cancer* **1998**, *76*, 563–570. [CrossRef]

65.   Guy, C.T.; Webster, M.A.; Schaller, M.; Parsons, T.J.; Cardiff, R.D.; Muller, W.J. Expression of the neu protooncogene in the mammary epithelium of transgenic mice induces metastatic disease. *Proc. Natl. Acad. Sci. USA* **1992**, *89*, 10578–10582. [CrossRef] [PubMed]

66.   Shah, N.; Antony, T.; Haddad, S.; Amenta, P.; Shirahata, A.; Thomas, T.J.; Thomas, T. Antitumor effects of bis(ethyl)polyamine analogs on mammary tumor development in FVB/NTgN (MMTVneu) transgenic mice. *Cancer Lett.* **1999**, *146*, 15–23. [CrossRef]

67.   Wolff, A.C.; Armstrong, D.K.; Fetting, J.H.; Carducci, M.K.; Riley, C.D.; Bender, J.F.; Casero, R.A., Jr.; Davidson, N.E. A Phase II study of the polyamine analog N^1,N^{11}-diethylnorspermine (DENSpm) daily for five days every 21 days in patients with previously treated metastatic breast cancer. *Clin. Cancer Res.* **2003**, *9*, 5922–5928. [PubMed]

68.   Balabhadrapathruni, S.; Santhakumaran, L.M.; Thomas, T.J.; Shirahata, A.; Gallo, M.A.; Thomas, T. Bis(ethyl)norspermine potentiates the apoptotic activity of the pure antiestrogen ICI 182780 in breast cancer cells. *Oncol. Rep.* **2005**, *13*, 101–108. [CrossRef] [PubMed]

69. Holst, C.M.; Staaf, J.; Jönsson, G.; Hegardt, C.; Oredsson, S.M. Molecular mechanisms underlying N^1,N^{11}-diethylnorspermine-induced apoptosis in a human breast cancer cell line. *Anticancer Drugs* **2008**, *19*, 871–883. [CrossRef] [PubMed]

70. Vijayanathan, V.; Venkiteswaran, S.; Nair, S.K.; Verma, A.; Thomas, T.J.; Zhu, B.T.; Thomas, T. Physiologic levels of 2-methoxyestradiol interfere with nongenomic signaling of 17β-estradiol in human breast cancer cells. *Clin. Cancer Res.* **2006**, *12*, 2038–2048. [CrossRef] [PubMed]

71. Pledgie-Tracy, A.; Billam, M.; Hacker, A.; Sobolewski, M.D.; Woster, P.M.; Zhang, Z.; Casero, R.A.; Davidson, N.E. The role of the polyamine catabolic enzymes SSAT and SMO in the synergistic effects of standard chemotherapeutic agents with a polyamine analogue in human breast cancer cell lines. *Cancer Chemother. Pharmacol.* **2010**, *65*, 1067–1081. [CrossRef] [PubMed]

72. Mahjoub, M.A.; Bakhshinejad, B.; Sadeghizadeh, M.; Babashah, S. Combination treatment with dendrosomal nanocurcumin and doxorubicin improves anticancer effects on breast cancer cells through modulating CXCR4/NF-κB/Smo regulatory network. *Mol. Biol. Rep.* **2017**, *44*, 341–351. [CrossRef] [PubMed]

73. Bunjobpol, W.; Dulloo, I.; Igarashi, K.; Concin, N.; Matsuo, K.; Sabapathy, K. Suppression of acetylpolyamine oxidase by selected AP-1 members regulates DNp73 abundance: Mechanistic insights for overcoming DNp73-mediated resistance to chemotherapeutic drugs. *Cell Death Differ.* **2014**, *21*, 1240–1249. [CrossRef] [PubMed]

74. Obakan, P.; Arısan, E.D.; Ozfiliz, P.; Çoker-Gurkan, A. Palavan-Unsal, N. Purvalanol A is a strong apoptotic inducer via activating polyamine catabolic pathway in MCF-7 estrogen receptor positive breast cancer cells. *Mol. Biol. Rep.* **2014**, *41*, 145–154. [CrossRef] [PubMed]

75. Pledgie, A.; Huang, Y.; Hacker, A.; Zhang, Z.; Woster, P.M.; Davidson, N.E.; Casero, R.A., Jr. Spermine oxidase SMO(PAOh1), Not N^1-acetylpolyamine oxidase PAO, is the primary source of cytotoxic H_2O_2 in polyamine analogue-treated human breast cancer cell lines. *J. Biol. Chem.* **2005**, *280*, 39843–39851. [CrossRef] [PubMed]

76. Nair, S.K.; Verma, A.; Thomas, T.J.; Chou, T.C.; Gallo, M.A.; Shirahata, A.; Thomas, T. Synergistic apoptosis of MCF-7 breast cancer cells by 2-methoxyestradiol and bis(ethyl)norspermine. *Cancer Lett.* **2007**, *250*, 311–322. [CrossRef] [PubMed]

77. Silva, T.M.; Fiuza, SM.; Marques, M.P.; Persson, L.; Oredsson, S. Increased breast cancer cell toxicity by palladination of the polyamine analogue N^1N^{11}-bis(ethyl)norspermine. *Amino Acids* **2014**, *46*, 339–352. [CrossRef] [PubMed]

78. Thomas, T.J.; Shah, N.; Faaland, C.A.; Gallo, M.A.; Yurkow, E.; Satyaswaroop, P.G.; Thomas, T. Effects of a bis(benzyl)spermine analog on MCF-7 breast cancer cells in culture and nude mice xenografts. *Oncol. Rep.* **1977**, *4*, 5–13. [CrossRef]

79. Casero, R.A., Jr.; Woster, P.M. Recent advances in the development of polyamine analogues as antitumor agents. *J. Med. Chem.* **2009**, *52*, 4551–4573. [CrossRef] [PubMed]

80. Jagu, E.; Pomel, S.; Pethe, S.; Loiseau, P.M.; Labruère, R. Polyamine-based analogs and conjugates as antikinetoplastid agents. *Eur. J. Med. Chem.* **2017**, *139*, 982–1015. [CrossRef] [PubMed]

81. Valasinas, A.; Reddy, V.K.; Blokhin, A.V.; Basu, H.S.; Bhattacharya, S.; Sarkar, A.; Marton, L.J.; Frydman, B. Long-chain polyamines (oligoamines) exhibit strong cytotoxicities against human prostate cancer cells. *Bioorg. Med. Chem.* **2003**, *11*, 4121–4131. [CrossRef]

82. Huang, Y.; Keen, J.C.; Hager, E.; Smith, R.; Hacker, A.; Frydman, B.; Valasinas, A.L.; Reddy, V.K.; Marton, L.J.; Casero, R.A., Jr.; et al. Regulation of polyamine analogue cytotoxicity by c-Jun in human MDA-MB-435 cancer cells. *Mol. Cancer Res.* **2004**, *2*, 81–88. [PubMed]

83. Zhu, Q.; Huang, Y.; Marton, L.J.; Woster, P.M.; Davidson, N.E.; Casero, R.A., Jr. Polyamine analogs modulate gene expression by inhibiting lysine-specific demethylase 1 (LSD1) and altering chromatin structure in human breast cancer cells. *Amino Acids* **2012**, *42*, 887–898. [CrossRef] [PubMed]

84. Huang, Y.; Pledgie, A.; Rubin, E.; Marton, L.J.; Woster, P.M.; Sukumar, S.; Casero, R.A., Jr.; Davidson, N.E. Role of p53/p21(Waf1/Cip1) in the regulation of polyamine analogue-induced growth inhibition and cell death in human breast cancer cells. *Cancer Biol. Ther.* **2005**, *4*, 1006–1013. [CrossRef] [PubMed]

85. Vujcic, S.; Halmekyto, M.; Diegelman, P.; Gan, G.; Kramer, D.L.; Janne, J.; Porter, C.W. Effects of conditional overexpression of spermidine/spermine N^1-acetyltransferase on polyamine pool dynamics, cell growth, and sensitivity to polyamine analogs. *J. Biol. Chem.* **2000**, *275*, 38319–38328. [CrossRef] [PubMed]

86. Akyol, Z.; Çoker-Gürkan, A.; Arisan, E.D.; Obakan-Yerlikaya, P.; Palavan-Ünsal, N. DENSpm overcame Bcl-2 mediated resistance against Paclitaxel treatment in MCF-7 breast cancer cells via activating polyamine catabolic machinery. *Biomed. Pharmacother.* **2016**, *84*, 2029–2041. [CrossRef] [PubMed]

87. Ha, H.C.; Woster, P.M.; Yager, J.D.; Casero, R.A., Jr. The role of polyamine catabolism in polyamine analogue-induced programmed cell death. *Proc. Natl. Acad. Sci. USA* **1997**, *94*, 11557–11562. [CrossRef] [PubMed]

88. Wang, X.; Feith, D.J.; Welsh, P.; Coleman, C.S.; Lopez, C.; Woster, P.M.; O'Brien, T.G.; Pegg, A.E. Studies of the mechanism by which increased spermidine/spermine N^1-acetyltransferase activity increases susceptibility to skin carcinogenesis. *Carcinogenesis* **2007**, *28*, 2404–2411. [CrossRef] [PubMed]

89. Goodwin, A.C.; Jadallah, S.; Toubaji, A.; Lecksell, K.; Hicks, J.L.; Kowalski, J.; Bova, G.S.; De Marzo, A.M.; Netto, G.J.; Casero, R.A., Jr. Increased spermine oxidase expression in human prostate cancer and prostatic intraepithelial neoplasia tissues. *Prostate* **2008**, *68*, 766–772. [CrossRef] [PubMed]

90. Huang, Y.; Greene, E.; Murray Stewart, T.; Goodwin, A.C.; Baylin, S.B.; Woster, P.M.; Casero, R.A., Jr. Inhibition of lysine-specific demethylase 1 by polyamine analogues results in reexpression of aberrantly silenced genes. *Proc. Natl. Acad. Sci. USA* **2007**, *104*, 8023–8028. [CrossRef] [PubMed]

91. Wang, Y.; Casero, R.A., Jr. Mammalian polyamine catabolism: A therapeutic target, a pathological problem, or both? *J. Biochem.* **2006**, *139*, 17–25. [CrossRef] [PubMed]

92. Zou, Z.; Chang, H.; Li, H.; Wang, S. Induction of reactive oxygen species: An emerging approach for cancer therapy. *Apoptosis* **2017**, *22*, 1321–1335. [CrossRef] [PubMed]

93. Thomas, R.M.; Thomas, T.; Wada, M.; Sigal, L.H.; Shirahata, A.; Thomas, T.J. Facilitation of the cellular uptake of a triplex-forming oligonucleotide by novel polyamine analogues: Structure-activity relationships. *Biochemistry* **1999**, *38*, 13328–13337. [CrossRef] [PubMed]

94. Xie, Y.; Murray-Stewart, T.; Wang, Y.; Yu, F.; Li, J.; Marton, L.J.; Casero, R.A., Jr.; Oupický, D. Self-immolative nanoparticles for simultaneous delivery of microRNA and targeting of polyamine metabolism in combination cancer therapy. *J. Control. Release* **2017**, *246*, 110–119. [CrossRef] [PubMed]

95. Ucal, S.; Häkkinen, M.R.; Alanne, A.L.; Alhonen, L.; Vepsäläinen, J.; Keinänen, T.A.; Hyvönen, M.T. Controlling of *N*-Alkylpolyamine Analogue Metabolism by Selective Deuteration. *Biochem. J.* **2018**, *475*, 663–676. [CrossRef] [PubMed]

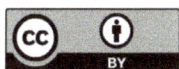

medical
sciences

MDPI

Review

Role of Polyamines in Immune Cell Functions

Rebecca S. Hesterberg [1,2], John L. Cleveland [3] and Pearlie K. Epling-Burnette [2,*]

1   University of South Florida Cancer Biology Graduate Program, University of South Florida,
    4202 East Fowler Ave, Tampa, FL 33620, USA; Rebecca.Hesterberg@moffitt.org
2   Department Immunology, PharmD, Moffitt Cancer Center & Research Institute, 12902 Magnolia Drive,
    23033 SRB, Tampa, FL 33612, USA
3   Department of Tumor Biology, Moffitt Cancer Center & Research Institute, 12902 Magnolia Drive,
    Tampa, FL 33612, USA; John.Cleveland@moffitt.org
*   Correspondence: Pearlie.Burnette@moffitt.org; Tel.: +1-813-745-6177

Received: 16 January 2018; Accepted: 2 March 2018; Published: 8 March 2018

Abstract: The immune system is remarkably responsive to a myriad of invading microorganisms and provides continuous surveillance against tissue damage and developing tumor cells. To achieve these diverse functions, multiple soluble and cellular components must react in an orchestrated cascade of events to control the specificity, magnitude and persistence of the immune response. Numerous catabolic and anabolic processes are involved in this process, and prominent roles for L-arginine and L-glutamine catabolism have been described, as these amino acids serve as precursors of nitric oxide, creatine, agmatine, tricarboxylic acid cycle intermediates, nucleotides and other amino acids, as well as for ornithine, which is used to synthesize putrescine and the polyamines spermidine and spermine. Polyamines have several purported roles and high levels of polyamines are manifest in tumor cells as well in autoreactive B- and T-cells in autoimmune diseases. In the tumor microenvironment, L-arginine catabolism by both tumor cells and suppressive myeloid cells is known to dampen cytotoxic T-cell functions suggesting there might be links between polyamines and T-cell suppression. Here, we review studies suggesting roles of polyamines in normal immune cell function and highlight their connections to autoimmunity and anti-tumor immune cell function.

Keywords: immunity; T-lymphocytes; B-lymphocytes; tumor immunity; metabolism; epigenetics; autoimmunity

1. Introduction

Metabolic regulation is a vital component of a coordinated immune response [1]. Dormant immune cells circulate in blood and tissues and morph into highly activated cells following antigen exposure. Activated immune cells act as sentinels throughout the body, and eradicate pathogens present in distinct ecosystems, in areas with diverse growth factors or low oxygen [2], and when nutrients are limiting [3], which can compromise their functional veracity. For a versatile and potent response, immune cells must make rapid and precise adaptations to these environmental changes [4]. To achieve its diverse functions, the immune system is comprised of heterogeneous populations of cells that are each capable of a broad range of responses. Importantly, all of these cells must adjust their metabolic activity to meet functional demands that include migration, proliferation and sometimes long-lasting persistence in these diverse environments [5,6].

Recent advances in understanding immunometabolism have shown that the energetic demands of unique T-cell subpopulations are linked to dynamic responses of the immune system. Most immune cells generate adenosine triphosphate (ATP) from glucose as their primary energy source, but drastic changes in metabolism are observed when transitioning from a quiescent to an activated state [7,8] and the complexity of metabolic circuits has confounded ascribing a particular function to one specific pathway or intermediate. Here, a focused discussion is provided that reviews the roles of an understudied metabolic pathway in immune cells, specifically that which controls polyamine homeostasis, in normal immune cell functions and immune-related diseases [9].

2. B-Cell Lymphopoiesis and Activation

As members of the adaptive immune system, T- and B-lymphocytes are fundamental components of an integrated immune response [10]. B-cell differentiation starts in the fetal liver and continues in the bone marrow during adult life [11]. Though both B- and T-cell populations are derived from a common lymphoid progenitor (CLP) in bone marrow [11–14], T-cells and B-cells differ by their mechanism of antigen recognition [15]. Specifically, B-cells express surface immunoglobulin (Ig) as a receptor for detecting circulating microorganisms. Antigen binding to Ig receptors activates B-cells and triggers their differentiation into plasma cells that produce and secrete copious amounts of soluble antibodies with distinct isotypes that selectively bind to the activating antigen. Further, a subset of antigen-activated B-cells differentiate into long-lasting memory B-cells, which allow for a more rapid response following re-exposure to cognate antigen [16].

The initial step in activating a B-cell response involves receptor-antigen interactions that occur in restricted areas of primary lymphoid organs such as the spleen, lymph node or tonsils [16]. On its surface, each B-cell expresses a single membrane bound Ig receptor (B-cell receptor, BCR) that is created through a unique process of somatic genomic recombination of immunoglobulin genes to form heterodimeric immunoglobulin receptors that results from the fusion of three separate gene segments, variable (V), diversity (D) and joining (J) genes (VDJ) that provide receptor diversity [15,17]. Both integrated T-cell and innate immune cell interactions are required for the activation of B-cells, which become progressively more antigen reactive via a process of hypermutation and class switching [16,18–20]. Precursor, immature and mature B-cells signal through the immunoglobulin receptor. Immature B-cells, expressing only membrane IgM heavy chain (mu) and the Ig_α and Ig_β, [21] undergo several selection events triggered by the recognition of self-molecules in bone marrow that prevent autoimmunity [9]. Since the V(D)J-BCR gene rearrangement process is stochastic, there is a random expression of self-reactive receptors that requires a systematic bioenergetic reprogramming to achieve clonal deletion or inactivation of self-reactive B-cells in circulation [18,22]. Autoreactive B-cells have been shown to increase glycolysis and oxygen consumption compared to normal antigen-activated B-cells [22,23]. Further, disabling glycolysis by treatment with the pyruvate dehydrogenase inhibitor dichloroacetate impairs antibody production both ex vivo and in vivo [22]. Moreover, B-cell specific deletion of the glucose transporter Glut1 or Myc revealed their role in B-lymphopoiesis, and that c-Myc is necessary for activation-induced expression of Glut1 [22,24]. Notably, overexpression and inhibitor studies have revealed that c-Myc directly and coordinately induces the transcription of ornithine decarboxylase (ODC), adenosylmethionine decarboxylase-1 (Amd1), spermidine synthase (Srm), and spermine synthase (Sms), four enzymes which direct polyamine biosynthesis [25]. Indeed, c-Myc itself is a transcription factor for ODC and Sms [26,27]. Ornithine decarboxylase functions as a dimer and is the rate-limiting enzyme in the pathway and converts ornithine to putrescine, which is then converted into spermidine and spermine. Ornithine decarboxylase is tightly controlled by rapid messenger RNA (mRNA) turnover, a very short protein half-life, as well as by antizyme that is translationally induced as polyamine levels rise and which directly binds to ODC and triggers its destruction by the proteasome [28]. Gene knockout studies in mice have established that ODC is essential for proper embryogenesis [29].

Increased expression of enzymes that direct polyamine production and polyamine levels occur after BCR activation [30]. Further, addition of spermine compromises activation-associated apoptosis, suggesting polyamines may be important in repressing the clonal deletion of B-cells after activation. Moreover, nitric oxide enhanced IgE class-switching by anti-trinitrophenyl (TNP) keyhole limpet hemocyanin-(KLH) is blocked in vivo by treatment with aminoguanidine, which inhibits serum diamine oxidase and prevents the conversion of extracellular polyamines into toxic products [31,32].

Thus, although there are scant reports directly linking polyamines to specific B-cell functions, the importance of Myc and the role that Myc plays in B-cell activation and development suggests direct links to polyamines.

3. The Role of Polyamines in T-Lymphopoiesis

T-cells express either an αβ or γδ T-cell receptor (TCR) that rearrange through non-homologous recombination of the V(D)J genes mediated by the activations of the recombination activating genes (*Rag*)1 and *Rag*2 [33,34] as described for B-cells [35]. Deletion of *Rag*1 or *Rag*2, whose expression is restricted to lymphocytes, leads to small lymphoid organs and to the complete loss of mature circulating T and B-cells in mice. Unlike B-cell development that largely occurs in the bone marrow, T-cells arise from a common lymphoid progenitor that migrates into the thymus [36,37] where environmental interactions with thymic epithelial cells [38], signaling via NOTCH1 [39,40] and TCR repertoire selection occurs at the population level through positive and negative selection processes similar to B-cells [33–35,41]. Most T-cells (95%) in the lymphoid compartment express αβTCRs [42], but are further delineated by surface expression of CD4 or CD8, which are required for major histocompatibility cluster (MHC)-class II and MHC-class I co-ligation, respectively [43,44]. The αβTCR receptor is also expressed on regulatory T-cells [45], on a minor population of natural killer (NK) T-cells [46], and on subtypes of intestinal intraepithelial lymphocytes (IELs) [47], which play regulatory roles in response to mucosal infections [48].

A major difference between B- and T-cells is the MHC-restricted nature of TCR antigen activation [43]. T-cells recognize their targets (e.g., virally infected cells) through interaction of small peptide fragments bound in the groove of an MHC molecule, which strengthens selectivity for self over non-self and protects against autoimmunity [43,44]. Professional antigen-presenting cells (APCs) such as B-cells, macrophages and dendritic cells (DC) express both MHC class I and MHC class II for activating CD4$^+$ and CD8$^+$ T-cells. Through receptor or phagocytosis-mediated antigen internalization, APCs process antigen into the correct fragment length for display by the MHC molecule [49]. These cells also express additional co-stimulatory signals including CD28, OX40 ligand, CD40L (Figure 1), which enhances the T-cell's response and provides a critical level of regulation [50–54]. Notably, the inducible co-stimulatory (ICOS) molecule, a member of the CD28 family, is essential for the T-cell mediated induction of immunoglobulin isotype class switching by activated B-cells [19]. Further T-cells undergo an educational process in the thymus mediated by Aire, a transcription factor expressed by medullary epithelial cells (mTECs) in the thymus, which induces the promiscuous expression of restricted peripheral tissue antigens (PTAs) [55] that trigger the clonal deletion of T-cells with potential self-reactivity before they can exit the thymus [56,57]. In part, this is due to the unique ability of Aire to recognize the hypomethylated amino-terminal tail of histone H3 [38,58], to bind to transcriptional sites of paused polymerases [59], and to control genes that direct mRNA splicing [57]. This process is critical to the formation of immunological tolerance, autoimmune prevention, and antitumor immunity [56,57,59,60].

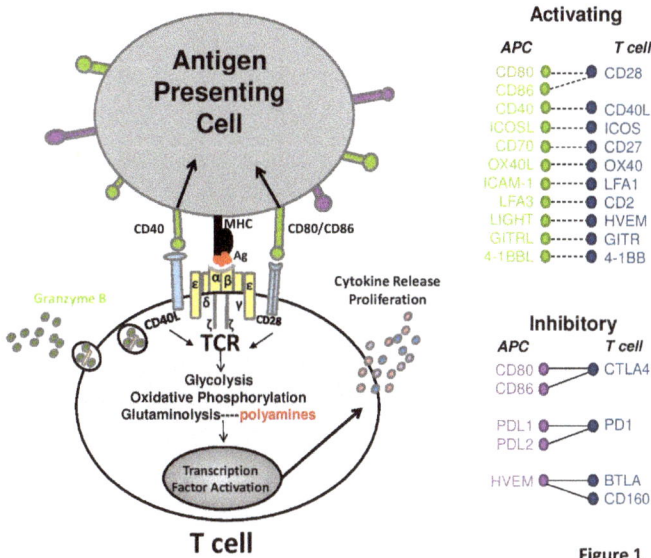

Figure 1. Stimulatory and inhibitory molecules expressed on T-cells. Diagram depicting the antigen presenting cell (APC) and T-cell interactions and activating receptors and ligands on these cells that govern the functional outcomes of T-cells such as cytotoxicity-associated granzyme B expression, cytokine release, and proliferation [61]. The T-cell receptor complex is composed of several proteins that are necessary for survival and signaling including T-cell receptor (TCR)α and TCRβ chains, CD3 signaling molecules δ/ε, CD3γ/ε and CD247 composed on the dimeric $\zeta\zeta$-chains or ζn (not shown). Co-stimulatory molecules on T-cells such as CD28, the founding member of the immunoglobulin (Ig) family of costimulatory receptors, are critical to amplify and sustain the signaling response. Activation leads to metabolic reprogramming to increase glycolysis, oxidative phosphorylation, and amino acid metabolism through glutaminolysis and ultimately to polyamine biosynthesis [62,63]. Additional receptors include CD40L (CD154), T-cell specific surface glycoprotein CD28, inducible T-cell costimulatory ICOS (CD278) which is a CD28-family molecule expressed on T-cells important for Th2 responses, Traf-linked tumor necrosis factor receptor family protein, CD27, which is important in T and B-cell memory formation and activation of natural killer (NK) cells [64–66], tumor necrosis factor receptor superfamily (TNFRSF), member 4 (TNFRSF4) also OX40 (CD134) expressed on activated T-cells [51–53], leukocyte-associated antigen-1 (LFA1) which is an integrin involved in T-cell migration [53,67], the adhesion molecule CD2 present on T-cells and NK cells (also known erythrocyte receptor and rosette receptor, LFA-2), herpesvirus entry mediator (HVEM) also known as tumor necrosis factor receptor superfamily member 14 (TNFRSF14), glucocorticoid-induced TNFR family related gene (GITR) a member of the TNFRSF [68], S-type lectin Galectin 9, T-cell immunoglobulin mucin domain 1 (TIM1) also known as hepatitis A virus cellular receptor 1 (HAVcr-1), and 4-1BB (CD137, TNFRS9). Corresponding receptors on APC are the classical costimulatory ligands CD80 (B7-1), CD86 (B7-2) that interact with CD28, TNFRS5 (CD40), human inducible costimulatory-ligand (ICOSL) [69], ligand for CD27 (CD70 also TNFSF7), OX40 ligand (OX40L), intercellular adhesion molecule 1 (ICAM-1, also CD54), leukocyte-associated antigen-3 (LFA3), HVEM counter-receptor lymphotoxin-like, exhibits inducible expression, and competes with herpes simplex virus glycoprotein D for HVEM, a receptor expressed by T-lymphocytes (LIGHT, also CD160), GITR ligand (GITRL), and 4-1BB ligand (4-1BBL). Inhibitory receptors and ligands are shown including cytotoxic T-lymphocyte antigen-4 (CTLA4) which interacts with CD80, CD86 that also recognizes CD28, programmed cell death protein 1 (PD1) receptor and its ligands PD-ligand 1 (PDL1) and PD-ligand 2 (PDL2), and B-and T-lymphocyte attenuator (BTLA) and CD160 [70] that both recognize HVEM. MHC: major histocompatibility complex, Ag: antigenic peptide.

Once released from the thymus, antigen-naive T-cells are primarily reliant on interleukin-7 (IL-7) which is critical for their growth and survival [71]. IL-7 directs the metabolic function of naïve cells by regulating basal glucose and amino acid metabolism via activation of Janus kinase (JAK3) and phosphorylation of signal transducer and activator of transcription-5 (STAT5) and PI3K/Akt/mTOR that promotes the surface expression of Glut1 and transport of glucose [72–74]. T-cells interact with peptide-loaded APCs in peripheral lymphoid organs, such as the spleen and lymph nodes, which stimulates the activation of their effector functions. Activation then triggers a complex cascade of signaling events (Figure 2) that leads to changes in metabolism [4–6]. Based on the cytokine milieu, CD4+ effector cells can differentiate into distinct subsets including T helper (Th)1, Th2, Th17, as well as FoxP3+ CD4+ regulatory T-cells (Tregs) which are all metabolically distinct [7,75]. Differential regulation of mammalian target of rapamycin mTOR, protein kinase B(Akt)-mediated phosphorylation of the tuberous sclerosis complex (TSC1/TSC2), and Ras family GTPase Rheb are critical in regulating this process [76–79]. Most notably, suppression of TOR complex 1 (mTORC1) pharmacologically and through genetic depletion of mTOR in T-cells leads to a predominance of Treg differentiation [80]. Functional specificity of mTOR is determined by its interacting proteins. The mTORC1 complex contains a small GTPase Rheb, a regulatory-associated protein of mTOR (raptor), the G protein β-subunit-like protein (GβL, also known as mLST8) and substrate 40 kDa (PRAS40) whereas, mTORC2 contains mTOR, and GβL with the rapamycin-insensitive companion of mTOR (rictor) and mammalian stress-activated protein kinase interacting protein-1 (mSin1) [81]. Signaling events such as activated AMP-activated protein kinase (AMPK) [82,83] that differentially antagonize the activation of mTORC1, polarize T-cell differentiation toward Tregs and simulate lipid oxidation [23]. Several surface markers such as L-selectin (CD62L) are also critical for metabolic reprogramming since they regulate homing and migration of T-cells into and out of lymphoid organs [84]. Although they express classical αβTCRs, NKT-cells function independent of MHC class I or II via interactions with a glycolipid antigen in the context of CD1d, a non-canonical MHC molecule. Based on the current literature, several of these fundamental events appear controlled by polyamines and/or are linked to key signaling molecules like mTOR or Myc (Figure 2) that control polyamine homeostasis.

Required role of polyamines in proper erythrocyte differentiation have been shown in studies with alpha-difluoromethylornithine (DFMO), a suicide inhibitor of ODC [91,92], but the impact of polyamines on lymphocyte development is largely unknown. Given established roles for putrescine (1,4-diaminobutane), spermidine and spermine in cell proliferation, DNA and RNA synthesis [93,94], as well as in protein translation in both cell free systems and in activated lymphocytes [62], polyamines are highly likely to play key roles in T-cell or B-cell development, particularly in scenarios where exogenous polyamines are limiting and there is compensatory mechanisms induced by polyamine uptake through designated energy-dependent transporters [95–101].

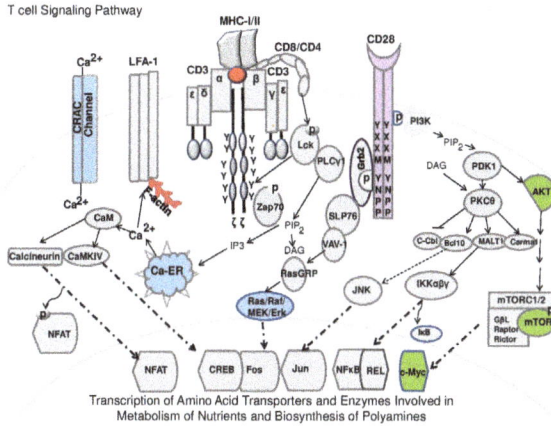

Figure 2. Proximal T-cell signaling cascade. Proximal signaling pathways downstream of the T-cell receptor (TCR)-antigen presenting cell (APC)ignaling complex (as described in Figure 1) are responsible for the cascade of events leading to metabolic reprogramming including the transcription of amino acid transporter and enzymes involved in metabolism of nutrients and biosynthesis of polyamines [85]. Phosphorylation of the immunoreceptor tyrosine-based activation motifs (ITAMs) on the cytoplasmic side of the TCR/CD3 complex engage numerous cascading interactions largely mediated by phosphorylation, dephosphorylation or ubiquitinylation resulting in cellular activation [61]. The initiating signal is generated by lymphocyte protein tyrosine kinase (Lck) and other proto-oncogene tyrosine-protein kinase (Src) family tyrosine kinases including the zeta-chain associated protein kinase (Zap-70) that is recruited to the TCR/CD3 complex. Costimulation through leukocyte-associated antigen-1 (LFA1) which is an integrin involved in T-cell migration or CD28 interaction with CD80 (B7-1) or CD86 (B7-2) (see also Figure 1) activates the phosphorylation of the YXXM or YNPP signaling motifs [86] which regulates glucose metabolism. CD28 leads to stable recruitment of the adaptor protein Grb2/GADS along with interleukin-2-indicible T-cell kinase (Itk), Lck, and phosphatidylinositide 3 kinase (PI3K) heterodimer p85/p110 and SLP76. These interactions promote the activation of VAV-1, RasGRP, and the Ras/Raf/MEK/Erk pathway downstream of phosphorylated SLP-76 and Zap-70 modulating the TCR signal strength [86]. A complement of transcription factors nuclear factor of activated T-cells (NFAT), cAMP response element-binding protein (CREB), Fox family transcription factor c-Fos, Jun (when in combination with c-Fos forms the AP-1 early response transcription factor complex, nuclear factor kappa-light-chain-enhancer of activated B cells (NFκB), an NFκB family member c-Rel, and c-Myc which coordinately regulate gene expression. Activation of CD28 leads to the phosphorylation of PI3K, phosphatidylinositol-3,4 bisphosphate (PIP2) and phosphoinositide-dependent kinase 1 (PDK1) [87] which integrates the TCR and CD28 signaling to induce the NFκB pathway including protein kinase C-theta (PKC-θ), and inhibits the ubiquitin ligase c-Cbl [88] leading to activation of Bcl10, Malt1, Carma1 (CBM) complex leading to IKK$\alpha\beta\gamma$ activation of NFκB and REL [87]. In addition to PKC-θ, phosphorylation of Akt is critical for the regulation of mTORC1 and mTORC2 complexes of mTOR that bind GβL and raptor or rictor, respectively [79,81]. This is a critical step in c-Myc-dependent transcriptional regulation that stimulates dramatic changes in metabolism including glucose, amino acid, nucleotide and polyamine biosynthesis [63,89]. Divalent cations such as calcium (Ca^{2+}) are induce downstream of phospholipase C γ1, PIP2, and indo inositol-1,4,5 triphosphate (IP3) which mobilizes the release of intracellular Ca^{2+} stores from the endoplasmic reticulum (Ca^{2+}-ER) a potential metabolic switch that suppresses intratumoral T-cell function [90]. Sustained signaling then promotes the influx of extracellular Ca^{2+} into the cells through calcium release-activated Ca^{2+} (CRAC) channels. Calcium-calmodulin interactions (Ca^{2+}/CaM) then activates the phosphatase calcineurin and calcium/calmodulin-dependent protein kinase type IV calmodulin (CaMKIV), which dephosphorylates the cytoplasmic subunits of nuclear factor of activated T-cells (NFAT) exposing a nuclear localization signal resulting in nuclear transport and phosphorylates CREB, respectively.

Conditional gene targeting in T-cells is accomplished using the lymphocyte-specific protein tyrosine kinase Lck or CD4-gene promoter fused Cre recombinases [39,102]. Expression of genes under the control of the *Lck* proximal promoter initiates conditional inactivation of genes early in T-cell development prior to the expression of T-cell lineage markers [103] versus CD4-Cre which directs gene expression after transition from the CD4⁺/CD8⁺ double-positive cell leading to gene deletion in both mature CD4⁺ and CD8⁺ single lineage T-cells in the periphery [39,43]. Although cell-specific *Odc* deletion in T-cells or B-cells has yet to be reported, several studies have assessed the effects of regulators of the polyamine pathway. The mTOR serine/threonine protein kinase senses the nutrient state and exists as two distinct protein complexes, mTORC1 and mTORC2. Cell growth (mass) is regulated by mTORC2 via c-Myc and, in turn, c-Myc coordinately induces polyamine biosynthetic enzymes through direct transcriptional regulation and through other mechanisms of regulation [26,27,63]. Notably, T-cells lacking *c-Myc* in *LckCre; c-Myc*$^{fl/fl}$ mice are severely defective in their proliferative response and fail to undergo progression through the double positive (CD4+/CD8+) stage, which is likely due to failed proliferation by early pre-TCR signaling [104]. Further, deletion of Mnt, a Myc antagonist, triggers apoptosis of thymic T-cells and blocks T-cell development [105]. As a target of Myc [25], select depletion of *Odc* in T-cells is needed to assess the importance of polyamines on thymic development.

4. Role of Polyamines in Antigen Activated T-Cells

Given that ODC enzymatic activity is significantly increased after T-cell activation, polyamine production is an important part of normal T-cell function [82,92,93]. Though other ODC-regulating proteins have been reported, c-Myc is the major regulator of enzymes involved in polyamine biosynthesis in T-cells [25,87]. Indeed, mice deficient in another transcriptional regulator of ODC, c-Fos, have been shown to have normal peripheral T-cells, further demonstrating that c-Myc is the master regulator of T-cell-associated polyamines [106,107].

Two of the amino acid precursors for ornithine, glutamine and arginine, are required for T-cell activation [108,109] downstream of TCR signaling events, including mTOR, Myc and mitogen-activated protein kinases/extracellular signal-regulated kinases (MAPK/ERK) [63,109] that are linked through integrated signaling (Figure 2). Polyamines are likely produced downstream of either arginine or glutamine due to the increase in ODC enzymatic activity [63,110,111]. Mass spectrometry-based global metabolomics and integrated transcriptome analyses have been used to map the changes in metabolic intermediates after TCR-stimulation [112]. Notably, proteins that regulate the arginine and proline pathways are enriched in TCR-stimulated CD4⁺ T-cells, and metabolic tracing studies have shown that TCR activation triggers flux of L-arginine Arg into ornithine, putrescine, and agmatine, and to lower levels of spermidine and proline. Catabolism of Arg into polyamines in CD4⁺ T-cells is regulated by mitochondrial arginase-2 (ARG2) as arginase-1 is not expressed in these cells. Interestingly, dietary supplementation of Arg during activation is associated with enhance mitochondrial oxidative phosphorylation (OXPHOS) and mitochondrial spare respiratory capacity (SRC) [113–115]. The morphology and numbers of mitochondria are critical determinants for SRC and in T-cells, for a functional memory response following secondary antigenic challenge [113–115]. Notably, in vivo Arg supplementation of transgenic mice bearing a TCR receptor that specifically recognizes the hemagglutinin antigen (HA 110–119 peptide) increases intracellular Arg levels and the survival of memory T-cells [112].

Although polyamines have not yet been shown to be involved in the memory response, the role of polyamines in survival in other cells suggests that proper polyamine pools may be necessary for this response [25,116,117]. Further, similar to phenotypes observed in other cell types, polyamines are required for T-cell proliferation manifest after TCR stimulation [63,118]. Accordingly, though the mechanism (s) is unclear, polyamine depletion during initial T-cell activation in vitro has been shown to impair cytotoxic function (CTL) against target cells [119–124].

5. Role of Polyamines and Anti-Tumor Immunity

Polyamines are essential components of T-cell and B-cell activation, where for example they are necessary for the effector functions and high rates of proliferation of T-cells [63,119–124]. However, polyamines play much different roles in other cell types of the immune system (Figure 3).

Surprisingly, several studies have demonstrated that ODC inhibition [133–136], and/or treatment with polyamine transport inhibitors (PTIs) significantly reduces rates of tumor growth and that this is due to increase in anti-tumor immunity. Further, the anti-tumor response is linked to T-cell anti-tumor activity, as the beneficial effects observed following treatment with ODC inhibitors and PTIs are reversed in Rag$^{-/-}$ mice lacking both T and B-cells, and in athymic nude mice that lack only T-cells consistent with activation of T-cells after polyamine depletion in tumor models [134,137]. Moreover, polyamine inhibition increases CD8$^+$ T-cell infiltration into the tumor bed [116,134,137]. Though CD8$^+$ T-cells isolated from a similar B16F10 melanoma model lack cytotoxic functions in vitro [136], it is clear that systemic polyamine inhibition of tumor-bearing mice restores T-cell anti-tumor immunity.

In the tumor microenvironment, cell populations suppress the immune response and contribute to tumor escape from immune surveillance [138]. These cells also use polyamines to invoke their suppressive activations and to support their metabolism (Figure 3). Suppressive myeloid cells are evident in many infectious diseases, including leishmaniasis [139], toxoplasmosis [140], candidiasis [141], and human immunodeficiency virus (HIV)-infected individuals [142], and are significantly elevated in tumor-bearing animals [128,143]. Comparable suppressive cells have been identified in both mouse models and human cancers including melanoma, breast cancer, pancreatic, non-small cell lung and leukemia [143].

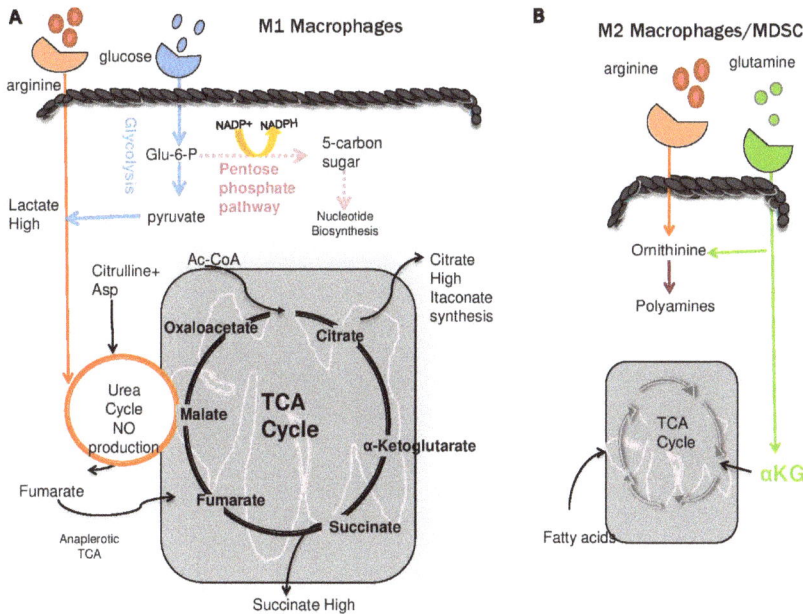

Figure 3. *Cont.*

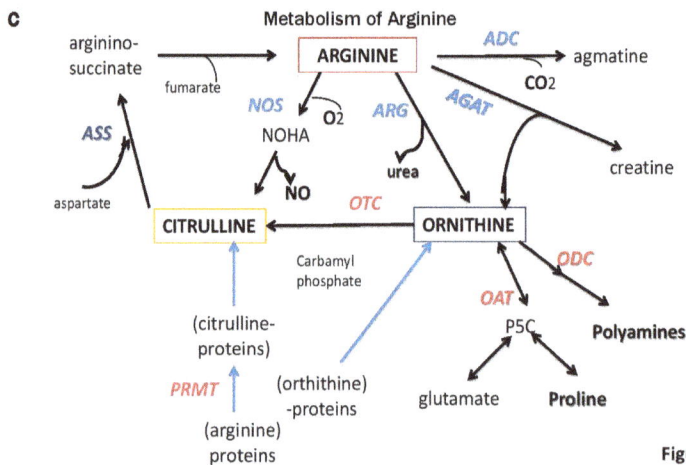

Figure 3. Bioenergetics of macrophage subsets. Monocyte-derived macrophages can be differentially polarized by the cytokine milieu [125,126]. (**A**) M1 macrophages originate from cells in the bone marrow and develop in inflammatory environments. Nitric oxide (NO) is the major byproduct of these cells arising from the reaction of arginine with oxygen through the actions of inducible nitric oxide synthase (iNOS) which produces citrulline and NO (see detailed pathway Figure 3C). Citrulline is then exported and re-imported to re-generate arginine and sustain NO production. A product of the degradation of arginine through this cycle is fumarate which is derived from the conversion of argininosuccinate to arginine (see Figure 3C). M1 macrophages are also critical for the production of cytokines and chemokines and for the production of itaconate which acts as an anti-microbial cellular metabolite. Succinate, a proinflammatory molecule that controls IL-1β expression, accumulates and stabilizes the oxygen sensing pathway regulated by hyposia-inducible factor 1-alpha (HIF1α) [125,127]. (**B**) Unlike M1 macrophages, the polarized M2 subtype reduces their ability to make NO and instead hydrolyzes imported arginine into ornithine and urea through the urea cycle (detailes in Figure 3C). M2 macrophages are therefore suppressive by competing for both arginine and glutamine that is necessary for effector T-cell functions [63,89,112]. To fuel their functions, including proliferation, M2 macrophages use fatty acids oxidation (FAO) which supports oxidative phosphorylation and electron transport through the tricarboxylic acid (TCA) cycle. Also present in the suppressive tumor microenvironment is a population of bone marrow derived immature myeloid cells known as myeloid derived suppressor cells [128,129]. While bioenergetics for these cells needs further analysis, they retain NO production and FAO, TCA and deplete arginine and glutamine [130] from the microenvironment. (**C**) Also detailed is the metabolism of arginine, L-citrulline and L-ornithine to produce fumarate from conversion of argininosuccinate. Citrulline plus aspartate generates argininosuccinate via the actions of argininosuccinate synthetase (ASS) in the cytosol and ornithine is converted to citrulline by carbamylphosphate plus ornithine via the enzymatic activity of ornithine transcarbamoylase (OTC). Additional enzymes and reactions include those metabolized by ODC: ornithinine decarboxylase, ARG: arginase 1 or arginase 2, ADC: arginine decarboxylase which is the biosynthetic enzyme for agmatate [131], OAT: ornithine aminotransferase, NOS: nitric oxide synthase, PRMT: protein arginine methyltransferases which is important for epigenetic regulation [132], and AGAT: L-arginine:glycine amidinotransferase which is the enzyme that catalyzes the transfer of an amidino group from L-arginine to produce L-ornithine and guanidinoacetate and acts as the immediate precursor of creatine.

Suppressive myeloid cells, specifically myeloid-derived suppressor cells (MDSCs), monocyte-derived M2 macrophages and some dendritic cells (DCs), can be present in high numbers in the tumor microenvironment. Based on the cytokine milieu, monocyte-derived macrophages can be polarized into M1 or M2 macrophages [116,144,145]. M2 macrophages do not make nitric oxide (NO), a major byproduct of M1 macrophages, and use arginase to hydrolyze imported arginine into ornithine and urea which depletes arginine in the tumor microenvironment, compromising intratumoral T-cell functions and survival [129,143,146]. Myeloid-derived suppressor cells (MDSCs) retain the ability to produce NO and high levels of reactive oxygen species (ROS) leading to nitration of tyrosine residues of the TCR which disrupts its interaction with the peptide-MHC complex during antigen presentation [49] (Figure 1). The suppressive functions of M2 macrophages relies on higher basal mitochondrial oxygen consumption rates driven by fatty acid oxidation (FAO) [147] and, accordingly, the development of M2 macrophages is blocked by inhibiting mitochondrial OXPHOS and FAO (Figure 3). Further, unlike M1 macrophages, M2 macrophages require glutaminolysis for proliferation and ODC inhibition through difluoromethylornithine (DFMO) or polyamine transport inhibitor treatment of tumor-bearing mice significantly reduces intratumoral suppressive MDSCs [116,134,137] which should improve Arg availability for T-cells that is necessary for their proliferation and persistence [108,112,148,149]. Polyamine inhibition also increases TNFα and IL-1 cytokine production by tumor infiltrating macrophages, suggesting reprogramming of macrophages into the M1 phenotype that augments presentation of tumor-associated antigens, increases citrulline export and import, and further supports the TCA cycle through arginine-derived fumarate [116,136]. Recently, it has been shown that arginine-derived polyamines produced by DCs induce IDO1 expression within the cell through Src kinase, which results in a more immunosuppressive phenotype [150]. This can also be exacerbated by bystander MDSCs that provide more polyamines in the extracellular milieu freely available to DCs. Inhibition of ODC by DFMO reduces this signaling network and promotes DCs to an immune stimulatory phenotype [150]. Thus, it appears that although polyamines are required for normal CD8[+] T-cell functions, the net effects of polyamine depletion on suppressive myeloid cells is to increase anti-tumor CD8[+] T-cell activity by restoring a more conducive tumor microenvironment.

6. Polyamines in Autoimmune Disease

Autoimmune diseases are provoked by abnormal, unchecked immune responses against normal host tissue, and are driven self-reactive TCRs and BCRs in the thymus and bone marrow. Further, suppressive immune populations including myeloid cells, regulatory T-cells (Tregs) and IELs, are necessary to establish peripheral tolerance against self-reactive effector T- and B-cells that escape negative selection [6,151,152]. Autoimmunity can arise in almost every peripheral tissue in the body, for example multiple sclerosis in the brain, thyroiditis and Graves's disease in the thyroid, rheumatoid arthritis and ankylosing spondylitis in the joints, psoriasis, eczema and scleroderma in the skin, diabetes in the pancreas, and celiac disease, ulcerative colitis, and Crohn's Disease that occur in the intestine. Interestingly, circulating polyamine levels are increased in patients with autoimmune diseases [153,154], polyamines have the ability to form nuclear aggregates [155–157] and it has been suggested that nuclear polyamine aggregates interact with DNA, RNA, or other macro-molecular structures to stabilize autoantigens. Strikingly, the most common autoimmune B-cell responses are generated to macromolecules such as double stranded DNA or single stranded DNA [158,159]. Abnormal polyamine structures have been noted in patients with systemic lupus erythematosus (SLE), and rheumatoid arthritis that are characterized by anti-nuclear antibodies consistent with this hypothesis.

7. Concluding Remarks

Recent studies have provided key mechanistic insights into how polyamines may regulate cell fate and proliferation. First, it has been shown that decreasing polyamine pools with the ODC inhibitor DFMO reduces pools of the methyl donor S-adenosylmethionine (SAM, an activated form of methionine) [160]. This appears to occur via effects of polyamines on harnessing the translation of SAM decarboxylase (SAMDC/AMD1) [161,162], which converts SAM to decarboxylated SAM (dcSAM) [163]. Thus, reductions in polyamine pools lead to increases in dcSAM and corresponding reductions in SAM pools. Notably, methylation of DNA and histone tails requires the transfer of the methyl group derived from SAM, and these epigenetic changes are required for changing the pattern of peripheral tissue antigens during negative selection [38,60]. Furthermore, unbiased metabolomic analyses of colon tumor cells revealed that treatment with DFMO also leads to profound reductions in thymidine and thymidine monophosphate (TMP), and that inhibitory effects of DFMO on growth can be overcome by treatment with exogenous thymidine [160,161]. Collectively, these findings suggest a model whereby BCR- and TCR-dependent activation of c-Myc coordinately induces polyamine biosynthesis, and where polyamines then regulate B-cell and T-cell growth, fate, and effector functions via both epigenetic and metabolic control.

Author Contributions: R.H. conceptualized and wrote the manuscript, J.L.C. edited the manuscript and contributed to concept development, and P.K.E.-B. contributed to concept development, writing, editing, and created the display items.

Conflicts of Interest: The authors receive support from Celgene Corporation through a grant to the Moffitt Cancer Center & Research Institute. The content of this manuscript is not influenced by this association.

References

1. Green, D.R. Metabolism and immunity: The old and the new. *Semin. Immunol.* **2012**, *24*, 383. [CrossRef] [PubMed]
2. Halligan, D.N.; Murphy, S.J.; Taylor, C.T. The hypoxia-inducible factor (HIF) couples immunity with metabolism. *Semin. Immunol.* **2016**, *28*, 469–477. [CrossRef] [PubMed]
3. Scharping, N.E.; Delgoffe, G.M. Tumor Microenvironment Metabolism: A New Checkpoint for Anti-Tumor Immunity. *Vaccines (Basel)* **2016**, *4*, 46. [CrossRef] [PubMed]
4. Pearce, E.L.; Pearce, E.J. Metabolic pathways in immune cell activation and quiescence. *Immunity* **2013**, *38*, 633–643. [CrossRef] [PubMed]
5. Michalek, R.D.; Rathmell, J.C. The metabolic life and times of a T-cell. *Immunol. Rev.* **2010**, *236*, 190–202. [CrossRef] [PubMed]
6. Gerriets, V.A.; Rathmell, J.C. Metabolic pathways in T cell fate and function. *Trends Immunol.* **2012**, *33*, 168–173. [CrossRef] [PubMed]
7. Olenchock, B.A.; Rathmell, J.C.; Vander Heiden, M.G. Biochemical Underpinnings of Immune Cell Metab.olic Phenotypes. *Immunity* **2017**, *46*, 703–713. [CrossRef] [PubMed]
8. O'Neill, L.A.; Kishton, R.J.; Rathmell, J. A guide to immunometabolism for immunologists. *Nat. Rev. Immunol.* **2016**, *16*, 553–565. [CrossRef] [PubMed]
9. Goodnow, C.C.; Sprent, J.; Fazekas de St Groth, B.; Vinuesa, C.G. Cellular and genetic mechanisms of self tolerance and autoimmunity. *Nature* **2005**, *435*, 590–597. [CrossRef] [PubMed]
10. Flajnik, M.F.; Kasahara, M. Origin and evolution of the adaptive immune system: Genetic events and selective pressures. *Nat. Rev. Genet.* **2010**, *11*, 47–59. [CrossRef] [PubMed]
11. Kondo, M.; Weissman, I.L.; Akashi, K. Identification of clonogenic common lymphoid progenitors in mouse bone marrow. *Cell* **1997**, *91*, 661–672. [CrossRef]
12. Akashi, K.; Traver, D.; Kondo, M.; Weissman, I.L. Lymphoid development from hematopoietic stem cells. *Int. J. Hematol.* **1999**, *69*, 217–226. [PubMed]
13. Akashi, K.; Traver, D.; Miyamoto, T.; Weissman, I.L. A clonogenic common myeloid progenitor that gives rise to all myeloid lineages. *Nature* **2000**, *404*, 193–197. [CrossRef] [PubMed]
14. Morrison, S.J.; Hemmati, H.D.; Wandycz, A.M.; Weissman, I.L. The purification and characterization of fetal liver hematopoietic stem cells. *Proc. Natl. Acad. Sci. USA* **1995**, *92*, 10302–10306. [CrossRef] [PubMed]
15. Venkitaraman, A.R.; Williams, G.T.; Dariavach, P.; Neuberger, M.S. The B-cell antigen receptor of the five immunoglobulin classes. *Nature* **1991**, *352*, 777–781. [CrossRef] [PubMed]
16. Pape, K.A.; Catron, D.M.; Itano, A.A.; Jenkins, M.K. The humoral immune response is initiated in lymph nodes by B cells that acquire soluble antigen directly in the follicles. *Immunity* **2007**, *26*, 491–502. [CrossRef] [PubMed]
17. Tonegawa, S. Somatic generation of antibody diversity. *Nature* **1983**, *302*, 575–581. [CrossRef] [PubMed]

18. Garside, P.; Ingulli, E.; Merica, R.R.; Johnson, J.G.; Noelle, R.J.; Jenkins, M.K. Visualization of specific B and T lymphocyte interactions in the lymph node. *Science* **1998**, *281*, 96–99. [CrossRef] [PubMed]
19. McAdam, A.J.; Greenwald, R.J.; Levin, M.A.; Chernova, T.; Malenkovich, N.; Ling, V.; Freeman, G.J.; Sharpe, A.H. ICOS is critical for CD40-mediated antibody class switching. *Nature* **2001**, *409*, 102–105. [CrossRef] [PubMed]
20. Honjo, T.; Kinoshita, K.; Muramatsu, M. Molecular mechanism of class switch recombination: Linkage with somatic hypermutation. *Annu. Rev. Immunol.* **2002**, *20*, 165–196. [CrossRef] [PubMed]
21. Gong, S.; Nussenzweig, M.C. Regulation of an early developmental checkpoint in the B cell pathway by Ig β. *Science* **1996**, *272*, 411–414. [CrossRef] [PubMed]
22. Caro-Maldonado, A.; Wang, R.; Nichols, A.G.; Kuraoka, M.; Milasta, S.; Sun, L.D.; Gavin, A.L.; Abel, E.D.; Kelsoe, G.; Green, D.R.; et al. Metabolic reprogramming is required for antibody production that is suppressed in anergic but exaggerated in chronically BAFF-exposed B cells. *J. Immunol.* **2014**, *192*, 3626–3636. [CrossRef] [PubMed]
23. Jones, R.G.; Pearce, E.J. MenTORing Immunity: mTOR Signaling in the Development and Function of Tissue-Resident Immune Cells. *Immunity* **2017**, *46*, 730–742. [CrossRef] [PubMed]
24. Siska, P.J.; van der Windt, G.J.; Kishton, R.J.; Cohen, S.; Eisner, W.; MacIver, N.J.; Kater, A.P.; Weinberg, J.B.; Rathmell, J.C. Suppression of Glut1 and Glucose Metabolism by Decreased Akt/mTORC1 Signaling Drives T Cell Impairment in B Cell Leukemia. *J. Immunol.* **2016**, *197*, 2532–2540. [CrossRef] [PubMed]
25. Nilsson, J.A.; Keller, U.B.; Baudino, T.A.; Yang, C.; Norton, S.; Old, J.A.; Nilsson, L.M.; Neale, G.; Kramer, D.L.; Porter, C.W.; et al. Targeting ornithine decarboxylase in Myc-induced lymphomagenesis prevents tumor formation. *Cancer Cell* **2005**, *7*, 433–444. [CrossRef] [PubMed]
26. Bello-Fernandez, C.; Packham, G.; Cleveland, J.L. The ornithine decarboxylase gene is a transcriptional target of c-Myc. *Proc. Natl. Acad. Sci. USA* **1993**, *90*, 7804–7808. [CrossRef] [PubMed]
27. Tessem, M.B.; Bertilsson, H.; Angelsen, A.; Bathen, T.F.; Drablos, F.; Rye, M.B. A Balanced Tissue Composition Reveals New Metabolic and Gene Expression Markers in Prostate Cancer. *PLoS ONE* **2016**, *11*, e0153727. [CrossRef] [PubMed]
28. Pegg, A.E. Regulation of ornithine decarboxylase. *J. Biol. Chem.* **2006**, *281*, 14529–14532. [CrossRef] [PubMed]
29. Pendeville, H.; Carpino, N.; Marine, J.C.; Takahashi, Y.; Muller, M.; Martial, J.A.; Cleveland, J.L. The ornithine decarboxylase gene is essential for cell survival during early murine development. *Mol. Cell. Biol.* **2001**, *21*, 6549–6558. [CrossRef] [PubMed]
30. Nitta, T.; Igarashi, K.; Yamashita, A.; Yamamoto, M.; Yamamoto, N. Involvement of polyamines in B cell receptor-mediated apoptosis: Spermine functions as a negative modulator. *Exp. Cell Res.* **2001**, *265*, 174–183. [CrossRef] [PubMed]
31. Ohmori, H.; Egusa, H.; Ueura, N.; Matsumoto, Y.; Kanayama, N.; Hikida, M. Selective augmenting effects of nitric oxide on antigen-specific IgE response in mice. *Immunopharmacology* **2000**, *46*, 55–63. [CrossRef]
32. Kramer, D.L.; Diegelman, P.; Jell, J.; Vujcic, S.; Merali, S.; Porter, C.W. Polyamine acetylation modulates polyamine metabolic flux, a prelude to broader metabolic consequences. *J. Biol. Chem.* **2008**, *283*, 4241–4251. [CrossRef] [PubMed]
33. Akamatsu, Y.; Monroe, R.; Dudley, D.D.; Elkin, S.K.; Gartner, F.; Talukder, S.R.; Takahama, Y.; Alt, F.W.; Bassing, C.H.; Oettinger, M.A. Deletion of the RAG2 C terminus leads to impaired lymphoid development in mice. *Proc. Natl. Acad. Sci. USA* **2003**, *100*, 1209–1214. [CrossRef] [PubMed]
34. Mombaerts, P.; Iacomini, J.; Johnson, R.S.; Herrup, K.; Tonegawa, S.; Papaioannou, V.E. RAG-1-deficient mice have no mature B and T lymphocytes. *Cell* **1992**, *68*, 869–877. [CrossRef]
35. Zhu, C.; Roth, D.B. Characterization of coding ends in thymocytes of scid mice: Implications for the mechanism of V(D)J recombination. *Immunity* **1995**, *2*, 101–112. [CrossRef]
36. Zlotoff, D.A.; Sambandam, A.; Logan, T.D.; Bell, J.J.; Schwarz, B.A.; Bhandoola, A. CCR7 and CCR9 together recruit hematopoietic progenitors to the adult thymus. *Blood* **2010**, *115*, 1897–1905. [CrossRef] [PubMed]
37. Zlotoff, D.A.; Zhang, S.L.; De Obaldia, M.E.; Hess, P.R.; Todd, S.P.; Logan, T.D.; Bhandoola, A. Delivery of progenitors to the thymus limits T-lineage reconstitution after bone marrow transplantation. *Blood* **2011**, *118*, 1962–1970. [CrossRef] [PubMed]
38. Meredith, M.; Zemmour, D.; Mathis, D.; Benoist, C. Aire controls gene expression in the thymic epithelium with ordered stochasticity. *Nat. Immunol.* **2015**, *16*, 942–949. [CrossRef] [PubMed]

39. Wolfer, A.; Wilson, A.; Nemir, M.; MacDonald, H.R.; Radtke, F. Inactivation of Notch1 impairs VDJβ rearrangement and allows pre-TCR-independent survival of early αβ Lineage Thymocytes. *Immunity* **2002**, *16*, 869–879. [CrossRef]

40. Tanigaki, K.; Tsuji, M.; Yamamoto, N.; Han, H.; Tsukada, J.; Inoue, H.; Kubo, M.; Honjo, T. Regulation of αβ/γδ T cell lineage commitment and peripheral T cell responses by Notch/RBP-J signaling. *Immunity* **2004**, *20*, 611–622. [CrossRef]

41. Mombaerts, P.; Clarke, A.R.; Rudnicki, M.A.; Iacomini, J.; Itohara, S.; Lafaille, J.J.; Wang, L.; Ichikawa, Y.; Jaenisch, R.; Hooper, M.L.; et al. Mutations in T-cell antigen receptor genes α and β block thymocyte development at different stages. *Nature* **1992**, *360*, 225–231. [CrossRef] [PubMed]

42. Haas, W.; Pereira, P.; Tonegawa, S. Gamma/delta cells. *Annu. Rev. Immunol.* **1993**, *11*, 637–685. [CrossRef] [PubMed]

43. Germain, R.N. T-cell development and the CD4-CD8 lineage decision. *Nat. Rev. Immunol.* **2002**, *2*, 309–322. [CrossRef] [PubMed]

44. Gao, G.F.; Tormo, J.; Gerth, U.C.; Wyer, J.R.; McMichael, A.J.; Stuart, D.I.; Bell, J.I.; Jones, E.Y.; Jakobsen, B.K. Crystal structure of the complex between human CD8αα and HLA-A2. *Nature* **1997**, *387*, 630–634. [CrossRef] [PubMed]

45. Josefowicz, S.Z.; Rudensky, A. Control of regulatory T cell lineage commitment and maintenance. *Immunity* **2009**, *30*, 616–625. [CrossRef] [PubMed]

46. Bendelac, A.; Savage, P.B.; Teyton, L. The biology of NKT cells. *Annu. Rev. Immunol.* **2007**, *25*, 297–336. [CrossRef] [PubMed]

47. McDonald, B.D.; Bunker, J.J.; Ishizuka, I.E.; Jabri, B.; Bendelac, A. Elevated T cell receptor signaling identifies a thymic precursor to the TCRαβ+ CD4−CD8β− intraepithelial lymphocyte lineage. *Immunity* **2014**, *41*, 219–229. [CrossRef] [PubMed]

48. Cheroutre, H. IELs: Enforcing law and order in the court of the intestinal epithelium. *Immunol. Rev.* **2005**, *206*, 114–131. [CrossRef] [PubMed]

49. Nagaraj, S.; Gupta, K.; Pisarev, V.; Kinarsky, L.; Sherman, S.; Kang, L.; Herber, D.L.; Schneck, J.; Gabrilovich, D.I. Altered recognition of antigen is a mechanism of CD8+ T cell tolerance in cancer. *Nat. Med.* **2007**, *13*, 828–835. [CrossRef] [PubMed]

50. Bertram, E.M.; Dawicki, W.; Watts, T.H. Role of T cell costimulation in anti-viral immunity. *Semin. Immunol.* **2004**, *16*, 185–196. [CrossRef] [PubMed]

51. Croft, M. Costimulation of T cells by OX40, 4-1BB, and CD27. *Cytokine Growth Factor Rev.* **2003**, *14*, 265–273. [CrossRef]

52. Gramaglia, I.; Weinberg, A.D.; Lemon, M.; Croft, M. Ox-40 ligand: A potent costimulatory molecule for sustaining primary CD4 T cell responses. *J. Immunol.* **1998**, *161*, 6510–6517. [PubMed]

53. Rogers, P.R.; Croft, M. CD28, Ox-40, LFA-1, and CD4 modulation of Th1/Th2 differentiation is directly dependent on the dose of antigen. *J. Immunol.* **2000**, *164*, 2955–2963. [CrossRef] [PubMed]

54. Cantrell, D.A. Transgenic analysis of thymocyte signal transduction. *Nat. Rev. Immunol.* **2002**, *2*, 20–27. [CrossRef] [PubMed]

55. Derbinski, J.; Schulte, A.; Kyewski, B.; Klein, L. Promiscuous gene expression in medullary thymic epithelial cells mirrors the peripheral self. *Nat. Immunol.* **2001**, *2*, 1032–1039. [CrossRef] [PubMed]

56. Abramson, J.; Giraud, M.; Benoist, C.; Mathis, D. Aire's partners in the molecular control of immunological tolerance. *Cell* **2010**, *140*, 123–135. [CrossRef] [PubMed]

57. Danan-Gotthold, M.; Guyon, C.; Giraud, M.; Levanon, E.Y.; Abramson, J. Extensive RNA editing and splicing increase immune self-representation diversity in medullary thymic epithelial cells. *Genome Biol.* **2016**, *17*, 219. [CrossRef] [PubMed]

58. Koh, A.S.; Kingston, R.E.; Benoist, C.; Mathis, D. Global relevance of Aire binding to hypomethylated lysine-4 of histone-3. *Proc. Natl. Acad. Sci. USA* **2010**, *107*, 13016–13021. [CrossRef] [PubMed]

59. Giraud, M.; Yoshida, H.; Abramson, J.; Rahl, P.B.; Young, R.A.; Mathis, D.; Benoist, C. Aire unleashes stalled RNA polymerase to induce ectopic gene expression in thymic epithelial cells. *Proc. Natl. Acad. Sci. USA* **2012**, *109*, 535–540. [CrossRef] [PubMed]

60. Herzig, Y.; Nevo, S.; Bornstein, C.; Brezis, M.R.; Ben-Hur, S.; Shkedy, A.; Eisenberg-Bord, M.; Levi, B.; Delacher, M.; Goldfarb, Y.; et al. Transcriptional programs that control expression of the autoimmune regulator gene Aire. *Nat. Immunol.* **2017**, *18*, 161–172. [CrossRef] [PubMed]

61. Chen, L.; Flies, D.B. Molecular mechanisms of T cell co-stimulation and co-inhibition. *Nat. Rev. Immunol.* **2013**, *13*, 227–242. [CrossRef] [PubMed]

62. Ito, K.; Igarashi, K. Polyamine regulation of the synthesis of thymidine kinase in bovine lymphocytes. *Arch. Biochem. Biophys.* **1990**, *278*, 277–283. [CrossRef]

63. Wang, R.; Dillon, C.P.; Shi, L.Z.; Milasta, S.; Carter, R.; Finkelstein, D.; McCormick, L.L.; Fitzgerald, P.; Chi, H.; Munger, J.; et al. The transcription factor Myc controls metabolic reprogramming upon T lymphocyte activation. *Immunity* **2011**, *35*, 871–882. [CrossRef] [PubMed]

64. Hendriks, J.; Gravestein, L.A.; Tesselaar, K.; van Lier, R.A.; Schumacher, T.N.; Borst, J. CD27 is required for generation and long-term maintenance of T cell immunity. *Nat. Immunol.* **2000**, *1*, 433–440. [CrossRef] [PubMed]

65. Agematsu, K.; Hokibara, S.; Nagumo, H.; Komiyama, A. CD27: A memory B-cell marker. *Immunol. Today* **2000**, *21*, 204–206. [CrossRef]

66. Takeda, K.; Oshima, H.; Hayakawa, Y.; Akiba, H.; Atsuta, M.; Kobata, T.; Kobayashi, K.; Ito, M.; Yagita, H.; Okumura, K. CD27-mediated activation of murine NK cells. *J. Immunol.* **2000**, *164*, 1741–1745. [CrossRef] [PubMed]

67. Smith, A.; Stanley, P.; Jones, K.; Svensson, L.; McDowall, A.; Hogg, N. The role of the integrin LFA-1 in T-lymphocyte migration. *Immunol. Rev.* **2007**, *218*, 135–146. [CrossRef] [PubMed]

68. Nocentini, G.; Giunchi, L.; Ronchetti, S.; Krausz, L.T.; Bartoli, A.; Moraca, R.; Migliorati, G.; Riccardi, C. A new member of the tumor necrosis factor/nerve growth factor receptor family inhibits T cell receptor-induced apoptosis. *Proc. Natl. Acad. Sci. USA* **1997**, *94*, 6216–6221. [CrossRef] [PubMed]

69. Khayyamian, S.; Hutloff, A.; Buchner, K.; Grafe, M.; Henn, V.; Kroczek, R.A.; Mages, H.W. ICOS-ligand, expressed on human endothelial cells, costimulates Th1 and Th2 cytokine secretion by memory CD4+ T cells. *Proc. Natl. Acad. Sci. USA* **2002**, *99*, 6198–6203. [CrossRef] [PubMed]

70. Cai, G.; Freeman, G.J. The CD160, BTLA, LIGHT/HVEM pathway: A bidirectional switch regulating T-cell activation. *Immunol. Rev.* **2009**, *229*, 244–258. [CrossRef] [PubMed]

71. Tan, J.T.; Dudl, E.; LeRoy, E.; Murray, R.; Sprent, J.; Weinberg, K.I.; Surh, C.D. IL-7 is critical for homeostatic proliferation and survival of naive T cells. *Proc. Natl. Acad. Sci. USA* **2001**, *98*, 8732–8737. [CrossRef] [PubMed]

72. Wofford, J.A.; Wieman, H.L.; Jacobs, S.R.; Zhao, Y.; Rathmell, J.C. IL-7 promotes Glut1 trafficking and glucose uptake via STAT5-mediated activation of Akt to support T-cell survival. *Blood* **2008**, *111*, 2101–2111. [CrossRef] [PubMed]

73. Pallard, C.; Stegmann, A.P.; van Kleffens, T.; Smart, F.; Venkitaraman, A.; Spits, H. Distinct roles of the phosphatidylinositol 3-kinase and STAT5 pathways in IL-7-mediated development of human thymocyte precursors. *Immunity* **1999**, *10*, 525–535. [CrossRef]

74. Jacobs, S.R.; Michalek, R.D.; Rathmell, J.C. IL-7 is essential for homeostatic control of T cell metabolism in vivo. *J. Immunol.* **2010**, *184*, 3461–3469. [CrossRef] [PubMed]

75. MacIver, N.J.; Michalek, R.D.; Rathmell, J.C. Metabolic regulation of T lymphocytes. *Annu. Rev. Immunol.* **2013**, *31*, 259–283. [CrossRef] [PubMed]

76. Inoki, K.; Li, Y.; Zhu, T.; Wu, J.; Guan, K.L. TSC2 is phosphorylated and inhibited by Akt and suppresses mTOR signalling. *Nat. Cell Biol.* **2002**, *4*, 648–657. [CrossRef] [PubMed]

77. Inoki, K.; Li, Y.; Xu, T.; Guan, K.L. Rheb GTPase is a direct target of TSC2 GAP activity and regulates mTOR signaling. *Genes Dev.* **2003**, *17*, 1829–1834. [CrossRef] [PubMed]

78. Inoki, K.; Zhu, T.; Guan, K.L. TSC2 mediates cellular energy response to control cell growth and survival. *Cell* **2003**, *115*, 577–590. [CrossRef]

79. Iwata, T.N.; Ramirez, J.A.; Tsang, M.; Park, H.; Margineantu, D.H.; Hockenbery, D.M.; Iritani, B.M. Conditional Disruption of Raptor Reveals an Essential Role for mTORC1 in B Cell Development, Survival, and Metabolism. *J. Immunol.* **2016**, *197*, 2250–2260. [CrossRef] [PubMed]

80. Delgoffe, G.M.; Kole, T.P.; Zheng, Y.; Zarek, P.E.; Matthews, K.L.; Xiao, B.; Worley, P.F.; Kozma, S.C.; Powell, J.D. The mTOR kinase differentially regulates effector and regulatory T cell lineage commitment. *Immunity* **2009**, *30*, 832–844. [CrossRef] [PubMed]

81. Guertin, D.A.; Stevens, D.M.; Thoreen, C.C.; Burds, A.A.; Kalaany, N.Y.; Moffat, J.; Brown, M.; Fitzgerald, K.J.; Sabatini, D.M. Ablation in mice of the mTORC components *raptor*, *rictor*, or *mLST8* reveals that mTORC2 is required for signaling to Akt-FOXO and PKCalpha, but not S6K1. *Dev. Cell* **2006**, *11*, 859–871. [CrossRef] [PubMed]

82. Shaw, R.J.; Kosmatka, M.; Bardeesy, N.; Hurley, R.L.; Witters, L.A.; DePinho, R.A.; Cantley, L.C. The tumor suppressor LKB1 kinase directly activates AMP-activated kinase and regulates apoptosis in response to energy stress. *Proc. Natl. Acad. Sci. USA* **2004**, *101*, 3329–3335. [CrossRef] [PubMed]

83. Woods, A.; Johnstone, S.R.; Dickerson, K.; Leiper, F.C.; Fryer, L.G.; Neumann, D.; Schlattner, U.; Wallimann, T.; Carlson, M.; Carling, D. LKB1 is the upstream kinase in the AMP-activated protein kinase cascade. *Curr. Biol.* **2003**, *13*, 2004–2008. [CrossRef] [PubMed]

84. Van der Windt, G.J.; Pearce, E.L. Metabolic switching and fuel choice during T-cell differentiation and memory development. *Immunol. Rev.* **2012**, *249*, 27–42. [CrossRef] [PubMed]

85. Frauwirth, K.A.; Riley, J.L.; Harris, M.H.; Parry, R.V.; Rathmell, J.C.; Plas, D.R.; Elstrom, R.L.; June, C.H.; Thompson, C.B. The CD28 signaling pathway regulates glucose metabolism. *Immunity* **2002**, *16*, 769–777. [CrossRef]

86. Boomer, J.S.; Green, J.M. An enigmatic tail of CD28 signaling. *Cold Spring Harb. Perspect. Biol.* **2010**, *2*, a002436. [CrossRef] [PubMed]

87. Park, S.G.; Schulze-Luehrman, J.; Hayden, M.S.; Hashimoto, N.; Ogawa, W.; Kasuga, M.; Ghosh, S. The kinase PDK1 integrates T cell antigen receptor and CD28 coreceptor signaling to induce NF-κB and activate T cells. *Nat. Immunol.* **2009**, *10*, 158–166. [CrossRef] [PubMed]

88. Balagopalan, L.; Barr, V.A.; Sommers, C.L.; Barda-Saad, M.; Goyal, A.; Isakowitz, M.S.; Samelson, L.E. c-Cbl-mediated regulation of LAT-nucleated signaling complexes. *Mol. Cell. Biol.* **2007**, *27*, 8622–8636. [CrossRef] [PubMed]

89. Verbist, K.C.; Guy, C.S.; Milasta, S.; Liedmann, S.; Kaminski, M.M.; Wang, R.; Green, D.R. Metabolic maintenance of cell asymmetry following division in activated T lymphocytes. *Nature* **2016**, *532*, 389–393. [CrossRef] [PubMed]

90. Ho, P.C.; Bihuniak, J.D.; Macintyre, A.N.; Staron, M.; Liu, X.; Amezquita, R.; Tsui, Y.C.; Cui, G.; Micevic, G.; Perales, J.C.; et al. Phosphoenolpyruvate Is a Metabolic Checkpoint of Anti-tumor T Cell Responses. *Cell* **2015**, *162*, 1217–1228. [CrossRef] [PubMed]

91. Maeda, T.; Wakasawa, T.; Shima, Y.; Tsuboi, I.; Aizawa, S.; Tamai, I. Role of polyamines derived from arginine in differentiation and proliferation of human blood cells. *Biol. Pharm. Bull.* **2006**, *29*, 234–239. [CrossRef] [PubMed]

92. Shima, Y.; Maeda, T.; Aizawa, S.; Tsuboi, I.; Kobayashi, D.; Kato, R.; Tamai, I. L-arginine import via cationic amino acid transporter CAT1 is essential for both differentiation and proliferation of erythrocytes. *Blood* **2006**, *107*, 1352–1356. [CrossRef] [PubMed]

93. Bachrach, U.; Persky, S. Interaction of oxidized polyamines with DNA. V. Inhibition of nucleic acid synthesis. *Biochim. Biophys. Acta* **1969**, *179*, 484–493. [CrossRef]

94. Francke, B. Cell-free synthesis of herpes simplex virus DNA: The influence of polyamines. *Biochemistry* **1978**, *17*, 5494–5499. [CrossRef] [PubMed]

95. Leveque, J.; Burtin, F.; Catros-Quemener, V.; Havouis, R.; Moulinoux, J.P. The gastrointestinal polyamine source depletion enhances DFMO induced polyamine depletion in MCF-7 human breast cancer cells in vivo. *Anticancer Res.* **1998**, *18*, 2663–2668. [PubMed]

96. Hessels, J.; Kingma, A.W.; Ferwerda, H.; Keij, J.; van den Berg, G.A.; Muskiet, F.A. Microbial flora in the gastrointestinal tract abolishes cytostatic effects of α-difluoromethylornithine in vivo. *Int. J. Cancer* **1989**, *43*, 1155–1164. [CrossRef] [PubMed]

97. Sugiyama, S.; Vassylyev, D.G.; Matsushima, M.; Kashiwagi, K.; Igarashi, K.; Morikawa, K. Crystal structure of PotD, the primary receptor of the polyamine transport system in *Escherichia coli*. *J. Biol. Chem.* **1996**, *271*, 9519–9525. [CrossRef] [PubMed]

98. Tomitori, H.; Kashiwagi, K.; Asakawa, T.; Kakinuma, Y.; Michael, A.J.; Igarashi, K. Multiple polyamine transport systems on the vacuolar membrane in yeast. *Biochem. J.* **2001**, *353*, 681–688. [CrossRef] [PubMed]

99. Satriano, J.; Isome, M.; Casero, R.A., Jr.; Thomson, S.C.; Blantz, R.C. Polyamine transport system mediates agmatine transport in mammalian cells. *Am. J. Physiol. Cell Physiol.* **2001**, *281*, C329–C334. [CrossRef] [PubMed]

100. Sakata, K.; Kashiwagi, K.; Igarashi, K. Properties of a polyamine transporter regulated by antizyme. *Biochem. J.* **2000**, *347*, 297–303. [CrossRef] [PubMed]

101. Uemura, T.; Stringer, D.E.; Blohm-Mangone, K.A.; Gerner, E.W. Polyamine transport is mediated by both endocytic and solute carrier transport mechanisms in the gastrointestinal tract. *Am. J. Physiol. Gastrointest. Liver Physiol.* **2010**, *299*, G517–G522. [CrossRef] [PubMed]

102. Wolfer, A.; Bakker, T.; Wilson, A.; Nicolas, M.; Ioannidis, V.; Littman, D.R.; Lee, P.P.; Wilson, C.B.; Held, W.; MacDonald, H.R.; et al. Inactivation of Notch 1 in immature thymocytes does not perturb CD4 or CD8T cell development. *Nat. Immunol.* **2001**, *2*, 235–241. [CrossRef] [PubMed]

103. Wildin, R.S.; Garvin, A.M.; Pawar, S.; Lewis, D.B.; Abraham, K.M.; Forbush, K.A.; Ziegler, S.F.; Allen, J.M.; Perlmutter, R.M. Developmental regulation of lck gene expression in T lymphocytes. *J. Exp. Med.* **1991**, *173*, 383–393. [CrossRef] [PubMed]

104. Dose, M.; Khan, I.; Guo, Z.; Kovalovsky, D.; Krueger, A.; von Boehmer, H.; Khazaie, K.; Gounari, F. c-Myc mediates pre-TCR-induced proliferation but not developmental progression. *Blood* **2006**, *108*, 2669–2677. [CrossRef] [PubMed]

105. Dezfouli, S.; Bakke, A.; Huang, J.; Wynshaw-Boris, A.; Hurlin, P.J. Inflammatory disease and lymphomagenesis caused by deletion of the Myc antagonist Mnt in T cells. *Mol. Cell. Biol.* **2006**, *26*, 2080–2092. [CrossRef] [PubMed]

106. Jain, J.; Nalefski, E.A.; McCaffrey, P.G.; Johnson, R.S.; Spiegelman, B.M.; Papaioannou, V.; Rao, A. Normal peripheral T-cell function in c-Fos-deficient mice. *Mol. Cell. Biol.* **1994**, *14*, 1566–1574. [CrossRef] [PubMed]

107. Wrighton, C.; Busslinger, M. Direct transcriptional stimulation of the ornithine decarboxylase gene by Fos in PC12 cells but not in fibroblasts. *Mol. Cell. Biol.* **1993**, *13*, 4657–4669. [CrossRef] [PubMed]

108. Choi, B.S.; Martinez-Falero, I.C.; Corset, C.; Munder, M.; Modolell, M.; Muller, I.; Kropf, P. Differential impact of L-arginine deprivation on the activation and effector functions of T cells and macrophages. *J. Leukoc. Biol.* **2009**, *85*, 268–277. [CrossRef] [PubMed]

109. Carr, E.L.; Kelman, A.; Wu, G.S.; Gopaul, R.; Senkevitch, E.; Aghvanyan, A.; Turay, A.M.; Frauwirth, K.A. Glutamine uptake and metabolism are coordinately regulated by ERK/MAPK during T lymphocyte activation. *J. Immunol.* **2010**, *185*, 1037–1044. [CrossRef] [PubMed]

110. Hunt, N.H.; Fragonas, J.C. Effects of anti-oxidants on ornithine decarboxylase in mitogenically-activated T lymphocytes. *Biochim. Biophys. Acta* **1992**, *1133*, 261–267. [CrossRef]

111. Widjaja, C.E.; Olvera, J.G.; Metz, P.J.; Phan, A.T.; Savas, J.N.; de Bruin, G.; Leestemaker, Y.; Berkers, C.R.; de Jong, A.; Florea, B.I.; et al. Proteasome activity regulates CD8+ T lymphocyte metabolism and fate specification. *J. Clin. Investig.* **2017**, *127*, 3609–3623. [CrossRef] [PubMed]

112. Geiger, R.; Rieckmann, J.C.; Wolf, T.; Basso, C.; Feng, Y.; Fuhrer, T.; Kogadeeva, M.; Picotti, P.; Meissner, F.; Mann, M.; et al. L-Arginine Modulates T Cell Metab.olism and Enhances Survival and Anti-tumor Activity. *Cell* **2016**, *167*, 829–842. [CrossRef] [PubMed]

113. Buck, M.D.; O'Sullivan, D.; Klein Geltink, R.I.; Curtis, J.D.; Chang, C.H.; Sanin, D.E.; Qiu, J.; Kretz, O.; Braas, D.; van der Windt, G.J.; et al. Mitochondrial Dynamics Controls T Cell Fate through Metabolic Programming. *Cell* **2016**, *166*, 63–76. [CrossRef] [PubMed]

114. Klein Geltink, R.I.; O'Sullivan, D.; Corrado, M.; Bremser, A.; Buck, M.D.; Buescher, J.M.; Firat, E.; Zhu, X.; Niedermann, G.; Caputa, G.; et al. Mitochondrial Priming by CD28. *Cell* **2017**, *171*, 385–397. [CrossRef] [PubMed]

115. Pearce, E.L.; Poffenberger, M.C.; Chang, C.H.; Jones, R.G. Fueling immunity: Insights into metabolism and lymphocyte function. *Science* **2013**, *342*, 1242454. [CrossRef] [PubMed]

116. Alexander, E.T.; Minton, A.; Peters, M.C.; Phanstiel, O.t.; Gilmour, S.K. A novel polyamine blockade therapy activates an anti-tumor immune response. *Oncotarget* **2017**, *8*, 84140–84152. [CrossRef] [PubMed]

117. Mandal, S.; Mandal, A.; Johansson, H.E.; Orjalo, A.V.; Park, M.H. Depletion of cellular polyamines, spermidine and spermine, causes a total arrest in translation and growth in mammalian cells. *Proc. Natl. Acad. Sci. USA* **2013**, *110*, 2169–2174. [CrossRef] [PubMed]

118. Gnanaprakasam, J.N.; Wang, R. MYC in Regulating Immunity: Metabolism and Beyond. *Genes (Basel)* **2017**, *8*, 88. [CrossRef] [PubMed]

119. Ehrke, M.J.; Porter, C.W.; Eppolito, C.; Mihich, E. Selective modulation by alpha-difluoromethylornithine of T-lymphocyte and antibody-mediated cytotoxic responses to mouse tumor allografts. *Cancer Res.* **1986**, *46*, 2798–2803. [PubMed]

120. Bowlin, T.L.; McKown, B.J.; Sunkara, P.S. Increased ornithine decarboxylase activity and polyamine biosynthesis are required for optimal cytolytic T lymphocyte induction. *Cell. Immunol.* **1987**, *105*, 110–117. [CrossRef]

121. Bowlin, T.L.; McKown, B.J.; Schroeder, K.K. Methyl-acetylenicputrescine (MAP), an inhibitor of polyamine biosynthesis, reduces the frequency and cytolytic activity of alloantigen-induced LyT 2.2 positive lymphocytes in vivo. *Int. J. Immunopharmacol.* **1989**, *11*, 259–265. [CrossRef]

122. Bowlin, T.L.; Rosenberger, A.L.; McKown, B.J. α-difluoromethylornithine, an inhibitor of polyamine biosynthesis, augments cyclosporin A inhibition of cytolytic T lymphocyte induction. *Clin. Exp. Immunol.* **1989**, *77*, 151–156. [PubMed]

123. Schall, R.P.; Sekar, J.; Tandon, P.M.; Susskind, B.M. Difluoromethylornithine (DFMO) arrests murine CTL development in the late, pre-effector stage. *Immunopharmacology* **1991**, *21*, 129–143. [CrossRef]

124. Bowlin, T.L.; Davis, G.F.; McKown, B.J. Inhibition of alloantigen-induced cytolytic T lymphocytes in vitro with (2R,5R)-6-heptyne-2,5-diamine, an irreversible inhibitor of ornithine decarboxylase. *Cell. Immunol.* **1988**, *111*, 443–450. [CrossRef]

125. Lampropoulou, V.; Sergushichev, A.; Bambouskova, M.; Nair, S.; Vincent, E.E.; Loginicheva, E.; Cervantes-Barragan, L.; Ma, X.; Huang, S.C.; Griss, T.; et al. Itaconate Links Inhibition of Succinate Dehydrogenase with Macrophage Metabolic Remodeling and Regulation of Inflammation. *Cell Metab.* **2016**, *24*, 158–166. [CrossRef] [PubMed]

126. Murray, P.J.; Rathmell, J.; Pearce, E. SnapShot: Immunometabolism. *Cell Metab.* **2015**, *22*, 190.e1. [CrossRef] [PubMed]

127. Kelly, B.; O'Neill, L.A. Metabolic reprogramming in macrophages and dendritic cells in innate immunity. *Cell Res.* **2015**, *25*, 771–784. [CrossRef] [PubMed]

128. Nagaraj, S.; Youn, J.I.; Gabrilovich, D.I. Reciprocal relationship between myeloid-derived suppressor cells and T cells. *J. Immunol.* **2013**, *191*, 17–23. [CrossRef] [PubMed]

129. Youn, J.I.; Collazo, M.; Shalova, I.N.; Biswas, S.K.; Gabrilovich, D.I. Characterization of the nature of granulocytic myeloid-derived suppressor cells in tumor-bearing mice. *J. Leukoc. Biol.* **2012**, *91*, 167–181. [CrossRef] [PubMed]

130. Rodriguez, P.C.; Ochoa, A.C.; Al-Khami, A.A. Arginine Metabolism in Myeloid Cells Shapes Innate and Adaptive Immunity. *Front. Immunol.* **2017**, *8*, 93. [CrossRef] [PubMed]

131. Zhu, M.Y.; Iyo, A.; Piletz, J.E.; Regunathan, S. Expression of human arginine decarboxylase, the biosynthetic enzyme for agmatine. *Biochim. Biophys. Acta* **2004**, *1670*, 156–164. [CrossRef] [PubMed]

132. Fuhrmann, J.; Thompson, P.R. Protein Arginine Methylation and Citrullination in Epigenetic Regulation. *ACS Chem. Biol.* **2016**, *11*, 654–668. [CrossRef] [PubMed]

133. Bronte, V.; Zanovello, P. Regulation of immune responses by L-arginine metabolism. *Nat. Rev. Immunol.* **2005**, *5*, 641–654. [CrossRef] [PubMed]

134. Ye, C.; Geng, Z.; Dominguez, D.; Chen, S.; Fan, J.; Qin, L.; Long, A.; Zhang, Y.; Kuzel, T.M.; Zhang, B. Targeting Ornithine Decarboxylase by α-Difluoromethylornithine Inhibits Tumor Growth by Impairing Myeloid-Derived Suppressor Cells. *J. Immunol.* **2016**, *196*, 915–923. [CrossRef] [PubMed]

135. Ziv, Y.; Fazio, V.W.; Kitago, K.; Gupta, M.K.; Sawady, J.; Nishioka, K. Effect of tamoxifen on 1,2-dimethylhydrazine-HCl-induced colon carcinogenesis in rats. *Anticancer Res.* **1997**, *17*, 803–810. [PubMed]

136. Bowlin, T.L.; Hoeper, B.J.; Rosenberger, A.L.; Davis, G.F.; Sunkara, P.S. Effects of three irreversible inhibitors of ornithine decarboxylase on macrophage-mediated tumoricidal activity and antitumor activity in B16F1 tumor-bearing mice. *Cancer Res.* **1990**, *50*, 4510–4514. [PubMed]

137. Hayes, C.S.; Shicora, A.C.; Keough, M.P.; Snook, A.E.; Burns, M.R.; Gilmour, S.K. Polyamine-blocking therapy reverses immunosuppression in the tumor microenvironment. *Cancer Immunol. Res.* **2014**, *2*, 274–285. [CrossRef] [PubMed]

138. Nagaraj, S.; Schrum, A.G.; Cho, H.I.; Celis, E.; Gabrilovich, D.I. Mechanism of T cell tolerance induced by myeloid-derived suppressor cells. *J. Immunol.* **2010**, *184*, 3106–3116. [CrossRef] [PubMed]

139. Sunderkotter, C.; Nikolic, T.; Dillon, M.J.; Van Rooijen, N.; Stehling, M.; Drevets, D.A.; Leenen, P.J. Subpopulations of mouse blood monocytes differ in maturation stage and inflammatory response. *J. Immunol.* **2004**, *172*, 4410–4417. [CrossRef] [PubMed]

140. Voisin, M.B.; Buzoni-Gatel, D.; Bout, D.; Velge-Roussel, F. Both expansion of regulatory GR1+ CD11b+ myeloid cells and anergy of T lymphocytes participate in hyporesponsiveness of the lung-associated immune system during acute toxoplasmosis. *Infect. Immun.* **2004**, *72*, 5487–5492. [CrossRef] [PubMed]

141. Mencacci, A.; Montagnoli, C.; Bacci, A.; Cenci, E.; Pitzurra, L.; Spreca, A.; Kopf, M.; Sharpe, A.H.; Romani, L. CD80+Gr-1+ myeloid cells inhibit development of antifungal Th1 immunity in mice with candidiasis. *J. Immunol.* **2002**, *169*, 3180–3190. [CrossRef] [PubMed]

142. Garg, A.; Spector, S.A. HIV type 1 gp120-induced expansion of myeloid derived suppressor cells is dependent on interleukin 6 and suppresses immunity. *J. Infect. Dis.* **2014**, *209*, 441–451. [CrossRef] [PubMed]

143. Kumar, V.; Patel, S.; Tcyganov, E.; Gabrilovich, D.I. The Nature of Myeloid-Derived Suppressor Cells in the Tumor Microenvironment. *Trends Immunol.* **2016**, *37*, 208–220. [CrossRef] [PubMed]

144. Niino, D.; Komohara, Y.; Murayama, T.; Aoki, R.; Kimura, Y.; Hashikawa, K.; Kiyasu, J.; Takeuchi, M.; Suefuji, N.; Sugita, Y.; et al. Ratio of M2 macrophage expression is closely associated with poor prognosis for Angioimmunoblastic T-cell lymphoma (AITL). *Pathol. Int.* **2010**, *60*, 278–283. [CrossRef] [PubMed]

145. Wang, Y.C.; He, F.; Feng, F.; Liu, X.W.; Dong, G.Y.; Qin, H.Y.; Hu, X.B.; Zheng, M.H.; Liang, L.; Feng, L.; et al. Notch signaling determines the M1 versus M2 polarization of macrophages in antitumor immune responses. *Cancer Res.* **2010**, *70*, 4840–4849. [CrossRef] [PubMed]

146. Nagaraj, S.; Nelson, A.; Youn, J.I.; Cheng, P.; Quiceno, D.; Gabrilovich, D.I. Antigen-specific CD4(+) T cells regulate function of myeloid-derived suppressor cells in cancer via retrograde MHC class II signaling. *Cancer Res.* **2012**, *72*, 928–938. [CrossRef] [PubMed]

147. Mills, E.L.; Kelly, B.; Logan, A.; Costa, A.S.H.; Varma, M.; Bryant, C.E.; Tourlomousis, P.; Dabritz, J.H.M.; Gottlieb, E.; Latorre, I.; et al. Succinate Dehydrogenase Supports Metabolic Repurposing of Mitochondria to Drive Inflammatory Macrophages. *Cell* **2016**, *167*, 457–470. [CrossRef] [PubMed]

148. Cao, Y.; Wang, Q.; Du, Y.; Liu, F.; Zhang, Y.; Feng, Y.; Jin, F. L-arginine and docetaxel synergistically enhance anti-tumor immunity by modifying the immune status of tumor-bearing mice. *Int. Immunopharmacol.* **2016**, *35*, 7–14. [CrossRef] [PubMed]

149. He, X.; Lin, H.; Yuan, L.; Li, B. Combination therapy with L-arginine and α-PD-L1 antibody boosts immune response against osteosarcoma in immunocompetent mice. *Cancer Biol. Ther.* **2017**, *18*, 94–100. [CrossRef] [PubMed]

150. Mondanelli, G.; Bianchi, R.; Pallotta, M.T.; Orabona, C.; Albini, E.; Iacono, A.; Belladonna, M.L.; Vacca, C.; Fallarino, F.; Macchiarulo, A.; et al. A Relay Pathway between Arginine and Tryptophan Metabolism Confers Immunosuppressive Properties on Dendritic Cells. *Immunity* **2017**, *46*, 233–244. [CrossRef] [PubMed]

151. Rubin, R.L.; Burlingame, R.W. Drug-induced autoimmunity: A disorder at the interface between metabolism and immunity. *Biochem. Soc. Trans.* **1991**, *19*, 153–159. [CrossRef] [PubMed]

152. Rathmell, J.C. Apoptosis and B cell tolerance. *Curr. Dir. Autoimmun.* **2003**, *6*, 38–60. [PubMed]

153. Teti, D.; Visalli, M.; McNair, H. Analysis of polyamines as markers of (patho)physiological conditions. *J. Chromatogr. B Anal. Technol. Biomed. Life Sci.* **2002**, *781*, 107–149. [CrossRef]

154. Karouzakis, E.; Gay, R.E.; Gay, S.; Neidhart, M. Increased recycling of polyamines is associated with global DNA hypomethylation in rheumatoid arthritis synovial fibroblasts. *Arthritis Rheumatol.* **2012**, *64*, 1809–1817. [CrossRef] [PubMed]

155. Pignata, S.; Di Luccia, A.; Lamanda, R.; Menchise, A.; D'Agostino, L. Interaction of putrescine with nuclear oligopeptides in the enterocyte-like Caco-2 cells. *Digestion* **1999**, *60*, 255–261. [CrossRef] [PubMed]

156. D'Agostino, L.; Di Luccia, A. Polyamines interact with DNA as molecular aggregates. *Eur. J. Biochem.* **2002**, *269*, 4317–4325. [CrossRef] [PubMed]

157. D'Agostino, L.; di Pietro, M.; Di Luccia, A. Nuclear aggregates of polyamines are supramolecular structures that play a crucial role in genomic DNA protection and conformation. *FEBS J.* **2005**, *272*, 3777–3787. [CrossRef] [PubMed]

158. Riboldi, P.; Gerosa, M.; Moroni, G.; Radice, A.; Allegri, F.; Sinico, A.; Tincani, A.; Meroni, P.L. Anti-DNA antibodies: A diagnostic and prognostic tool for systemic lupus erythematosus? *Autoimmunity* **2005**, *38*, 39–45. [CrossRef] [PubMed]

159. Fineschi, S.; Borghi, M.O.; Riboldi, P.; Gariglio, M.; Buzio, C.; Landolfo, S.; Cebecauer, L.; Tuchynova, A.; Rovensky, J.; Meroni, P.L. Prevalence of autoantibodies against structure specific recognition protein 1 in systemic lupus erythematosus. *Lupus* **2004**, *13*, 463–468. [CrossRef] [PubMed]

160. Casero, R.A., Jr. Say what? The activity of the polyamine biosynthesis inhibitor difluoromethylornithine in chemoprevention is a result of reduced thymidine pools? *Cancer Discov.* **2013**, *3*, 975–977. [CrossRef] [PubMed]

161. Ruan, H.; Hill, J.R.; Fatemie-Nainie, S.; Morris, D.R. Cell-specific translational regulation of *S*-adenosylmethionine decarboxylase mRNA. Influence of the structure of the 5′ transcript leader on regulation by the upstream open reading frame. *J. Biol. Chem.* **1994**, *269*, 17905–17910. [PubMed]

162. Ruan, H.; Shantz, L.M.; Pegg, A.E.; Morris, D.R. The upstream open reading frame of the mRNA encoding *S*-adenosylmethionine decarboxylase is a polyamine-responsive translational control element. *J. Biol. Chem.* **1996**, *271*, 29576–29582. [CrossRef] [PubMed]

163. Bale, S.; Lopez, M.M.; Makhatadze, G.I.; Fang, Q.; Pegg, A.E.; Ealick, S.E. Structural basis for putrescine activation of human *S*-adenosylmethionine decarboxylase. *Biochemistry* **2008**, *47*, 13404–13417. [CrossRef] [PubMed]

![medical sciences]

MDPI

Review

Skeletal Muscle Pathophysiology: The Emerging Role of Spermine Oxidase and Spermidine

Manuela Cervelli [1],*, Alessia Leonetti [1], Guglielmo Duranti [2], Stefania Sabatini [2], Roberta Ceci [2] and Paolo Mariottini [1]

[1] Department of Science, Università degli Studi di Roma "Roma Tre", 00146 Rome, Italy; alessia.leonetti@uniroma3.it (A.L.); paolo.mariottini@uniroma3.it (P.M.)

[2] Department of of Movement Human and Health Sciences, Unit of Biology, Genetics and Biochemistry, Università degli Studi di Roma "Foro Italico", Piazza Lauro De Bosis 15, 00135 Rome, Italy; guglielmo.duranti@uniroma4.it (G.D.); stefania.sabatini@uniroma4.it (S.S.); roberta.ceci@uniroma4.it (R.C.)

* Correspondence: manuela.cervelli@uniroma3.it

Received: 30 November 2017; Accepted: 9 February 2018; Published: 14 February 2018

Abstract: Skeletal muscle comprises approximately 40% of the total body mass. Preserving muscle health and function is essential for the entire body in order to counteract chronic diseases such as type II diabetes, cardiovascular diseases, and cancer. Prolonged physical inactivity, particularly among the elderly, causes muscle atrophy, a pathological state with adverse outcomes such as poor quality of life, physical disability, and high mortality. In murine skeletal muscle C2C12 cells, increased expression of the spermine oxidase (SMOX) enzyme has been found during cell differentiation. Notably, SMOX overexpression increases muscle fiber size, while SMOX reduction was enough to induce muscle atrophy in multiple murine models. Of note, the SMOX reaction product spermidine appears to be involved in skeletal muscle atrophy/hypertrophy. It is effective in reactivating autophagy, ameliorating the myopathic defects of collagen VI-null mice. Moreover, spermidine treatment, if combined with exercise, can affect D-gal-induced aging-related skeletal muscle atrophy. This review hypothesizes a role for SMOX during skeletal muscle differentiation and outlines its role and that of spermidine in muscle atrophy. The identification of new molecular pathways involved in the maintenance of skeletal muscle health could be beneficial in developing novel therapeutic lead compounds to treat muscle atrophy.

Keywords: aging; atrophy; autophagy; oxidative stress; polyamines; skeletal muscle; spermidine; spermine oxidase; transgenic mouse

---

1. Introduction

Skeletal muscle is the largest tissue of the human body and represents 40–50% of body weight, varying according to physiological and pathological conditions [1]. Skeletal muscle atrophy is a frequent and disabling condition. It involves different molecular mechanisms that occur during pathogenesis, hence knowledge of the cellular signal pathways that mediate muscle atrophy is still limited [2].

Polyamines (PAs) are essential for normal cell growth, proliferation, and differentiation, and the tissue levels of individual PAs are maintained and buffered via complex regulatory mechanisms [3,4]. The functional roles of the natural PAs, putrescine (Put), spermidine (Spd), and spermine (Spm), are under active investigation in broad research areas, from neuroscience to cancer and biochemistry [5–9]. Two key enzymes, ornithine decarboxylase and S-adenosylmethionine decarboxylase, control PA biosynthesis. PA catabolism is finely regulated by the enzymes N^1-acetyltransferase, polyamine oxidase, and spermine oxidase (SMOX) (Figure 1).

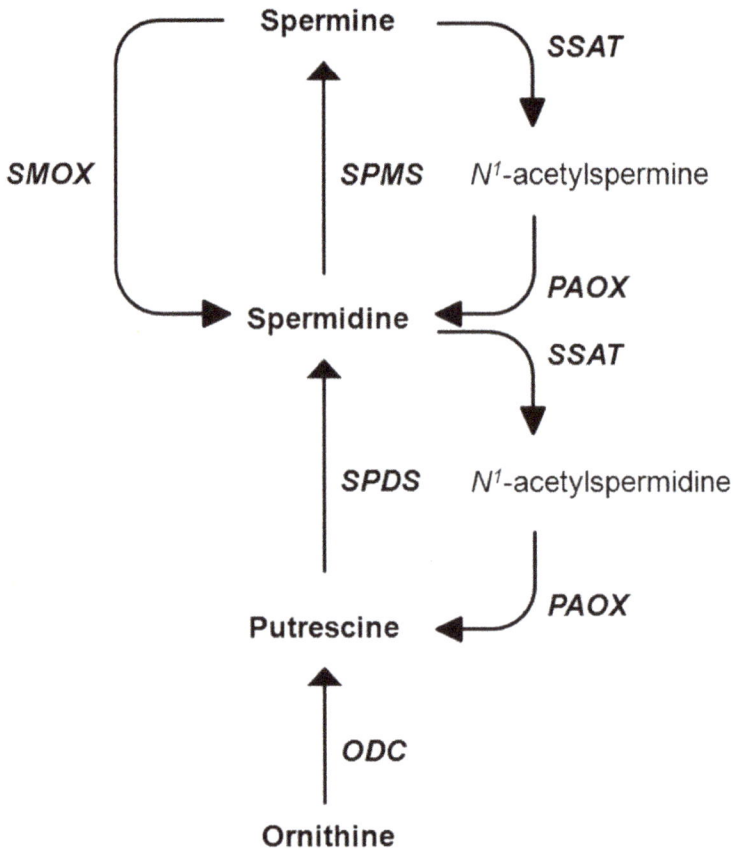

Figure 1. Polyamine metabolism. Schematic representation of mammalian polyamine metabolism showing enzyme network and substrate interconversion pathways. ODC: ornithine decarboxylase; SSAT: spermidine/spermine N^1-acetyltransferase; PAOX: polyamine oxidase; SMOX: spermine oxidase; SPMS: spermine synthase; SPDS: spermidine synthase.

Polyamine levels can be altered by physiological stimuli or by inhibitors or inducers of PA metabolic enzymes. Animal models with genetically altered PA synthesis or catabolism offer a versatile way to affect PA homoeostasis in different tissues and for a prolonged period. These models help to define the physiological importance of PAs by offering tools to develop treatment therapies for pathophysiological conditions derived from deregulated PA homeostasis [10]. Given the role of PAs in the development of many tissues, numerous studies have examined the relation between PAs and skeletal muscle atrophy and hypertrophy. The importance of PAs in muscle disease is highlighted by an alteration of their levels in muscular fibers undergoing degeneration and regeneration [10]. Altered PA levels are present in muscles of patients suffering from Duchenne muscular dystrophy, a pathology characterized by muscle fiber atrophy [11]. Limb girdle dystrophy patients, who display weakness and wasting of the muscles in the arms and legs, show higher levels of PAs in skeletal muscle [11,12]. Similarly, in mouse and hamster models of muscle dystrophy, PA content is altered compared to control muscles [12]. Although the strong association between PA levels and muscle mass is evident, the potential mechanism by which PAs regulate muscle growth is still unclear [10]. This review focuses on the role of SMOX in skeletal muscle pathophysiology, underlining its role in myogenesis and muscle atrophy. The use of a Total-SMOX animal model could help to find new therapies able to counteract physiological atrophy due to aging and pathology.

2. Skeletal Muscle Differentiation

Skeletal muscle differentiation (myogenesis) is a finely regulated process involving a cascade of muscle-specific genes whose expression is coordinated in a timely manner to cell cycle withdrawal and synthesis of muscle contractile proteins. During embryogenesis, muscle fibers are established through a highly ordered multistep process leading from mononucleated-undifferentiated cells (myoblasts) to polynuclear cells (myotubes) [13]. The process is predominantly regulated by myogenic regulatory factors (MRFs) of the basic helix-loop-helix family of transcription factors including MyoD, Myf5, myogenin, and MRF4, together with other transcription factors such as paired box 3 (Pax3) and paired box 7 (Pax7) [14]. MyoD and Myf5, expressed before the onset of myogenic differentiation, promote proliferation and differentiation of myogenic progenitor cells into myoblasts [15], while myogenin plays an important role in the differentiation of myoblasts into myotubes and MRF4 participates in differentiation and cell fate determination [16] (Figure 2).

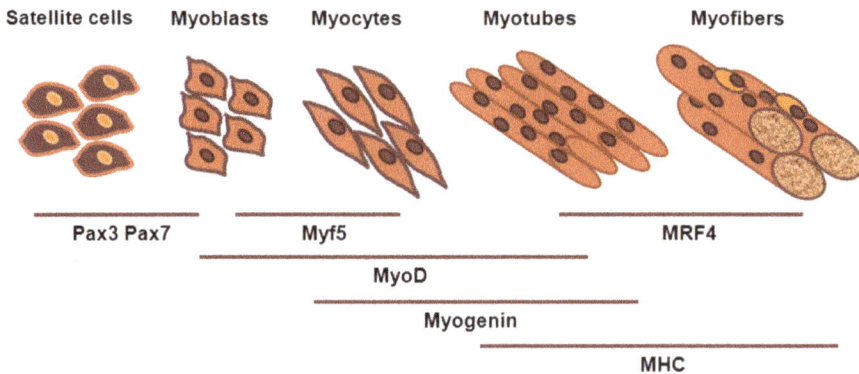

Figure 2. Schematic illustration of the skeletal muscle differentiation process. During myogenesis, Pax3 and Pax7 are activated in quiescent progenitors. Then, progenitor cells differentiate into proliferating muscle precursor cells (myoblasts) and Myf5, MyoD, and myogenin expression stimulates myoblasts to differentiate into myotubes. The terminal stage of differentiation is mediated by the activation of genes responsible for muscle fiber (myofiber) architecture and functionality such as MHCs. Pax: paired box; Myf5: myogenic factor 5; MyoD: myogenic factor 3; MRF4: myogenic regulatory factor 4; MHC: myosin heavy chain.

It must be pointed out that exiting from the cell cycle is a critical regulatory event for successful myogenic differentiation. The elevation of MyoD expression in proliferating myoblasts induces transcriptional upregulation of cell-cycle inhibitors, such as the cyclin-dependent kinase inhibitor p21, that play a fundamental role in establishing the post-mitotic state in skeletal muscle, completing the differentiation process [17,18]. After cell cycle arrest, late differentiation markers such as myosin heavy chain (MHC) are induced [19] (Figure 2).

During adulthood, constant muscle remodeling is possible due to the presence of a specialized population of myogenic progenitors, the satellite cells, placed underneath the myofiber basal lamina. These cells, mitotically quiescent, act as muscle stem cells, allowing the repair and maintenance of myofibers. Upon damage or stress, satellite cells divide asymmetrically: some of them reconstitute the pool of quiescent satellite stem cells and others differentiate into myoblasts. Finally, they fuse together to form new myotubes or fuse with damaged myotubes to repair them [19]. The mouse C2C12 muscle cell line is a useful model system to study the differentiation process. In vitro, myogenic differentiation can be obtained by serum deprivation from myoblast cultures. In high serum, myoblasts proliferate, while after serum removal, they go into an early differentiation stage. Later, cells begin to fuse, forming multinucleated myotubes positive for the characteristic muscle-specific protein MHC [20–22].

Many molecules are highly regulated during the phenotypic conversion of rapidly dividing C2C12 myoblasts into fully differentiated post-mitotic myotubes. Among them, SMOX, the most recently characterized polyamine catabolic enzyme [23], has been shown to be modulated during C2C12 differentiation [24].

3. Spermine Oxidase and Muscle Tissue

In animal cells, SMOX can be considered a multitasking enzyme, primarily involved in controlling PA metabolism, and its substrate Spm and reaction product Spd are ubiquitous polycations that have several important control functions in cells, ranging from basic DNA synthesis to regulation of cell proliferation and differentiation [25].

The SMOX gene is highly expressed in muscle tissue, as demonstrated by Cervelli et al. [26], who reported a high level of both transcript and enzymatic activity. Interestingly, in C2C12 myoblast cultures, after serum removal, SMOX messenger RNA (mRNA) activity was downregulated in the early stage of differentiation. Successively, SMOX was induced in a time-dependent manner, at the level of both transcription and enzymatic activity. Overall, the pattern of SMOX expression profile in the C2C12 muscle cell line showed high levels in proliferating myoblasts followed by low levels in differentiating myoblasts and high levels again in myocytes/myotubes [24].

The relatively high expression levels of SMOX in proliferating cells could promote cell expansion and/or survival. It is known that SMOX can be rapidly induced in response to stress and is responsible for H_2O_2 production in different cell lines and malignant tissues [27]. Thus, a decrease in its activity, and possibly a reduction in the level of H_2O_2, could allow for subsequent cell differentiation, with fully differentiated cells resuming its expression and activity. A failure of SMOX downregulation might result in defective differentiation and potentially contribute to neoplastic transformation. It is interesting that SMOX expression levels are high in several tumor cell types [28,29]. Hence, there needs to be investigation of mechanisms through which SMOX expression can be re-established, as occurs at the end of the differentiation process. Interestingly, a recent study suggested that muscle atrophy induced by limb immobilization, fasting, muscle denervation, and aging could be related to a significant reduction of SMOX expression, while on the other hand, when SMOX was overexpressed, muscle fiber size increased. Furthermore, SMOX overexpression resulted in a decrease of the genes that promote muscle atrophy and, conversely, an increase in the expression of genes that help to maintain muscle mass [30]. Overall, these results highlight the importance of SMOX in the pathophysiology of skeletal muscle. Remarkably, involvement of p21 has been observed in atrophy conditions: the induction of p21 actively promotes muscle atrophy, leading to reduced SMOX expression; on the other hand, a relatively low level of p21 allows a higher level of SMOX expression, which helps to maintain basal skeletal muscle gene expression and fiber size [30]. It has been reported that p21 plays an important role in the in vivo healing process in muscular injury [31], and moreover, it has been shown that p21 protein is modulated during C2C12 differentiation: it increases at cell cycle exit and entrance into differentiation and declines in a time-dependent manner thereafter [32].

In Figure 3, a hypothetical scheme shows the possible relation between SMOX and p21 during myogenesis.

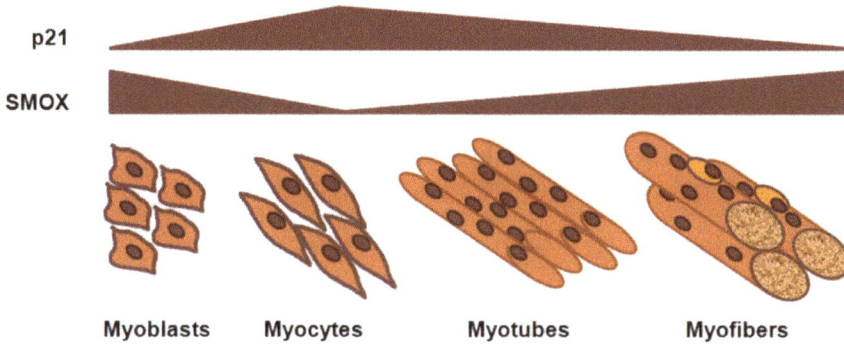

Figure 3. Hypothetical representation of SMOX regulation during myogenesis. During the early stage of muscle differentiation, p21 expression increases, leading to reduced SMOX expression. During the differentiation process, p21 declines in a time-dependent manner, allowing SMOX expression.

Spermine oxidase catalyzes the direct back-conversion of Spm to Spd, 3-aminopropanal (3-AP), which is non-enzymatically converted to acrolein, and hydrogen peroxide (H_2O_2) (Figure 4), and each of these products can affect muscle tissue in various ways, as described in the following sections.

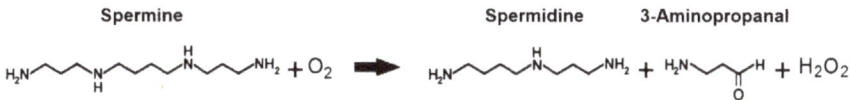

Figure 4. Enzymatic reaction catalyzed by SMOX. The physiological substrate spermine (Spm) is oxidized by the SMOX enzyme into spermidine (Spd) with the production of 3-aminopropanal and hydrogen peroxide.

4. Hydrogen Peroxide and Muscle Tissue

The reactive oxygen species (ROS) H_2O_2 produced by SMOX is a two-electron, nonradical oxidant molecule that is stable and freely diffuses within and between cells, and therefore acts as a signaling molecule. Numerous studies on myogenesis have indicated that muscle development is particularly sensitive to environmental and endogenous H_2O_2 levels. On the whole, H_2O_2 is directly engaged in modulating the expression of several enzymes related to the cellular redox state and plays a wide range of important roles in a variety of cells in a concentration-dependent manner [33–36]. It is to be noted that at low concentrations, H_2O_2 can activate various enzymes, such as phosphatases, modulating cell signaling; on the other hand, at high concentrations, it causes oxidative stress, leading to irreversible cell damage [37].

Some studies, conducted by treating cells with H_2O_2 exogenously, provide evidence for its negative effect on myoblast differentiation. Treatment with H_2O_2 slowed differentiation [38] and reduced myogenin and MHC protein content and creatine kinase activity, as well as troponin I gene transcription, in a dose-dependent manner [39]. The inhibition of myotube formation was reversible when the powerful antioxidant *N*-acetylcysteine had been previously added to the culture medium [39]. It was also reported that H_2O_2 administration markedly reduced *Myf5*, *MRF4* gene, and myogenin expression [40]. Another study showed that during the early stages of C2C12 myoblast differentiation, mildly toxic treatment with H_2O_2 resulted in the depletion of glutathione (GSH), the main thiol antioxidant, leading to further intracellular accumulation of ROS. The oxidative environment favored the activation of nuclear factor-kappa B (NF-κB), a redox-sensitive transcription factor, thus contributing to the lower expression of MyoD and impaired myogenesis [41].

On the other hand, other studies have shown a crucial role for endogenous ROS concentrations during skeletal muscle differentiation. It has been reported that an increase in skeletal muscle NADPH oxidase isoform 2 (NOX2) activity during differentiation leads to a rise in superoxide anion O_2^-, a molecule quickly converted by dismutation into H_2O_2. In line with the positive role of ROS, this increase seems to be crucial to muscle differentiation via NF-κB/inducible nitric oxide synthase (iNOS) pathway activation [42] and to the promotion of skeletal muscle precursor cell proliferation [43]. Furthermore, it has been observed that enhancement of the endogenous H_2O_2 level can regulate the cellular GSH redox balance from the very early stage to the fully differentiated myotube formation stage [44]. On the whole, these data clearly show that myogenesis is a process very sensitive to the intracellular redox environment, and at the same time supports the notion that H_2O_2 can have different outcomes depending on its intracellular concentration.

5. Acrolein and Muscle Tissue

Numerous conditions, such as oxidative stress, inflammation, alcoholic myopathy, and renal failure in which acrolein is increased, have been associated with muscle deterioration and dysfunction, resulting in disease. Moreover, increased muscle catabolism has been linked to an exogenous source of acrolein due to cigarette smoking [45]. Exposure of skeletal myotubes to cigarette smoke stimulates muscle catabolism via increased oxidative stress, activation of p38 mitogen-activated protein kinases (p38MAPK), and upregulation of muscle-specific E3 ubiquitin ligases [45]. In addition, acrolein treatment was sufficient to induce an increase of free radicals, activation of p38 MAPK, up-regulation of the muscle-specific E3 ligases atrogin-1 and MuRF1, degradation of MHC, and atrophy of myotubes. Inhibition of p38MAPK by SB203580 abolished acrolein-induced muscle catabolism [46]. These studies demonstrated that acrolein is able to activate a signaling cascade, inducing muscle catabolism in skeletal myotubes. Acrolein can also react with amino acid residues in proteins, consequently modifying protein function and inducing apoptosis [47] or tissue damage such as brain infarction [48,49]. Notwithstanding that high levels of acrolein within cells have been linked to toxicity, it has recently been demonstrated that SMOX plays a central role in the formation of bile canalicular lumen in liver cells by activating the protein kinase B (AKT) pathway through acrolein production [50].

6. Spermidine and Muscle Tissue

Considering the new emerging role of SMOX in muscle physiology, it is of primary importance to have full knowledge of this enzyme and the role of its catabolic product Spd. In several model systems such as yeast, flies, worms, and human immune cells, Spd levels decrease during aging, and by giving Spd as a dietary supplement, lifespan is extended [51]. In fact, a Spd-rich diet postpones age-related phenomena, such as the progressive decline of locomotor activity in flies [52]. In this variety of model organisms, Spd was found to suppress several aging-associated parameters, such as overproduction of ROS and the level of necrotic cell death [51]. Recently, Spd has been of great interest in the prevention or treatment of muscle diseases, since it may play a role in skeletal muscle atrophy/hypertrophy [53]. Aging brings a loss of skeletal muscle, and it has been shown that Spd cellular concentrations decrease during age progression [51]. Autophagy contributes to age-related degeneration processes, since it has been proven to decrease with aging. In physiological conditions, autophagy has a critical role, acting as a cell housekeeper by degrading damaged or unnecessary organelles and allowing the recovery of metabolites under nutrient starvation [54,55]. Deregulated autophagy contributes to neurodegenerative disorders, as well as liver, heart, and muscle diseases [56]. Variations in autophagic flux have been demonstrated to affect muscle homeostasis and body metabolic state [56]. Different studies have analyzed the correlation between aging, Spd, and autophagy and led to the identification of Spd as a strong and specific inducer of autophagy [56]. Through the autophagy mechanism, Spd extends lifespan by triggering epigenetic deacetylation of histone H3 through inhibition of histone acetyltransferases, suppressing oxidative stress and necrosis [51]. This mechanism provides protection from the aging process for several tissues, including heart, brain, and skeletal muscle, thereby endorsing longevity [1]. In an aged population, regular and proper exercise is used as a stimulus for muscle adaptation, attenuating the loss of skeletal muscle [57]. Autophagy attenuation that occurs with aging has been shown to be reduced with exercise training, thus endurance exercise enhances autophagic signaling in aged mice [1]. The coupled use of a specific autophagic inducer such as Spd and regular exercise could activate autophagy, establishing a correct health level for the maintenance of skeletal muscle [1].

Extended bed rest or post-surgery immobilization, resulting in skeletal muscle unloading, leads to skeletal muscle atrophy. In addition, exposure to a microgravity environment during extended space flights brings skeletal muscle disuse/atrophy similar to what is observed in prolonged bed rest or post-surgery recovery [53]. Abukhalaf et al. [53] reported that skeletal muscle polyamine levels appear to be influenced by microgravity (hind limb suspension) in a fiber-type-specific manner. Unloading-induced atrophy was accompanied by a dramatic decrease in Spd level (68%) in slow-twitch (type I) muscle fibers, but a slight increase (14%) in fast-twitch (type II) ones, when comparing unloaded and control animals [53]. No significant changes were observed in Spm levels in either type of muscle fibers [53]. On the other hand, individuals suffering from myasthenia gravis display high Spd and Spm levels, associated with muscle weakness and severe atrophy [58]. Further contrasting results were obtained by Bongers et al. [30], who did not detect any change of Spd levels in limb immobilization, suggesting that enzymes other than SMOX or transporters are able to maintain Spd levels when SMOX activity is reduced. Taking into account all these works, it can be seen that the direct role of polyamines, and in particular Spd, needs to be clarified in detail, but it is necessary to refer to the specific type of muscle fiber in light of the results obtained by Abukhalaf et al. [53].

A frequently used model to induce aging-related damage in vivo is administration of D-galactose (D-gal) [59], which causes accumulation of ROS with final oxidative stress [60].

Excessive apoptosis and deficient autophagy may cause skeletal muscle atrophy due to pathological events or the aging process, thus promoting cell death and disease progression [1]. Recently, AMP-activated protein kinase (AMPK) has been found to phosphorylate forkhead box O3 (FOXO3a) on Ser588 [61]. The FOXO3a signaling pathway induces the transcriptional activation of autophagy-related genes that oversee protein degradation. FOXO3a was found to be necessary for the reduction of skeletal muscle atrophy induced by D-gal and to maintain proliferation in aging skeletal muscle cells by Spd and exercise. These results suggest that exercise and Spd may share mediators that act on similar pathways in varying degrees, generating a synergistic effect for delaying skeletal muscle senescence [1]. Therefore, Spd, by activating the AMPK-dependent autophagy pathway, may decrease endoplasmic reticulum stress and reduce apoptosis [62].

Skeletal muscle atrophy is a state that not only characterizes a physiological condition in the aged population or in extended bed rest, but also is present in several pathologies with different etiologies.

Spermidine was found to ameliorate myopathic defects in the animal model of Ullrich congenital muscular dystrophy (UCMD) and Bethlem myopathy (BM) (*col6a1*$^{-/-}$ mice) by reactivating autophagy in skeletal muscle [56]. In this model, there is overactivation of AKT, which causes defective autophagy by activating the mechanistic target of rapamycin, which inhibits the transcription of genes under FOXO [56]. Ineffective autophagy brings an accumulation of damaging organelles in the myofibers, which degenerate with time. However, this process is reversible through dietary and pharmacological approaches [63]. The beneficial effects of Spd administration in *col6a1*$^{-/-}$ mice are linked to its ability to reactivate autophagic flux [64], as demonstrated by a significant increase of LC3B and autophagosome formation [56]. Spermidine seems to act on AKT [56]. AKT kinase negatively regulates, by phosphorylation, the activity of FOXO transcriptional factors. Low levels of phosphorylated AKT activate FOXO [65]. Spermidine is able to reduce AKT phosphorylation, which triggers the translocation of FOXO transcriptional factors into the nucleus, promoting autophagy [56]. BM and UCMD patients display respiratory insufficiency due to a loss of diaphragm function [66], which is also the most affected muscle in *col6a1*$^{-/-}$ mice, where it shows a high incidence of apoptotic myofibers [63]. Spermidine was able to rescue the aspects of both BM and UCMD.

FOXO proteins are key transcription factors regulated by different post-translational modifications, and among these, inhibitory acetylation by the histone acetyltransferase EP300 has been described in detail [67]. Since Spd also acts as a histone acetyltransferase inhibitor, and one of its known targets is EP300 [68], it is probable that Spd controls FOXO activity at multiple levels and determines a permissive condition for its activity by the action of AMPK, AKT, and EP300 (Figure 5).

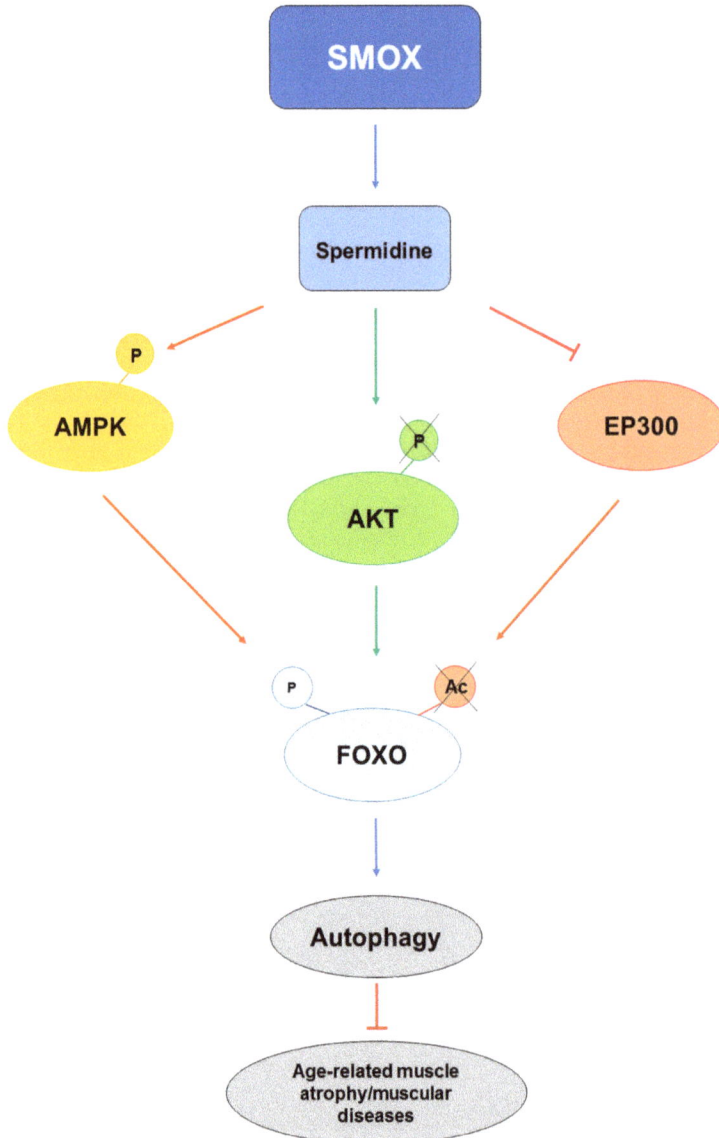

Figure 5. Schematic representation of spermidine effects of on forkhead box (FOXO) activity. The effect of spermidine on the attenuation of age-related skeletal muscle atrophy and diseases, through regulating autophagy via 5′ AMP-activated protein kinase (AMPK)/Protein Kinase B (AKT)/E1A binding protein p300 (EP300)-FOXO signal pathways.

7. Polyamines and Muscle Diseases

The involvement of Spd, and of PAs in general, has also been demonstrated in amyotrophic lateral sclerosis (ALS) and Duchenne muscular dystrophy. Results from ALS patients and SOD1^{G93A} mice, an animal model for ALS pathology, showed deregulated PA levels in plasma, skeletal muscle, and cerebral cortex. The many roles of PAs include ALS-relevant processes such as protection against stress induced by ROS [69], modulation of glutamate ion channel receptors [70], and induction of autophagy [51]. Similarly, muscular fibrosis after denervation displays a constant rise in PA concentration [71].

Duchenne muscular dystrophy is a neuromuscular disease caused by mutation(s) in the dystrophin gene. It is characterized by skeletal muscle and cardiopulmonary complications, resulting in shorter life expectancy. The mouse model that better represents this disease is a double-mutant mouse (*dmd/utrn* double mutant (*mdx-dm*)), where both dystrophin and utrophin genes are mutated [72]. It is hypothesized that the rapid deterioration of muscles is correlated to reduced action of androgens [73]. In a recent study, GTx-026 a nonsteroidal selective-androgen receptor modulator that selectively builds muscle and bone, was used to treat *mdx-dm* mice, and it was able to increase muscle mass, function, and survival [72]. While GTx-026 failed to reverse the genes altered by dystrophin knockdown, it regulated the expression of several genes of the PA pathway. Interestingly, GTx-026 significantly increased SMOX by two- to threefold. Moreover, genes belonging to the Spd pathway were also upregulated by GTx-026, confirming the crucial role of Spd in enhancing cellular lifespan and proposing SMOX as a key gene in muscle diseases [72].

8. The Transgenic Mouse Line: *Total-Smox*

Spermine oxidase is a significant positive regulator of muscle gene expression and fiber size. Recently, a Cre/loxP-based double transgenic mouse line overexpressing the Smox gene in all organs was engineered and named Total-Smox. Transgenic Green Fluorescent Protein (GFP)-Smox (formerly JoSMOX) mice described by Cervelli et al. [6] were crossed with Total-CRE mice reported by [74], to obtain the Total-Smox genetic line (Figure 6). This new experimental model was further genetically stabilized by back-crossing Total-Smox mice 10 times with C57BL/6 mice. SMOX overexpression in transgenic individuals can be detected in all tissues by β-Galoctiside (3-Gal) staining (Figure 6) and reverse-transcriptase/PCR analysis. As expected, SMOX enzyme activity was higher in all the organs of Total-Smox mice in comparison to their syngenic littermates, including skeletal muscle and heart, as demonstrated by [75].

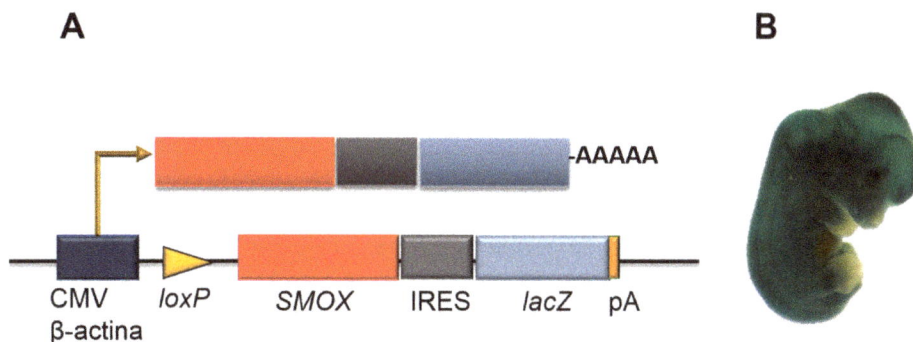

Figure 6. *Total-Smox* mouse line. (**A**) Scheme of the genetic construct of the *Total-Smox* mouse line upon recombination of the loxP sites by Cre recombinase. The β-actin/Cytomegalovirus (CMV) fusion promoter drives the ubiquitous expression of the *SMOX* gene and the *lacZ* reporter gene. IRES: internal ribosome entry site. (**B**) LacZ staining of *Total-Smox* embryo.

In the literature, it is known that muscle diseases and aging-related pathologies display high oxidative stress by accumulating H_2O_2 in the muscles [76]. In the Total-Smox mouse model, chronic H_2O_2 production due to SMOX overexpression leads to an imbalanced cellular redox state in both types of muscle tissue. Skeletal muscle displays lower oxidative damage compared to the heart, evoking a different redox adaptation through upregulation of the enzymatic antioxidant system [75]. This tissue shows a significant decline in the ratio of GSH reduced and oxidized due to a decrease in the total amount of GSH. SMOX overexpression increases glutathione S-transferase activity, since it is an enzyme induced under oxidative stress and is responsible for the detoxification of molecules through the formation of S-conjugates [75]. SMOX overexpression in the skeletal muscle of Total-Smox mice also causes a concomitant increase of catalase, another detoxifying enzyme, indicating that there is a counteracting action against the high H_2O_2 production. Moreover, the amount of Spd has been found to be elevated in Total-Smox mice compared to control animals, while the levels of Put and Spm were not altered. Spermidine has been proven to be essential in muscle physiology by promoting autophagy, counteracting aging, and extending lifespan [1].

Considering the important role of Spd in many pathologies and that Spd increases in the *Total-Smox* mice model while no changes are reported for the other PAs, it would be interesting to cross this mouse line with animal models of different diseases such as Ullrich congenital muscular dystrophy and Bethlem myopathy (*col6a1$^{-/-}$* mice), amyotrophic lateral sclerosis (SOD1^{G93A} mice), and Duchenne muscular dystrophy (*mdx-dm* mice), to evaluate the possibility of rescuing the phenotype.

9. Conclusions and Future Perspectives

This review points out that not only does SMOX take part in cancer and neurological disorders, but it is also involved in skeletal muscle pathophysiology. New evidence for the physiological roles of the PA pathway in atrophy could represent a major area for future research [77]. The prevalence rate of sarcopenia is up to 33% in elderly people, and the number is expected to increase due to lifestyle, dietary habits, and aging [78]. It has been known that PA levels and related enzyme activities decline during aging; however, until recently, it was not clear how alterations in PA metabolism could affect the aging process [79]. Several experiments have highlighted the role of Spd as an autophagy inducer that ameliorates age-related muscle atrophy [80]. Different mouse models of disease have brought new insights as to the possibility of Spd as a therapeutic food option. This approach could be used to design innovative treatments, with Spd alone or in combination with other approaches, leading to clinical trials. In perspective, this nutraceutical-based, autophagy-inducing approach could be applied to different therapies, and could also be extended to other inherited muscle pathologies involving defective activity of the autophagy machinery. Of note, Spd is the reaction product of SMOX, which in turn has been demonstrated to help in maintaining muscle mass and counteracting the expression of genes that promote muscle atrophy. Understanding how the SMOX/p21 axis and its product Spd influence muscle gene expression and how they are linked to autophagy and muscular atrophy during aging is a new field of investigation. Considering that SMOX is highly expressed in skeletal muscle and regulates the amount of PAs, it can become a key gene to target for treatment of diseases such as Duchenne muscular dystrophy.

In perspective, *Total-Smox* mice could be considered as a valuable genetic animal model to investigate the role of SMOX and Spd in muscular physiology, shedding light on new therapies to test in several muscle diseases.

Acknowledgments: The authors wish to thank Rosetta Ponzo for revision of the English text. This work was supported by grants from the Università degli Studi di Roma "Foro Italico" (DIP DSS10-15 and DIP DSS08-16) to S.S. and the Università degli Studi di Roma "Roma Tre" contribution to the laboratories (CAL/2016) to M.C. and P.M. and by the Ph.D. School (Department of Science) contribution 2017 to A.L.

Author Contributions: All authors contributed equally to the conception of this review.

Conflicts of Interest: The authors declare no conflict of interest.

References

1. Fan, J.; Yang, X.; Li, J.; Shu, Z.; Dai, J.; Liu, X.; Li, B.; Jia, S.; Kou, X.; Yang, Y.; et al. Spermidine coupled with exercise rescues skeletal muscle atrophy from D-gal-induced aging rats through enhanced autophagy and reduced apoptosis via AMPK-FOXO3a signal pathway. *Oncotarget* **2017**, *8*, 17475–17490. [CrossRef] [PubMed]

2. Ali, S.; Garcia, J.M. Sarcopenia, cachexia and aging: Diagnosis, mechanisms and therapeutic options-a mini-review. *Gerontology* **2014**, *60*, 294–305. [CrossRef] [PubMed]

3. Rea, G.; Bocedi, A.; Cervelli, M. What is the biological function of the polyamines? *IUBMB Life* **2004**, *56*, 167–169. [CrossRef] [PubMed]

4. Cervelli, M.; Angelucci, E.; Germani, F.; Amendola, R.; Mariottini, P. Inflammation, carcinogenesis and neurodegeneration studies in transgenic animal models for polyamine research. *Amino Acids* **2014**, *46*, 521–530. [CrossRef] [PubMed]

5. Cervelli, M.; Bellavia, G.; Fratini, E.; Amendola, R.; Polticelli, F.; Barba, M.; Rodolfo, F.; Signore, F.; Gucciardo, G.; Grillo, R.; et al. Spermine oxidase (SMO) activity in breast tumor tissues and biochemical analysis of the anticancer spermine analogues BENSpm and CPENSpm. *BMC Cancer* **2010**, *10*, 555–564. [CrossRef] [PubMed]

6. Cervelli, M.; Bellavia, G.; D'amelio, M.; Cavallucci, V.; Moreno, S.; Berger, J.; Nardacci, R.; Marcoli, M.; Maura, G.; Piacentini, M.; et al. A New Transgenic Mouse Model for Studying the Neurotoxicity of Spermine Oxidase Dosage in the Response to Excitotoxic Injury. *PLoS ONE* **2013**, *8*, e64810. [CrossRef] [PubMed]

7. Cervelli, M.; Pietropaoli, S.; Signore, F.; Amendola, R.; Mariottini, P. Polyamines metabolism and breast cancer: State of the art and perspectives. *Breast Cancer Res. Treat.* **2014**, *148*, 233–248. [CrossRef] [PubMed]

8. Cervelli, M.; Leonetti, A.; Cervoni, L.; Ohkubo, S.; Xhani, M.; Stano, P.; Federico, R.; Polticelli, F.; Mariottini, P.; Agostinelli, E. Stability of spermine oxidase to thermal and chemical denaturation: Comparison with bovine serum amine oxidase. *Amino Acids* **2016**, *48*, 2283–2291. [CrossRef] [PubMed]

9. Cervetto, C.; Vergani, L.; Passalacqua, M.; Ragazzoni, M.; Venturini, A.; Cecconi, F.; Berretta, N.; Mercuri, N.; D'Amelio, M.; Maura, G.; et al. Astrocyte-Dependent Vulnerability to Excitotoxicity in Spermine Oxidase-Overexpressing Mouse. *Neuromol. Med.* **2016**, *18*, 50–68. [CrossRef] [PubMed]

10. Lee, N.K.; MacLean, H.E. Polyamines, androgens, and skeletal muscle hypertrophy. *J. Cell. Physiol.* **2011**, *226*, 1453–1460. [CrossRef] [PubMed]

11. Kaminska, A.M.; Stern, L.Z.; Russell, D.H. Altered muscle polyamine levels in human neuromuscular diseases. *Ann. Neurol.* **1981**, *9*, 605–607. [CrossRef] [PubMed]

12. Rudman, D.; Kutner, M.H.; Chawla, R.K.; Goldsmith, M.A. Abnormal polyamine metabolism in hereditary muscular dystrophies: Effect of human growth hormone. *J. Clin. Investig.* **1980**, *65*, 95–102. [CrossRef] [PubMed]

13. Hernández-Hernández, J.M.; García-González, E.G.; Brun, C.E.; Rudnicki, M.A. The myogenic regulatory factors, determinants of muscle development, cell identity and regeneration. *Semin. Cell Dev. Biol.* **2017**, *72*, 10–18. [CrossRef] [PubMed]

14. Soleimani, V.D.; Punch, V.G.; Kawabe, Y.; Jones, A.E.; Palidwor, G.A.; Porter, C.J.; Cross, J.W.; Carvajal, J.J.; Kockx, C.E.; van IJcken, W.F.; et al. Transcriptional dominance of Pax7 in adult myogenesis is due to high-affinity recognition of homeodomain motifs. *Dev. Cell* **2012**, *22*, 1208–1220. [CrossRef] [PubMed]

15. Gianakopoulos, P.J.; Mehta, V.; Voronova, A.; Cao, Y.; Yao, Z.; Coutu, J.M.; Wang, X.; Waddington, M.S.; Tapscott, S.J.; Skerjanc, I.S. MyoD directly up-regulates premyogenic mesoderm factors during induction of skeletal myogenesis in stem cells. *J. Biol. Chem.* **2011**, *286*, 2517–2525. [CrossRef] [PubMed]

16. Kassar-Duchossoy, L.; Gayraud-Morel, B.; Gomès, D.; Rocancourt, D.; Buckingham, M.; Shinin, V.; Tajbakhsh, S. Mrf4 determines skeletal muscle identity in Myf5: Myod double-mutant mice. *Nature* **2004**, *431*, 466–471. [CrossRef] [PubMed]

17. Ishido, M.; Kami, K.; Masuhara, M. In vivo expression patterns of MyoD, p21, and Rb proteins in myonuclei and satellite cells of denervated rat skeletal muscle. *Am. J. Physiol. Cell Physiol.* **2004**, *287*, 484–493. [CrossRef] [PubMed]

18. Wei, Q.; Paterson, B.M. Regulation of MyoD function in the dividing myoblast. *FEBS Lett.* **2001**, *490*, 171–178. [CrossRef]

19. Zanou, N.; Gailly, P. Skeletal muscle hypertrophy and regeneration: Interplay between the myogenic regulatory factors (MRFs) and insulin-like growth factors (IGFs) pathways. *Cell. Mol. Life Sci.* **2013**, *70*, 4117–4130. [CrossRef] [PubMed]
20. Duranti, G.; La Rosa, P.; Dimauro, I.; Wannenes, F.; Bonini, S.; Sabatini, S.; Parisi, P.; Caporossi, D. Effects of salmeterol on skeletal muscle cells: Metabolic and pro-apoptotic features. *Med. Sci. Sports Exerc.* **2011**, *43*, 2259–2273. [CrossRef] [PubMed]
21. Ceci, R.; Duranti, G.; Rossi, A.; Savini, I.; Sabatini, S. Skeletal muscle differentiation: Role of dehydroepiandrosterone sulfate. *Horm. Metab. Res.* **2011**, *43*, 702–707. [CrossRef] [PubMed]
22. Duranti, G.; Ceci, R.; Sgrò, P.; Sabatini, S.; Di Luigi, L. Influence of the PDE5 inhibitor tadalafil on redox status and antioxidant defense system in C2C12 skeletal muscle cells. *Cell Stress Chaperones* **2017**, *22*, 389–396. [CrossRef] [PubMed]
23. Cervelli, M.; Amendola, R.; Polticelli, F.; Mariottini, P. Spermine oxidase: Ten years after. *Amino Acid* **2012**, *42*, 441–450. [CrossRef] [PubMed]
24. Cervelli, M.; Fratini, E.; Amendola, R.; Bianchi, M.; Signori, E.; Ferraro, E.; Lisi, A.; Federico, R.; Marcocci, L.; Mariottini, P. Increased spermine oxidase (SMO) activity as a novel differentiation marker of myogenic C2C12 cells. *Int. J. Biochem. Cell Biol.* **2009**, *41*, 934–944. [CrossRef] [PubMed]
25. Pegg, A.E. Functions of Polyamines in Mammals. *J. Biol. Chem.* **2016**, *291*, 14904–14912. [CrossRef] [PubMed]
26. Cervelli, M.; Bellini, A.; Bianchi, M.; Marcocci, L.; Nocera, S.; Polticelli, F.; Federico, R.; Amendola, R.; Mariottini, P. Mouse spermine oxidase gene splice variants. Nuclear subcellular localization of a novel active isoform. *Eur. J. Biochem.* **2004**, *271*, 760–770. [CrossRef] [PubMed]
27. Pledgie, A.; Huang, Y.; Hacker, A.; Zhang, Z.; Woster, P.M.; Davidson, N.E.; Casero, R.A., Jr. Spermine oxidase SMO(PAOh1), Not N^1-acetylpolyamine oxidase PAO, is the primary source of cytotoxic H_2O_2 in polyamine analogue-treated human breast cancer cell lines. *J. Biol. Chem.* **2005**, *280*, 39843–39851. [CrossRef] [PubMed]
28. Park, M.H.; Igarashi, K. Polyamines and their metabolites as diagnostic markers of human diseases. *Biomol. Ther. (Seoul)* **2013**, *21*, 1–9. [CrossRef] [PubMed]
29. Murray-Stewart, T.R.; Woster, P.M.; Casero, R.A., Jr. Targeting polyamine metabolism for cancer therapy and prevention. *Biochem. J.* **2016**, *473*, 2937–2953. [CrossRef] [PubMed]
30. Bongers, K.S.; Fox, D.K.; Kunkel, S.D.; Stebounova, L.V.; Murry, D.J.; Pufall, M.A.; Ebert, S.M.; Dyle, M.C.; Bullard, S.A.; Dierdorff, J.M.; et al. Spermine oxidase maintains basal skeletal muscle gene expression and fiber size and is strongly repressed by conditions that cause skeletal muscle atrophy. *Am. J. Physiol. Endocrinol. Metab.* **2015**, *308*, 144–158. [CrossRef] [PubMed]
31. Chinzei, N.; Hayashi, S.; Ueha, T.; Fujishiro, T.; Kanzaki, N.; Hashimoto, S.; Sakata, S.; Kihara, S.; Haneda, M.; Sakai, Y.; et al. p21 deficiency delays regeneration of skeletal muscular tissue. *PLoS ONE* **2015**, *10*, e0125765. [CrossRef] [PubMed]
32. Boonstra, K.; Bloemberg, D.; Quadrilatero, J. Caspase-2 is required for skeletal muscle differentiation and myogenesis. *Biochim. Biophys. Acta* **2018**, *1865*, 95–104. [CrossRef] [PubMed]
33. Higuchi, M.; Dusting, G.J.; Peshavariya, H.; Jiang, F.; Hsiao, S.T.; Chan, E.C.; Liu, G.S. Differentiation of human adipose-derived stem cells into fat involves reactive oxygen species and Forkhead box O1 mediated upregulation of antioxidant enzymes. *Stem Cells Dev.* **2013**, *22*, 878–888. [CrossRef] [PubMed]
34. Huang, L.S.; Jiang, P.; Feghali-Bostwick, C.; Reddy, S.P.; Garcia, J.G.N.; Natarajan, V. Lysocardiolipin acyltransferase regulates TGF-β mediated lung fibroblast differentiation. *Free Radic. Biol. Med.* **2017**, *112*, 162–173. [CrossRef] [PubMed]
35. Subramani, B.; Subbannagounder, S.; Ramanathanpullai, C.; Palanivel, S.; Ramasamy, R. Impaired redox environment modulates cardiogenic and ion-channel gene expression in cardiac-resident and non-resident mesenchymal stem cells. *Exp. Biol. Med.* **2017**, *242*, 645–656. [CrossRef] [PubMed]
36. Katz, A. Role of reactive oxygen species in regulation of glucose transport in skeletal muscle during exercise. *J. Physiol.* **2016**, *594*, 2787–2794. [CrossRef] [PubMed]
37. Kozakowska, M.; Pietraszek-Gremplewicz, K.; Jozkowicz, A.; Dulak, J. The role of oxidative stress in skeletal muscle injury and regeneration: Focus on antioxidant enzymes. *J. Muscle Res. Cell Motil.* **2015**, *36*, 377–393. [CrossRef] [PubMed]
38. Hansen, J.M.; Klass, M.; Harris, C.; Csete, M. A reducing redox environment promotes C2C12 myogenesis: Implications for regeneration in aged muscle. *Cell Biol. Int.* **2007**, *31*, 546–553. [CrossRef] [PubMed]

39. Langen, R.C.; Schols, A.M.; Kelders, M.C.; Van Der Velden, J.L.; Wouters, E.F.; Janssen-Heininger, Y.M. Tumor necrosis factor-α inhibits myogenesis through redox-dependent and independent pathways. *Am. J. Physiol. Cell Physiol.* **2002**, *283*, C714–C721. [CrossRef] [PubMed]

40. Furutani, Y.; Murakami, M.; Funaba, M. Differential responses to oxidative stress and calcium influx on expression of the transforming growth factor-β family in myoblasts and myotubes. *Cell Biochem. Funct.* **2009**, *27*, 578–582. [CrossRef] [PubMed]

41. Ardite, E.; Barbera, J.A.; Roca, J.; Fernández-Checa, J.C. Glutathione depletion impairs myogenic differentiation of murine skeletal muscle C2C12 cells through sustained NF-kappaB activation. *Am. J. Pathol.* **2004**, *165*, 719–728. [CrossRef]

42. Piao, Y.J.; Seo, Y.H.; Hong, F.; Kim, J.H.; Kim, Y.J.; Kang, M.H.; Kim, B.S.; Jo, S.A.; Jo, I.; Jue, D.M.; et al. Nox 2 stimulates muscle differentiation via NF-kappaB/iNOS pathway. *Free Radic. Biol. Med.* **2005**, *38*, 989–1001. [CrossRef] [PubMed]

43. Mofarrahi, M.; Brandes, R.P.; Gorlach, A.; Hanze, J.; Terada, L.S.; Quinn, M.T.; Mayaki, D.; Petrof, B.; Hussain, S.N. Regulation of proliferation of skeletal muscle precursor cells by NADPH oxidase. *Antioxid. Redox Signal.* **2008**, *10*, 559–574. [CrossRef] [PubMed]

44. Ding, Y.; Choi, K.J.; Kim, J.H.; Han, X.; Piao, Y.; Jeong, J.H.; Choe, W.; Kang, I.; Ha, J.; Forman, H.J.; et al. Endogenous hydrogen peroxide regulates glutathione redox via nuclear factor erythroid 2-related factor 2 downstream of phosphatidylinositol 3-kinase during muscle differentiation. *Am. J. Pathol.* **2008**, *172*, 1529–1541. [CrossRef] [PubMed]

45. Rom, O.; Kaisari, S.; Aizenbud, D.; Reznick, A.Z. Identification of possible cigarette smoke constituents responsible for muscle catabolism. *J. Muscle Res. Cell Motil.* **2012**, *33*, 199–208. [CrossRef] [PubMed]

46. Rom, O.; Kaisari, S.; Aizenbud, D.; Reznick, AZ. The effects of acetaldehyde and acrolein on muscle catabolism in C2 myotubes. *Free Radic. Biol. Med.* **2013**, *65*, 190–200. [CrossRef] [PubMed]

47. Nakamura, M.; Tomitori, H.; Suzuki, T.; Sakamoto, A.; Terui, Y.; Saiki, R.; Dohmae, N.; Igarashi, K.; Kashiwagi, K. Inactivation of GAPDH as one mechanism of acrolein toxicity. *Biochem. Biophys. Res. Commun.* **2013**, *430*, 1265–1271. [CrossRef] [PubMed]

48. Hirose, T.; Saiki, R.; Uemura, T.; Suzuki, T.; Dohmae, N.; Ito, S.; Takahashi, H.; Ishii, I.; Toida, T.; Kashiwagi, K.; et al. Increase in acrolein-conjugated immunoglobulins in saliva from patients with primary Sjögren's syndrome. *Clin. Chim. Acta* **2015**, *450*, 184–189. [CrossRef] [PubMed]

49. Uemura, T.; Suzuki, T.; Saiki, R.; Dohmae, N.; Ito, S.; Takahashi, H.; Toida, T.; Kashiwagi, K.; Igarashi, K.I. Activation of MMP-9 activity by acrolein in saliva from patients with primary Sjögren's syndrome and its mechanism. *Int. J. Biochem. Cell Biol.* **2017**, *88*, 84–91. [CrossRef] [PubMed]

50. Uemura, T.; Takasaka, T.; Igarashi, K.; Ikegaya, H. Spermine oxidase promotes bile canalicular lumen formation through acrolein production. *Sci. Rep.* **2017**, *7*, 14841. [CrossRef] [PubMed]

51. Eisenberg, T.; Knauer, H.; Schauer, A.; Büttner, S.; Ruckenstuhl, C.; Carmona-Gutierrez, D.; Ring, J.; Schroeder, S.; Magnes, C.; Antonacci, L.; et al. Induction of autophagy by spermidine promotes longevity. *Nat. Cell Biol.* **2009**, *11*, 1305–1314. [CrossRef] [PubMed]

52. Minois, N. Molecular basis of the 'anti-aging' effect of spermidine and other natural polyamines—A Mini Review. *Gerontology* **2014**, *60*, 319–326. [CrossRef] [PubMed]

53. Abukhalaf, I.K.; Von Deutsch, D.A.; Wineski, L.E.; Silvestrov, N.A.; Abera, S.A.; Sahlu, S.W.; Potter, D.E. Effect of hindlimb suspension and clenbuterol treatment on polyamine levels in skeletal muscle. *Pharmacology* **2002**, *65*, 145–154. [CrossRef] [PubMed]

54. Levine, B.; Kroemer, G. Autophagy in the pathogenesis of disease. *Cell* **2008**, *132*, 27–42. [CrossRef] [PubMed]

55. Rogov, V.; Deotsch, V.; Johansen, T.; Kirkin, V. Interactions between autophagy receptors and ubiquitin-like proteins form the molecular basis for selective autophagy. *Mol. Cell* **2014**, *53*, 167–178. [CrossRef] [PubMed]

56. Chrisam, M.; Pirozzi, M.; Castagnaro, S.; Blaauw, B.; Polishchuck, R.; Cecconi, F.; Grumati, P.; Bonaldo, P. Reactivation of autophagy by spermidine ameliorates the myopathic defects of collage, I-null mice. *Autophagy* **2015**, *11*, 2142–2152. [CrossRef] [PubMed]

57. Sanchez, A.M.; Bernardi, H.P.G.; Candau, R.B. Autophagy is essential to support skeletal muscle plasticity in response to endurance exercise. *Am. J. Physiol. Regul. Integr. Comp. Physiol.* **2014**, *307*, 956–969. [CrossRef] [PubMed]

58. Szathmáry, I.; Selmeci, L.; Szobor, A.; Molnár, J. Altered polyamine levels in skeletal muscle of patients with myasthenia gravis. *Clin. Neuropathol.* **1994**, *13*, 181–184. [PubMed]

59. Song, X.; Bao, M.; Li, D.; Li, Y.M. Advanced glycation in D-galactose induced mouse aging model. *Mech. Ageing Dev.* **1999**, *108*, 239–251. [CrossRef]

60. Du, Z.; Yang, Q.; Liu, L.; Li, S.; Zhao, J.; Hu, J.; Liu, C.; Qian, D.; Gao, C. NADPH oxidase 2-dependent oxidative stress, mitochondrial damage and apoptosis in the ventral cochlear nucleus of D-galactose-induced aging rats. *Neuroscience* **2015**, *286*, 281–292. [CrossRef] [PubMed]

61. Greer, E.L.; Oskoui, P.R.; Banko, M.R.; Maniar, J.M.; Gygi, M.P.; Gygi, S.P.; Brunet, A. The energy sensor AMP-activated protein kinase directly regulates the mammalian FOXO3 transcription factor. *J. Biol. Chem.* **2007**, *282*, 30107–30119. [CrossRef] [PubMed]

62. Tirupathi Pichiah, P.B.; Suriyakalaa, U.; Kamalakkannan, S.; Kokilavani, P.; Kalaiselvi, S.; SankarGanesh, D.; Gowri, J.; Archunan, G.; Cha, Y.S.; Achiraman, S. Spermidine may decrease ER stress in pancreatic beta cells and may reduce apoptosis via activating AMPK dependent autophagy pathway. *Med. Hypotheses* **2011**, *77*, 677–679. [CrossRef] [PubMed]

63. Bonaldo, P.; Braghetta, P.; Zanetti, M.; Piccolo, S.; Volpin, D.; Bressan, G.M. Collagen VI deficiency induces early onset myopathy in the mouse: An animal model for Bethlem myopathy. *Hum. Mol. Genet.* **1998**, *7*, 2135–2140. [CrossRef] [PubMed]

64. Grumati, P.; Coletto, L.; Sabatelli, P.; Cescon, M.; Angelin, A.; Bertaggia, E.; Blaauw, B.; Urciuolo, A.; Tiepolo, T.; Merlini, L.; et al. Autophagy is defective in collagen VI muscular dystrophies, and its reactivation rescues myofiber degeneration. *Nat. Med.* **2010**, *16*, 1313–1320. [CrossRef] [PubMed]

65. Mammucari, C.; Milan, G.; Romanello, V.; Masiero, E.; Rudolf, R.; Del Piccolo, P.; Burden, S.J.; Di Lisi, R.; Sandri, C.; Zhao, J.; et al. FoxO3 controls autophagy in skeletal muscle in vivo. *Cell Metab.* **2007**, *6*, 458–471. [CrossRef] [PubMed]

66. Bernardi, P.; Bonaldo, P. Dysfunction of mitochondria and sarcoplasmic reticulum in the pathogenesis of collagen VI muscular dystrophies. *Ann. N. Y. Acad. Sci.* **2008**, *1147*, 303–311. [CrossRef] [PubMed]

67. Bertaggia, E.; Coletto, L.; Sandri, M. Posttranslational modifications control FoxO3 activity during denervation. *AJP Cell Physiol.* **2012**, *302*, 587–596. [CrossRef] [PubMed]

68. Pietrocola, F.; Lachkar, S.; Enot, D.P.; Niso-Santano, M.; Bravo-San Pedro, J.M.; Sica, V.; Izzo, V.; Maiuri, M.C.; Madeo, F.; Marino, G.; et al. Spermidine induces autophagy by inhibiting the acetyltransferase EP300. *Cell Death Differ.* **2014**, *22*, 509–516. [CrossRef] [PubMed]

69. Virgili, M.; Crochemore, C.; Peña-Altamira, E.; Contestabile, A. Regional and temporal alterations of ODC/polyamine system during ALS-like neurodegenerative motor syndrome in G93A transgenic mice. *Neurochem. Int.* **2006**, *48*, 201–207. [CrossRef] [PubMed]

70. Sirrieh, R.E.; MacLean, D.M.; Jayaraman, V. Subtype-dependent N-methyl-D-aspartate receptor amino-terminal domain conformations and modulation by spermine. *J. Biol. Chem.* **2015**, *290*, 12812–12820. [CrossRef] [PubMed]

71. Patin, F.; Corcia, P.; Vourc'h, P.; Nadal-Desbarats, L.; Baranek, T.; Goossens, J.F.; Bruno, C. Omics to Explore Amyotrophic Lateral Sclerosis Evolution: The Central Role of Arginine and Proline Metabolism. *Mol. Neurobiol.* **2017**, *54*, 5361–5374. [CrossRef] [PubMed]

72. Ponnusamy, S.; Sullivan, R.D.; You, D.; Zafar, N.; He Yang, C.; Thiyagarajan, T.; Johnson, D.L.; Barrett, M.L.; Koehler, N.J.; Star, M.; et al. Androgen receptor agonists increase lean mass, improve cardiopulmonary functions and extend survival in preclinical models of Duchenne muscular dystrophy. *Hum. Mol. Genet.* **2017**, *26*, 2526–2540. [CrossRef] [PubMed]

73. Cozzoli, A.; Capogrosso, R.F.; Sblendorio, V.T.; Dinardo, M.M.; Jagerschmidt, C.; Namour, F.; Camerino, G.M.; De Luca, A. GLPG0492, a novel selective androgen receptor modulator, improves muscle performance in the exercised-mdx mouse model of muscular dystrophy. *Pharmacol. Res.* **2013**, *72*, 9–24. [CrossRef] [PubMed]

74. Schwenk, F.; Baron, U.; Rajewsky, K. A cre-transgenic mouse strain for the ubiquitous deletion of loxP-flanked gene segments including deletion in germ cells. *Nucleic Acids Res.* **1995**, *23*, 5080–5081. [CrossRef] [PubMed]

75. Ceci, R.; Duranti, G.; Leonetti, A.; Pietropaoli, S.; Spinozzi, F.; Marcocci, L.; Amendola, R.; Cecconi, F.; Sabatini, S.; Mariottini, P.; et al. Adaptive responses of heart and skeletal muscle to spermine oxidase overexpression: Evaluation of a new transgenic mouse model. *Free Radic. Biol. Med.* **2017**, *103*, 216–225. [CrossRef] [PubMed]

76. Ji, L.L. Exercise at old age: Does it increase or alleviate oxidative stress? *Ann. N. Y. Acad. Sci.* **2001**, *928*, 236–247. [CrossRef] [PubMed]

77. Lang, F.; Aravamudhan, S.; Nolte, H.; Tuerk, C.; Hölper, S.; Müller, S.; Krüger, M. Dynamic changes in the skeletal muscle proteome during denervation-induced atrophy. *Dis. Models Mech.* **2017**, *10*, 881–896. [CrossRef] [PubMed]

78. Cruz-Jentoft, A.J.; Landi, F.; Schneider, S.M.; Zuniga, C.; Arai, H.; Boirie, Y.; Chen, LK.; Fielding, R.A.; Martin, F.C.; Michel, J.P.; et al. Prevalence of and interventions for sarcopenia in ageing adults: A systematic review. Report of the International Sarcopenia Initiative (EWGSOP and IWGS). *Age Ageing* **2014**, *43*, 748–759. [CrossRef] [PubMed]

79. Cerrada-Gimenez, M.; Pietilä, M.; Loimas, S.; Pirinen, E.; Hyvönen, M.T.; Keinänen, T.A.; Alhonen, L. Continuous oxidative stress due to activation of polyamine catabolism accelerates aging and protects against hepatotoxic insults. *Transgenic Res.* **2011**, *20*, 387–396. [CrossRef] [PubMed]

80. Tancini, B.; Urbanelli, L.; Magini, A.; Polchi, A.; Emiliani, C. Extending lifespan through autophagy stimulation: A future perspective. *JGG* **2017**, *65*, 110–123.

Review

Alpha-Difluoromethylornithine, an Irreversible Inhibitor of Polyamine Biosynthesis, as a Therapeutic Strategy against Hyperproliferative and Infectious Diseases

Nicole LoGiudice [†], Linh Le [†], Irene Abuan [†], Yvette Leizorek [†] and Sigrid C. Roberts *

Pacific University School of Pharmacy, Hillsboro, OR 97123, USA; logi8930@pacificu.edu (N.L.);
le1143@pacificu.edu (L.L.); abua7932@pacificu.edu (I.A.); leiz1835@pacificu.edu (Y.L.)
* Correspondence: sroberts@pacificu.edu; Tel.: +1-503-352-7289
† These authors contributed equally to this work.

Received: 9 January 2018; Accepted: 5 February 2018; Published: 8 February 2018

Abstract: The fluorinated ornithine analog α-difluoromethylornithine (DFMO, eflornithine, ornidyl) is an irreversible suicide inhibitor of ornithine decarboxylase (ODC), the first and rate-limiting enzyme of polyamine biosynthesis. The ubiquitous and essential polyamines have many functions, but are primarily important for rapidly proliferating cells. Thus, ODC is potentially a drug target for any disease state where rapid growth is a key process leading to pathology. The compound was originally discovered as an anticancer drug, but its effectiveness was disappointing. However, DFMO was successfully developed to treat African sleeping sickness and is currently one of few clinically used drugs to combat this neglected tropical disease. The other Food and Drug Administration (FDA) approved application for DFMO is as an active ingredient in the hair removal cream Vaniqa. In recent years, renewed interest in DFMO for hyperproliferative diseases has led to increased research and promising preclinical and clinical trials. This review explores the use of DFMO for the treatment of African sleeping sickness and hirsutism, as well as its potential as a chemopreventive and chemotherapeutic agent against colorectal cancer and neuroblastoma.

Keywords: polyamines; ornithine decarboxylase; difluoromethylornithine; eflornithine; DFMO; African sleeping sickness; hirsutism; colorectal cancer; neuroblastoma

---

1. Introduction

The fluorinated ornithine analog α-difluoromethylornithine (DFMO, eflornithine, ornidyl) (Figure 1) is an irreversible suicide inhibitor of ornithine decarboxylase (ODC), the first and rate-limiting enzyme of polyamine biosynthesis. The compound was developed over 40 years ago as a targeted ODC inhibitor. Polyamines are polycationic aliphatic amines that are ubiquitous and essential for virtually all eukaryotic and prokaryotic cells [1–4]. The three main polyamines are putrescine, spermidine, and spermine; however, other polyamines exist, and the polyamine biosynthetic pathway and polyamine composition vary between species [1,4]. Polyamines have numerous functions, including, but not limited to, replication, protein biosynthesis, transcription, signal transduction, cell cycle progression, apoptosis, and stress resistance [5–7]. However, cellular proliferation is emerging as one of the most prominent functions. While quiescent cells show low ODC activity and putrescine levels, actively dividing cells exhibit increased ODC activity [8–11]. Enzyme activity is high during late G1 phase before DNA synthesis begins and again in the G2/M transition phase [8–11]. An essential downstream function of the polyamine pathway is the hypusination and activation of eukaryotic initiation factor eIF5A, vital for protein synthesis and cellular proliferation [12].

Furthermore, multiple studies have found elevated polyamine levels in tumor cells and that inhibition of polyamine pathway enzymes interferes with cellular proliferation [1,11,13,14].

The activity of ODC is highly regulated with multiple mechanisms. ODC is rapidly turned over within the cell by an antizyme-mediated process based on intracellular polyamine concentrations [14]. In addition, regulation of transcription and translation of *ODC* affects the amount of enzyme being expressed [14]. It has been found that ODC activity is increased and polyamines are present at higher concentrations in some solid tumors [11,14]. This has led to ODC inhibition being investigated as a therapeutic strategy for various malignancies, although long-term inhibition of this enzyme is challenging due to the rapid turnover and tight regulation of ODC production.

This review will provide an overview of the mechanism of action and pharmacokinetic properties of DFMO and summarize its clinically approved use against African sleeping sickness and hirsutism. In addition, the promise of DFMO to combat colon cancer and neuroblastoma will be discussed.

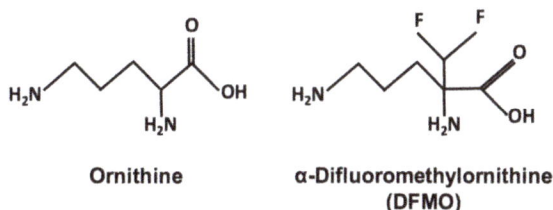

Figure 1. Structures of ornithine and the fluorinated ornithine analog α-difluoromethylornithine (DFMO).

2. Pharmacodynamic and Pharmacokinetic Properties of DFMO

DFMO was designed as an enzyme-activated irreversible inhibitor of ODC [15]. The compound is decarboxylated by the enzyme and covalently binds to ODC [16]. The irreversible inhibition of ODC results in depletion of putrescine and spermidine; however, only incomplete depletion of spermine occurs [11].

Biosynthesis of polyamines (Figure 2) starts with the conversion of L-arginine to ornithine. The rate-limiting step of polyamine biosynthesis is the conversion of ornithine to putrescine by ODC. Putrescine is further converted to spermidine and spermine by the sequential actions of spermidine synthase (SPDSYN) and spermine synthase (SPMSYN). The aminopropyl group for spermidine and spermine synthesis is donated from decarboxylated S-adenosylmethionine, which is formed from S-adenosylmethionine by the action of S-adenosylmethionine decarboxylase (ADOMETDC). In mammalian cells, a back-conversion pathway exists where spermine oxidase (SMO) converts spermine to spermidine and the combined actions of spermidine/spermine N^1-acetyltransferase (SSAT) and N-acetyl polyamine oxidase (APAO) convert spermine to spermidine and putrescine. Putrescine and acetylated polyamines can be exported, and polyamines can be imported by the cell.

Currently, the Food and Drug Administration (FDA) approves DFMO for female facial hirsutism and human African trypanosomiasis or sleeping sickness [10,17–22]. Vaniqa (Allergan, Irvine, CA, US) is a 13.9% DFMO cream available for hirsutism with a prescription [10,18,20]. In addition, DFMO has orphan drug status for a variety of cancers including: neuroblastoma, colon, gastric, and pancreatic cancer. DFMO is available as oral, intravenous, and topical formulations. The oral bioavailability of DFMO is approximately 50% for both the solution and tablet formulations [23]. Very little protein binding occurs, and DFMO can cross the blood brain barrier. The serum half-life is 1.5–5 h and 80% of it is excreted in the urine [17]. The rapid elimination from the blood makes high treatment doses necessary, and these poor pharmacokinetic properties render the drug less than ideal [24,25]. For the topical cream Vaniqa, less than 1% of the dose is absorbed systemically and the elimination half is 8 h [26,27].

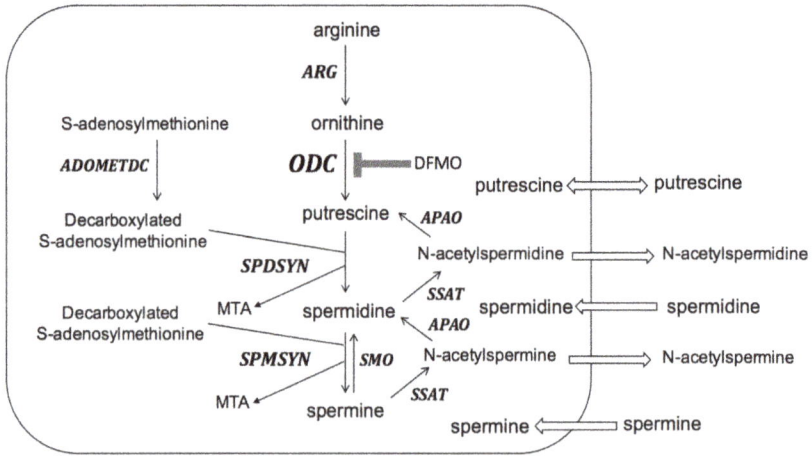

Figure 2. The polyamine pathway in mammalian cells. ARG: arginase; ODC: ornithine decarboxylase; SPDSYN: spermidine synthase; SPMSYN: spermine synthase; ADOMETDC: S-adenosylmethionine decarboxylase; SMO: spermine oxidase; SSAT: spermidine/spermine N1-acetyltransferase; APAO: N-acetyl polyamine oxidase; MTA: methylthioadenosine. Polyamine biosynthesis through ARG, ODC, SPDSYN, SPMSYN and ADOMETDC is shown. Back-conversion from spermine to spermidine occurs through the action of SMO and back-conversion from spermine to spermidine and putrescine is catalyzed by SSAT and APAO. Export of putrescine and acetylated polyamines and import of polyamines from the extracellular milieu are indicated. DFMO inhibition of ODC is shown.

The DFMO dose in colon cancer prevention trials is typically around 500 mg/m^2/day DFMO daily in tablet form over several years [28–31]. Higher oral doses are tested clinically for neuroblastoma treatment in children; 2000 mg/m^2/day DMFO alone or up to 9000 mg/m^2/day in combination therapies [32]. For the treatment of African sleeping sickness, DFMO is given intravenously at 400 mg/kg/day for 14 days as monotherapy or 800 mg/kg/day for seven days in combination with Nifurtimox [17,33]. The topical formulation of DFMO in Vaniqa is 13.9%.

At doses between 1–3 g/m^2/day, DFMO has minimal side effects including anemia, mild gastrointestinal upset, and reversible ototoxicity [10]. Because the drug has been found to reduce fetal body weight and may induce skeletal variations, DFMO is labeled pregnancy category C and should be used with caution [10]. The overall good safety profile has led to DFMO being considered as long-term chemoprevention or adjunct to chemotherapy.

3. African Sleeping Sickness

The parasites that give rise to African sleeping sickness are *Trypanosoma brucei rhodesiense* and *Trypanosoma brucei gambiense*. The bites of the tsetse flies transmit infectious trypanosomes to humans and livestock primarily seen in sub-Saharan Africa. The parasite invades the blood, lymph, and ultimately the central nervous system (CNS). The infection may present as acute or chronic depending on the subspecies. *T. b. rhodesiense* give rise to an acute disease state while *T. b. gambiense* are usually responsible for a chronic condition. If left untreated, both forms of the disease may lead to death commonly due to meningoencephalitis. There are no current vaccines available to prevent African sleeping sickness, and only four drugs approved for treatment: suramin, pentamidine, melarsoprol, and DFMO (eflornithine) [17,21,22].

Suramin and pentamidine are equally effective in treating early-stage gambiense disease but are less effective with the rhodesiense form [17,22]. Their effectiveness is limited to the early-stage of the disease mainly due to poor cerebrospinal fluid (CSF) penetration. For late-stage manifestations involving the CNS, melarsoprol is the only agent that can treat both subspecies, *T. b. rhodesiense* and *T. b. gambiense* [17,34]. Melarsoprol is a trivalent arsenical compound with weak CSF penetration but effective trypanocidal activity. The drawback to melarsoprol is the significant toxicity and irritation at the injection site. Serious reactions such as arsenic encephalopathy occurs in 10% of patients and are fatal in about half of those cases [17,34]. On top of the high risk for toxicity, a large percentage of non-responders have been observed suggesting the emergence of melarsoprol resistance [17,33,35].

The fourth drug, DFMO, is effective against both early- and late-stage gambiense disease and was approved by the FDA in 1990 [4,17,19,21,22,34]. DFMO has been shown to be the safer alternative to melarsoprol and became the preferred first-line treatment for late-stage gambiense disease [17,19,21,22,34]. The drug effectively crosses the blood–brain barrier and enters the CSF to irreversibly inhibit the parasite's ODC enzyme [17,19,21,22,34]. Interestingly, the affinity of DFMO to parasite and human ODC is similar and the drug inactivates both enzymes effectively [21]. However, the human enzyme is rapidly turned over (half-life less than an hour), while the parasites ODC has a much longer half-life (over 6 h), thus polyamine biosynthesis in trypanosomes is selectively more affected by the suicide inhibitor than the human host [3,36]. The inhibition of ODC by DFMO leads to the loss of polyamine biosynthesis crucial for parasite replication and survival. Genetic obliteration of ODC by gene deletion or knockdown confirmed that polyamine biosynthesis is essential for parasites [37,38]. Polyamines are especially important in *T. brucei* as, in a reaction unique to trypanosomatids, spermidine is conjugated with glutathione to produce trypanothione, which is essential for maintaining the intracellular redox balance and for defense against oxidative stress [2–4]. Furthermore, the polyamine biosynthetic pathway of *T. brucei* is significantly different from that of the mammalian host (Figure 3). Parasites do not contain a functional arginase to synthesize ornithine, the immediate substrate of ODC, do not synthesize or utilize spermine, and have no back-conversion pathway [4]. *T. brucei* parasites reside as extracellular parasites in the bloodstream. Because blood contains only nanomolar concentrations of polyamines and polyamine uptake is poor in trypanosomes, the parasites depend completely on endogenous biosynthesis [4]. This observation, together with the fact that ODC is a stable enzyme in parasites, explains the efficacy of DFMO for the treatment of *T. b. gambiense*. However, DFMO is not effective against *T. b. rhodesiense*; the reason for this is not well understood, but may possibly be due to the rapid disease progression or a higher ODC activity and turnover [3,39].

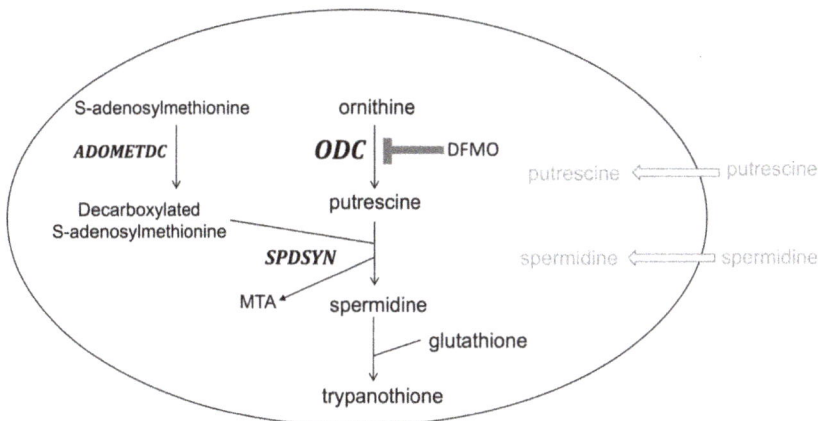

Figure 3. The polyamine pathway in *Trypanosoma brucei*. *Trypanosoma brucei* parasites lack ARG, SPMSYN, and a back-conversion pathway. Unique to trypanosomatids is the formation of trypanothione, a conjugate of spermidine and glutathione. Polyamine import is minimal as transport capacities are poor and levels of polyamines in blood are low. DFMO inhibition of ODC is shown.

Although DFMO was originally developed as an anti-cancer agent, its first clinical use was against African trypanosomes. Studies by Cyrus Bacchi and coworkers showed that the compound eliminated parasites in mice and that putrescine biosynthesis was reduced in parasites [40]. A first field trial in Sudan using oral DMFO showed effectiveness in 1985 [41] and subsequent studies using intravenous administration of DFMO were promising [25,42]. In fact, DFMO was so successful in treating patients, even in the comatose later stages of the disease, that the nickname "resurrection drug" was coined [19,43]. However, DFMO has high manufacturing costs and had to be administered at large doses (400 mg/kg/day, 4 times daily for 1–2 weeks) with an average cost per patient of $700 at the time [44]. Consequently, the use of DFMO in Africa waned in the 1990. Aventis (Schiltigheim, France), who had acquired the company originally manufacturing DFMO, sold left-over DFMO to Médecins Sans Frontières and gave the patent rights to the World Health Organization [19]. In an interesting development, DFMO was approved as the active ingredient in Vaniqa, a hair removal face cream, in 2000. The CBS show 60 min aired a segment on the use of DFMO for vanity contrasting this to the unavailability of the drug in Africa to save lives. In the end, the scandal resulted in Aventis producing and providing DFMO for use in Africa [19].

The standard dosing regimen of DFMO is intravenous infusions of 100 mg/kg every 6 h for a total of 14 days [17]. This dosing scheme is notably labor intensive, costly, and difficult to administer in developing countries with limited health care resources. In addition, although DFMO is tolerated well, some dose-dependent side effects were noted, such as reversible ototoxicity [18]. Attempts to make the dosing regimen shorter or to administer oral DFMO were not efficacious [17]. To make treatment more manageable, nifurtimox, a synthetic nitrofuran compound with antiprotozoal activity, was added in combination to DFMO therapy. The nifurtimox eflornithine combination therapy (NECT) demonstrated similar efficacy and significantly reduced the amount of DMFO, lessened adverse effects, and shortened treatment duration [17,21,45]. In combination therapy, DFMO is dosed 400 mg/kg/day by intravenous route every 12 h for seven days with concurrent nifurtimox 15 mg/kg/day by oral route every eight hours for 10 days [33]. NECT is listed as one of the essential medicines by the World Health Organization [17,21,45]. Relapse of the disease may occur more than 12 months after treatment; therefore, patient follow-up is recommended [45].

DFMO is also effective against a variety of other parasites [4]. For example, in *Leishmania* spp., DFMO has been found to reduce parasite burden in mice and hamsters [46–48] and an *ODC* gene deletion mutant was found to be virtually non-infective in mice [49]. Thus, DFMO may have potential against other parasitic diseases as well.

4. Vaniqa

Vaniqa is a DFMO 13.9% cream that is FDA approved since 2000 for unwanted facial hair in women. Unwanted hair growth, or hirsutism, is a significant problem in an estimated 40% of women in the US leading to decreased self-esteem, confidence, and increased social avoidance [20,50]. The hair follicle (HF), an organ composed of 20 different cell types, is highly proliferative and expresses high ODC activity, although the exact function of polyamines for HF is unknown [9]. Early DFMO trials in patients suffering from African sleeping sickness noted reversible hair loss [9,51], which may have spurred the development of DFMO for Vaniqa. Interestingly, the commonly noted side effect of reversible ototoxicity with DFMO is most likely due to impairment of hair cells of the organ of Corti in the inner ear being affected [18,52,53]. Transgenic mouse models have shown marked hair loss when DFMO was administered [54]. Other genetically manipulated mouse models have also confirmed a connection between polyamine metabolism and hair growth, although they show evidence of a more complicated relationship between putrescine and hair growth [9,55–57].

Clinical trials with Vaniqa demonstrated significant reduction in unwanted hair in the chin and upper lip area [20,50]. In one study, 58% in the DFMO group were rated as improved, compared to 34% in the control group [50]. Another trial included patient self-assessment, inquiring about the level of bother and discomfort in social gatherings or work due to the skin condition [20]. Again, DFMO resulted in significant benefits compared to the vehicle control [20]. One early study was performed with [^{14}C]-labeled DFMO to study percutaneous absorption and pharmacokinetics. Labeled DFMO was not detected in plasma and in only small amounts in urine, demonstrating very limited absorption of topical DFMO [27]. Vaniqa is tolerated well, with low rates of the most common side effects: acne, redness or stinging of the skin [10,50].

The cream must be applied twice daily, with marked improvements seen at week 8 and continued improvements seen at week 24 [50]. This treatment is not a permanent solution for hirsutism; Vaniqa must be continued indefinitely or hair will regrow within two months [50]. DFMO application also increases the effectiveness of laser therapy, which is commonly used for the removal of unwanted hair [50,58].

Various studies have looked at using Vaniqa for pseudofolliculitis barbae (persistent irritations from shaving) and chemoprevention for skin cancer [9,10]. Furthermore, clinical trials found that 10% DFMO ointment reduced the number of actinic keratosis with a concomitant reduction in spermidine levels in the treated skin [59,60].

5. Colon Cancer

Colorectal cancer (CRC) is the second leading cause of cancer-related deaths in the United States and the third most common cancer in men and in women (www.cdc.org). The slow development of CRC and common genetic alterations involved in disease progression warrant the use of chemopreventive measures to reduce its incidence. Although polyamine production is usually tightly controlled by multiple mechanisms in mammalian cells, dysregulation is often observed in cancer cells [13,28,61–64]. In general, elevated levels of polyamines are frequently associated with cancers, and higher polyamine concentrations were detected in colon polyps compared to normal mucosa [13,65,66]. Increased activity of ODC and higher levels of polyamines, mainly putrescine and spermidine, have been found in CRC cell lines, rodent models, and human tissue samples [13,28,61–64]. Thus, the ability of DFMO to suppress polyamine content in CRC is of particular interest, although the molecular mechanism by which polyamines contribute to cancer are not well understood [67].

Commonly mutated genes in colon cancer are *APC* tumor-suppressor gene, *KRAS* proto-oncogene, and *MYC* proto-oncogene (Figure 4). *APC* is mutated in familial adenomatous polyposis (FAP), an inherited CRC syndrome, and in over 80% of sporadic colon cancer [28,65]. In normal cells, APC inhibits the proto-oncogene and transcription factor MYC and loss of MYC inhibition results in overexpression of ODC and polyamine synthesis [28,65,68]. Activation of MYC is associated with almost 70% of CRC [13]. *KRAS* is another proto-oncogene and important intracellular signaling molecule. Mutation and activation of *KRAS* lead to increased ODC expression and decreased expression of spermidine/spermine N^1-acetyltransferase 1 (SSAT), resulting in decreased polyamine catabolism and export [65,68]. ODC activity was found to be upregulated in KRAS mutant cancer cells and DFMO treatment prevented tumor formation in nude mice injected with KRAS activated tumors [69]. The study found that DFMO-mediated prevention of tumorigenesis may occur via altered cell–cell communication [69]. Intriguingly, DFMO treatment prevented occurrence of *KRAS* mutations in a CRC mouse model, which suggests that DFMO is effective in suppressing colon carcinogenesis in the chemopreventive setting [70]. Similarly, animal models have shown that DFMO appears to be especially effective in preventing the transition of noninvasive to invasive cancers [13].

Using metabolomics to examine cellular changes, several metabolites were found to be significantly altered by DFMO treatment in the human colorectal adenocarcinoma cell line HT-29, including a decrease in putrescine, spermidine, methylthioadenosine, and N-acetylputrescine [71]. A separate metabolomic study found that DFMO treatment in ApcMin mice caused a significant reduction in thymidine pools, presumably due to the link of polyamine metabolism to S-adenosylmethionine pathways [72]. The authors conjecture that DFMO mediated ODC inhibition may enhance S-adenosylmethionine decarboxylase activity and thus exhaust levels of S-adenosylmethionine, resulting in impaired methionine cycling and thymidine pools. Indeed, the authors go as far as to imply that the reduction in thymidine pools may be the underlying mechanism of DFMO colon cancer prevention [72].

Depletion of polyamines leads to a reduction in transcription factors and pluripotency factors required for expression of several proteins and enzymes essential to cell cycle progression [67]. DFMO-treated HCT116 colon cancer cells showed lower levels of transcription factor HMGA2 and pluripotency factor LIN28 [67]. Growth inhibition was rescued by exogenous putrescine supplementation, confirming that the growth reducing effects of DFMO are due to polyamine depletion in the colon cancer cells [67]. Given the importance of the growth-associated factors in neoplasia, these findings suggest that polyamines are oncometabolites that influence tumorigenesis by regulating specific growth factors, and thus leading to the increase in tumor numbers and tumor growth rates [67].

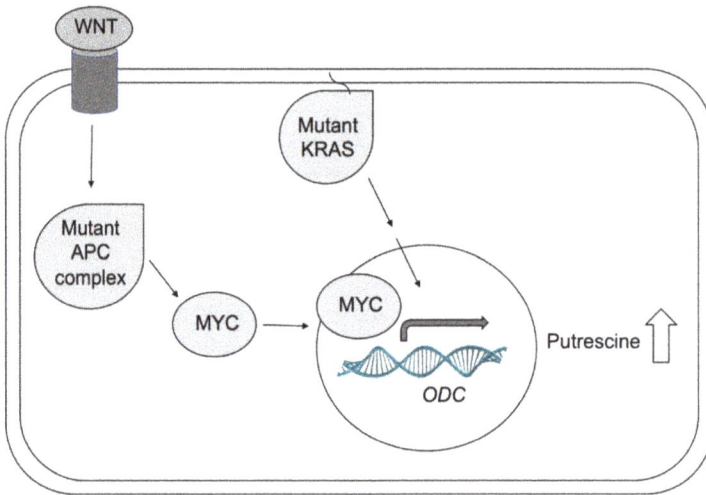

Figure 4. Simplified model of ODC activation in cancer. Activated WNT signaling and APC mutations lead to activation of MYC. The transcription factor MYC travels into the nucleus and induces transcription of several genes, including *ODC*. Mutated *KRAS* can also lead to increased *ODC* expression. In cancers, *APC*, *MYC*, or *KRAS* are often mutated, leading to increased ODC expression and activity and elevated levels of putrescine.

Although DFMO is not FDA-approved for the indication of CRC prevention yet, several clinical trials have shown that DFMO alone significantly alters metabolites in colon cancer cells [13,63–65,73]. Furthermore, the use of DFMO in combination with other agents for chemopreventive CRC therapy has shown promising results. As inflammation is typically associated with cancer and polyamines have been linked to inflammation-induced carcinogenesis, nonsteroidal anti-inflammatory drugs (NSAID) are potential candidates for combination therapy [28,65]. Cyclooxygenase 1 and 2 (COX-1 and COX-2), the target of NSAIDs, are critical enzymes in inflammatory pathways and both, COX-1 and COX-2, are induced by *APC* mutations [70]. In addition to being anti-inflammatory agents, NSAIDs have been shown to increase polyamine catabolism by inducing SSAT and polyamine export [28,62,65,68]. Thus, the combination of polyamine synthesis inhibitors and anti-inflammatory agents are particularly promising as cancer preventive agents. However, while NSAIDs have been associated with reduced colon cancer rates, they also increase the risk of gastrointestinal and/or cardiovascular side effects, therefore establishing safety profiles is important. Inhibitors of COX-1 and COX-2 (aspirin, sulindac, and piroxicam) and a selective inhibitor of COX-2 (celecoxib) have been advanced for combination therapy with DFMO [28,61,62,65,68]. Aspirin and piroxicam in combination with DFMO have been tested in animal studies and showed promising results [74,75]. A phase II clinical trial with aspirin and DFMO combination therapy to test for adenoma recurrence is underway [76]. The NSAIDs sulindac and celecoxib have been tested in animals and in clinical trials in combination with DFMO with encouraging outcomes [28–31,61,62,68,70,77,78]. When combined with sulindac, the effect of DFMO on reducing colon tumor numbers and total intestinal polyamine content was enhanced in a azoxymethane-induced colon cancer rat model [70]. Similar results were found in clinical trials [28,30,31,77,78]. A clinical trial with a low dose of 500 mg oral DMFO and 150 mg oral sulindac daily showed impressive efficacy in preventing adenomas in high risk patients with a history of resected adenomas [30,62,78]. The study found a 70% reduction in adenomas, a 92% decrease in advanced adenomas, and a 95% decrease in recurrence of more than one adenoma [30,62,78]. Furthermore, no increased cardiovascular events were observed and a follow up study that specifically investigated the risk of cardiovascular complications confirmed that the combination of DFMO and sulindac did not increase adverse cardiovascular outcomes in patients with normal risk for heart disease [31]. A currently ongoing clinical trial examines the safety and efficacy of the combination of oral 750 mg DFMO and 150 mg sulindac in patients with FAP [28]. A clinical trial testing celecoxib alone and in combination with DFMO found a higher reduction of adenoma count and burden for the combination than for celecoxib alone [29]. The combination treatment reduced adenoma count by 13% compared to 1% with celecoxib alone, and adenoma burden was reduced by 40% with the combination treatment compared to a 27% reduction with the NSAID alone [29]. Both sulindac and celecoxib monotherapy have been used off-label in FAP patients with some efficacy [28].

One important aspect to consider is that polyamines and polyamine precursors like arginine can also be taken up by diet. An increased intake in red meat, which is high in arginine, has been shown to decrease overall survival of colon cancer patients [65,77]. Thus, dietary recommendations may be a useful addition to any DFMO-based chemoprevention strategy [65,77].

There is evidence that genetic polymorphism in the *ODC* genomic locus has influence on the effectiveness of DFMO and NSAIDs [61,62,79]. A single nucleotide polymorphism (SNP) in the *ODC* promoter region has been associated with higher efficacy of aspirin for the risk reduction of adenoma recurrence in animals and in clinical trials [79,80]. It is likely that the variation in the promoter region makes *ODC* more susceptible for overexpression in some cancers and thus inhibition of polyamine synthesis by DFMO and/or NSAIDs is particularly effective in cancer prevention in patients exhibiting this polymorphism. The genetic marker may be a diagnostic tool to screen for high risk of cancer development and to identify patients that would receive the highest benefit from chemoprevention [80].

Another study examined rosuvastatin as an alternative to NSAIDs for combination therapy with DFMO in mice [81]. Statins, although mostly known for their inhibition of cholesterol synthesis, also have growth modulatory and anti-inflammatory effects and have been promoted as potential chemopreventive agents [81]. In addition, rosuvastatin appears to inhibit arginase activity and polyamine biosynthesis, thus DFMO and rosuvastatin may have an additive effect in decreasing polyamine levels [81]. Azoxymethane-induced colon cancer rat models were dosed individually and in combination with 500 ppm DFMO and 50 ppm rosuvastatin in food [81]. Rosuvastatin plus DFMO suppressed colon adenocarcinoma multiplicity by 76% compared to rosuvastatin monotherapy, 29%, and DFMO monotherapy, 46%, suggesting additive effects [81]. As expected, the treated colon tumors exhibited decreased polyamine content. Furthermore, they exhibited increased intratumoral natural killer (NK) cells expressing perforin plus interferon (IFN)-γ compared with untreated colon tumors [81]. This boost of innate immune cells in colon tumors may provide enhanced cytotoxic effects that can trigger lysis of the tumor cells, thus preventing progression of colon adenocarcinomas [81].

In these chemopreventive clinical trials, oral DFMO is typically given at 500 mg/m^2/day, with few side effects [13,29–31,64,78]. A particular concern with DFMO treatment is reversible ototoxicity [82,83]; however, only some dose-dependent reversible ototoxicity was observed while most trials reported no ototoxicity [13,29–31,64,78]. The low levels of both DFMO and NSAIDs required for cancer prevention compared to treatment correlate to the observed low levels of toxicity.

In summary, evidence from in vitro studies, animal models, and clinical trials demonstrate that DFMO mediated ODC inhibition causes metabolic changes, in particular polyamine depletion, which, in turn, decreases cellular proliferation and increases carcinogenesis suppression. Combination treatment of DFMO with NSAIDs and possibly other agents like rosuvastatin are even more effective and, although not yet FDA approved, are promising approaches for the chemoprevention of CRC.

6. Neuroblastoma

The term neuroblastoma is used to describe a range of tumors affecting immature nerve cells and accounts for 97% of these tumors. This cancer is fast spreading and can be found anywhere on the sympathetic nervous system, although 40% is found on the adrenal gland. The majority of patients affected by this cancer are children under four years of age. If diagnosed before the age of 1, patients have favorable outcomes. Patients older than one with advanced stage disease have a more negative, often lethal outcome stemming from progressive disease despite intensive multimodality therapy [32,84,85]. About 13% of pediatric cancer mortality can be attributed to neuroblastoma [86].

Genomic analyses in neuroblastoma tumors have shown that *MYC* mutations are present in about 40% of high risk tumors that show poor prognosis [32,87]. Evidence from a transgenic tyrosine hydroxylase (TH) promoter-driven mouse model (TH-MYCN) confirms that *MYC* is a driver gene as these mice develop peripheral neural tumors with complete penetrance in homozygous mice [88,89]. As observed in other cancers, and described above for colon cancer, the *ODC* gene is a target of MYC and neuroblastoma tumors often show high ODC activity [32,89]. Importantly, the ODC locus maps within 5 Mb of *MYC*, and both genes are often co-amplified [32]. Studies have linked inferior outcomes with higher expression of several polyamine biosynthetic enzymes and lower expression of catabolic enzymes in patients and animal models [32,84,90]. Indeed, *ODC* and other polyamine enzyme-encoding genes might be useful as diagnostic markers, independent of *MYC* [32]. High ODC expression has also been found in high risk tumors even without *MYC* mutations, confirming that ODC is a bona fide target [90].

Because MYC has proven notoriously 'undruggable' [91] and high tumor ODC activity has been linked to lower survival in cancer patients [32,90], inhibition of ODC by DFMO is a promising therapeutic strategy. Indeed, DFMO administration prevents tumor formation in the hemizygous TH-MYCN mouse and delays tumor onset in the homozygous mouse [90]. In vitro studies with neuroblastoma cell lines found that DFMO inhibited proliferation and reduced colony formation in a dose-dependent manner [32]. In human MYCN-amplified neuroblastoma cell lines, inhibition of ODC led to increased expression of cyclin-dependent kinase inhibitor p27Kip1, retinoblastoma protein (Rb) hypophosphorylation, and cell cycle arrest [92]. In TH-MYCN mice DFMO administration resulted in an increased expression of cyclin-dependent kinase inhibitor 1, p21^{Cip1}, a protein normally repressed in neuroblastoma, also resulting in cell cycle arrest and prevention of tumor cell migration and invasion of tumor cells [85,89]. New studies in the TH-MYCN mouse model suggest that polyamine depletion also has an effect on the tumor microenvironment and may modulate the immune response to prevent tumor proliferation [32].

The combination of DFMO with several other agents has shown even more promise. Early studies found that the reduction of polyamine biosynthesis by combined inhibition of ODC and ADOMETDC by DFMO and methylglyoxal-bis-guanylhydrazone (MGBG), respectively, was more potent than single agent treatment in reducing tumor formation in a rat prostate cancer model [93]. Compound SAM486 was developed based on the MGBG pharmacophore and has proven to be more efficacious than DFMO alone in extending survival in the TH-MYCN neuroblastoma mouse model [94]. DFMO treatment resulted in reduction of putrescine levels but not spermidine and spermine [90], while inhibition with DFMO and SAM486 showed reduced levels of all three polyamines [94]. Thus, the effectiveness of the drug combination can be rationalized by a more effective overall depletion of polyamines. Similar to observations in CRC chemopreventive trials, the addition of celecoxib also increased the effectiveness of DFMO in TH-MYCN mice [94]. This could be due to celecoxib's anti-inflammatory ability and/or its effect on polyamine catabolism. Furthermore, combination therapy with DFMO enhances the effect of the traditional chemotherapy agents, such as vincristine, cyclophosphamide, cisplatin, and topotecan in TH-MYCN mice [90,94].

Polyamine uptake is increased in DFMO treated cells and polyamine supplementation rescues the inhibitory effect of DFMO suggesting that scavenging of surrounding polyamines could negate the consequences of DFMO treatment [95–97]. It is likely that the early disappointing clinical trials in the 1970s and 1980s could at least in part be attributed to monotherapy with DFMO causing increased uptake of polyamines [1,98–103]. New approaches that combined the inhibition of polyamine synthesis and uptake, coined polyamine-blocking therapy (PBT), are effective in melanoma cell lines and in transgenic mice. Furthermore, this strategy resulted in an inhibition of the local tumor immunosuppressive response [99,100,104]. The dual impact of PBT on inhibition of polyamine dependent proliferation and as an immunostimulatory strategy may be particularly promising in anticancer therapy [99]. Indeed, recent studies confirmed that the combined strategy is also successful in neuroblastoma, as well as prostate, pancreatic, and breast cancer cell lines [102,105–107].

Numerous clinical trials have been initiated with children suffering from neuroblastoma. A completed phase I clinical trial investigated DFMO with and without the topoisomerase inhibitor etoposide with the main goal to assess drug safety [85]. Children with refractory or recurrent neuroblastoma were given between 500–1500 mg/m^2 DFMO orally twice a day. DFMO was given for three weeks as monotherapy followed by DFMO plus etoposide. No dose-limiting or drug related adverse side effects were observed. The study also evaluated the effect of the SNP rs2302616 in the *ODC* locus and found that this variant was associated with enhanced ODC susceptibility and responsiveness to DFMO therapy [85]. Two phase II clinical trials are currently underway to investigate the safety and efficacy of DFMO monotherapy to prevent recurrence of neuroblastoma that is in remission (NCT01586260 and NCT02395666). A similar Neuroblastoma Maintenance Therapy Trial is also ongoing (NCT 02679144). A phase I study is testing different doses of DFMO in combination with celecoxib, cyclophosphamide, and topotecan (NCT02030964) and another ongoing trial investigates the combination of DFMO with bortezimib for relapsed or refractory neuroblastoma (NCT 02139397). In yet another trial, DFMO will be used for two years as maintenance care after treatment with targeted therapy (NCT 02559778).

Although treatment with DFMO against cancer had been overall disappointing in the past, preclinical studies in the TH-MYCN neuroblastoma mouse model and clinical trials in children with this devastating disease are promising. Neuroblastoma may be especially responsive to DFMO treatment because *MYC* mutations and increased ODC activity are so prevalent in this type of cancer. Indeed, further genotyping a patient's tumor for *MYC* mutations, or ideally ODC overexpression, may improve clinical success of DFMO even more. In addition, not just high ODC activity but also genetic polymorphism within the *ODC* locus appears to be predictive of DFMO treatment success for both CRC prevention and neuroblastoma treatment. As with all new cancer strategies in the era of precision medicine, analyzing the molecular and genetic characteristics of a patient's tumor will be key to determine if any given drug strategy might be successful.

It should be noted that DFMO has also been advanced as a chemotherapeutic strategy for cancers other than colorectal cancer and neuroblastoma. For example, pancreatic cancer is an intriguing target, as the pancreas has the highest levels of spermidine of any human tissue [108,109]. Studies showed that DFMO is effective in preventing pancreatic cancer in genetically engineered KRAS mice [110] and furthermore that the combination therapy of ODC and polyamine transport inhibition in pancreatic cancer cell lines and pancreatic cancer mouse models inhibited cancer cell survival and prolonged the survival of tumor bearing mice [106,108]. Other areas of prolific research include but are not limited to skin, endometrial, and breast cancer [111–116]. Clinical trials with DFMO are ongoing for prostate cancer, skin cancer and precancerous conditions, gastric cancer, bladder cancer, the rare brain cancer anaplastic astrocytoma, cervical cancer, and esophageal cancer (www.clinicaltrials.gov). In summary, this resurgence of DFMO as an investigative cancer preventive or cancer treatment agent is remarkable.

7. Conclusions

DFMO targets a strange assortment of diseases, from parasitic infections to hirsutism to cancer. However, the common underlying molecular mechanism of all these afflictions is hyperproliferation. African trypanosomes are rapidly proliferating protozoan parasites, hirsutism is caused by aberrant and increased hair follicle growth, and the basic definition of cancer is uncontrolled proliferation. Polyamines have emerged as crucial metabolites for rapidly proliferating cells and DFMO inhibits ODC, the first and rate-limiting enzyme in this pathway. Although these ubiquitous and essential cations play many roles, their vital importance for cellular proliferation is now taking center stage in research.

The history of DFMO is intriguing as it was originally developed as an anticancer drug and is now experiencing a resurgence for exactly that purpose. Paradoxically though, the drug is currently not FDA approved for cancer prevention or treatment, but rather for an infectious disease and unwanted hair growth. Here, DFMO serves as an impressive example for drug development for neglected tropical diseases. With no economic incentives, DFMO production for Africa came to a halt and only the conscience of affluent countries reinvigorated its production. DFMO is also an intriguing drug as it has poor pharmacokinetic parameters, and modern drug development would probably not advance this chemical today. It is surprising that, despite much research, no other ODC inhibitors have been developed for clinical use, which elevates DFMO as an important and successful compound. Indeed, recently several clinical trials—with more underway—show that DFMO has a good safety profile and promising efficacy as chemopreventive or chemotherapeutic agent against colorectal cancer, neuroblastoma, and other hyperproliferative diseases.

In summary, DFMO has proven tremendously beneficial for numerous disease states and the recent surge in basic research and preclinical and clinical trials promises more future benefits.

Author Contributions: N.L. drafted the introduction, the section on pharmacodynamics and pharmacokinetic properties of DFMO, and the Vaniqa section; Y.L. drafted the section on neuroblastoma; I.A. drafted the section on colon cancer; L.L. drafted the section on African sleeping sickness; and S.R. was responsible for finalizing the manuscript.

Conflicts of Interest: The authors declare no conflict of interest.

References

1. Gerner, E.W.; Meyskens, F.L., Jr. Polyamines and cancer: Old molecules, new understanding. *Nat. Rev. Cancer* **2004**, *4*, 781–792. [CrossRef] [PubMed]
2. Heby, O.; Persson, L.; Rentala, M. Targeting the polyamine biosynthetic enzymes: A promising approach to therapy of African sleeping sickness, Chagas' disease, and leishmaniasis. *Amino Acids* **2007**, *33*, 359–366. [CrossRef] [PubMed]
3. Heby, O.; Roberts, S.C.; Ullman, B. Polyamine biosynthetic enzymes as drug targets in parasitic protozoa. *Biochem. Soc. Trans.* **2003**, *31*, 415–419. [CrossRef] [PubMed]
4. Roberts, S.; Ullman, B. Parasite Polyamines as Pharmaceutical Targets. *Curr. Pharm. Des.* **2017**, *23*, 3325–3341. [CrossRef] [PubMed]
5. Miller-Fleming, L.; Olin-Sandoval, V.; Campbell, K.; Ralser, M. Remaining Mysteries of Molecular Biology: The Role of Polyamines in the Cell. *J. Mol. Biol.* **2015**, *427*, 3389–3406. [CrossRef] [PubMed]
6. Igarashi, K.; Kashiwagi, K. Modulation of cellular function by polyamines. *Int. J. Biochem. Cell Biol.* **2010**, *42*, 39–51. [CrossRef] [PubMed]
7. Bachrach, U.; Wang, Y.C.; Tabib, A. Polyamines: New cues in cellular signal transduction. *News Physiol. Sci.* **2001**, *16*, 106–109. [CrossRef] [PubMed]
8. Rai, P.R.; Somani, R.R.; Kandpile, P.S. Ornithine Decarboxylase Inhibition: A strategy to combat various diseases. *Mini Rev. Med. Chem.* **2017**. [CrossRef] [PubMed]
9. Ramot, Y.; Pietila, M.; Giuliani, G.; Rinaldi, F.; Alhonen, L.; Paus, R. Polyamines and hair: A couple in search of perfection. *Exp. Dermatol.* **2010**, *19*, 784–790. [CrossRef] [PubMed]

10. Smith, K.J.; Skelton, H. α-Difluoromethylornithine, a polyamine inhibitor: Its potential role in controlling hair growth and in cancer treatment and chemo-prevention. *Int. J. Dermatol.* **2006**, *45*, 337–344. [CrossRef] [PubMed]

11. Wallace, H.M.; Fraser, A.V. Inhibitors of polyamine metabolism: Review article. *Amino Acids* **2004**, *26*, 353–365. [CrossRef] [PubMed]

12. Park, M.H.; Nishimura, K.; Zanelli, C.F.; Valentini, S.R. Functional significance of eIF5A and its hypusine modification in eukaryotes. *Amino Acids* **2010**, *38*, 491–500. [CrossRef] [PubMed]

13. Meyskens, F.L., Jr.; Gerner, E.W. Development of difluoromethylornithine (DFMO) as a chemoprevention agent. *Clin. Cancer Res.* **1999**, *5*, 945–951. [PubMed]

14. Pegg, A.E. Regulation of ornithine decarboxylase. *J. Biol. Chem.* **2006**, *281*, 14529–14532. [CrossRef] [PubMed]

15. Metcalf, B.W.; Bey, P.; Danzin, C.; Jung, M.J.; Casara, P.; Vevert, J.P. Catalytic Irreversible Inhibition of Mammalian Ornithine Decarboxylase (E.C.4.1.1.17) by Substrate and Product Analogs. *J. Am. Chem. Soc.* **1978**, *100*, 2551–2553. [CrossRef]

16. Pegg, A.E.; McGovern, K.A.; Wiest, L. Decarboxylation of α-difluoromethylornithine by ornithine decarboxylase. *Biochem. J.* **1987**, *241*, 305–307. [CrossRef] [PubMed]

17. Babokhov, P.; Sanyaolu, A.O.; Oyibo, W.A.; Fagbenro-Beyioku, A.F.; Iriemenam, N.C. A current analysis of chemotherapy strategies for the treatment of human African trypanosomiasis. *Pathog. Glob. Health* **2013**, *107*, 242–252. [CrossRef] [PubMed]

18. Coyne, P.E., Jr. The eflornithine story. *J. Am. Acad. Dermatol.* **2001**, *45*, 784–786. [CrossRef] [PubMed]

19. Ebikeme, C. The death and life of the resurrection drug. *PLoS Negl. Trop. Dis.* **2014**, *8*, e2910. [CrossRef] [PubMed]

20. Jackson, J.; Caro, J.J.; Caro, G.; Garfield, F.; Huber, F.; Zhou, W.; Lin, C.S.; Shander, D.; Schrode, K.; Eflornithine, H.S.G. The effect of eflornithine 13.9% cream on the bother and discomfort due to hirsutism. *Int. J. Dermatol.* **2007**, *46*, 976–981. [CrossRef] [PubMed]

21. Jacobs, R.T.; Nare, B.; Phillips, M.A. State of the art in African trypanosome drug discovery. *Curr. Top. Med. Chem.* **2011**, *11*, 1255–1274. [CrossRef] [PubMed]

22. Steverding, D. The development of drugs for treatment of sleeping sickness: A historical review. *Parasit Vectors* **2010**, *3*, 15. [CrossRef] [PubMed]

23. Carbone, P.P.; Douglas, J.A.; Thomas, J.; Tutsch, K.; Pomplun, M.; Hamielec, M.; Pauk, D. Bioavailability study of oral liquid and tablet forms of α-difluoromethylornithine. *Clin. Cancer Res.* **2000**, *6*, 3850–3854. [PubMed]

24. Legros, D.; Ollivier, G.; Gastellu-Etchegorry, M.; Paquet, C.; Burri, C.; Jannin, J.; Buscher, P. Treatment of human African trypanosomiasis—Present situation and needs for research and development. *Lancet Infect. Dis.* **2002**, *2*, 437–440. [CrossRef]

25. Milord, F.; Pepin, J.; Loko, L.; Ethier, L.; Mpia, B. Efficacy and toxicity of eflornithine for treatment of *Trypanosoma brucei gambiense* sleeping sickness. *Lancet* **1992**, *340*, 652–655. [CrossRef]

26. Jobanputra, K.S.; Rajpal, A.V.; Nagpur, N.G. Eflornithine. *Indian J. Dermatol. Venereol. Leprol.* **2007**, *73*, 365–366. [CrossRef] [PubMed]

27. Malhotra, B.; Noveck, R.; Behr, D.; Palmisano, M. Percutaneous absorption and pharmacokinetics of eflornithine HCl 13.9% cream in women with unwanted facial hair. *J. Clin. Pharmacol.* **2001**, *41*, 972–978. [CrossRef] [PubMed]

28. Burke, C.A.; Dekker, E.; Samadder, N.J.; Stoffel, E.; Cohen, A. Efficacy and safety of eflornithine (CPP-1X)/sulindac combination therapy versus each as monotherapy in patients with familial adenomatous polyposis (FAP): Design and rationale of a randomized, double-blind, Phase III trial. *BMC Gastroenterol.* **2016**, *16*, 87. [CrossRef] [PubMed]

29. Lynch, P.M.; Burke, C.A.; Phillips, R.; Morris, J.S.; Slack, R.; Wang, X.; Liu, J.; Patterson, S.; Sinicrope, F.A.; Rodriguez-Bigas, M.A.; et al. An international randomised trial of celecoxib versus celecoxib plus difluoromethylornithine in patients with familial adenomatous polyposis. *Gut* **2016**, *65*, 286–295. [CrossRef] [PubMed]

30. Meyskens, F.L., Jr.; McLaren, C.E.; Pelot, D.; Fujikawa-Brooks, S.; Carpenter, P.M.; Hawk, E.; Kelloff, G.; Lawson, M.J.; Kidao, J.; McCracken, J.; et al. Difluoromethylornithine plus sulindac for the prevention of sporadic colorectal adenomas: A randomized placebo-controlled, double-blind trial. *Cancer Prev. Res.* **2008**, *1*, 32–38. [CrossRef] [PubMed]

31. Zell, J.A.; Pelot, D.; Chen, W.P.; McLaren, C.E.; Gerner, E.W.; Meyskens, F.L. Risk of cardiovascular events in a randomized placebo-controlled, double-blind trial of difluoromethylornithine plus sulindac for the prevention of sporadic colorectal adenomas. *Cancer Prev. Res.* **2009**, *2*, 209–212. [CrossRef] [PubMed]

32. Bassiri, H.; Benavides, A.; Haber, M.; Gilmour, S.K.; Norris, M.D.; Hogarty, M.D. Translational development of difluoromethylornithine (DFMO) for the treatment of neuroblastoma. *Transl. Pediatr.* **2015**, *4*, 226–238. [PubMed]

33. Priotto, G.; Kasparian, S.; Mutombo, W.; Ngouama, D.; Ghorashian, S.; Arnold, U.; Ghabri, S.; Baudin, E.; Buard, V.; Kazadi-Kyanza, S.; et al. Nifurtimox-eflornithine combination therapy for second-stage African *Trypanosoma brucei gambiense* trypanosomiasis: A multicentre, randomised, phase III, non-inferiority trial. *Lancet* **2009**, *374*, 56–64. [CrossRef]

34. Priotto, G.; Pinoges, L.; Fursa, I.B.; Burke, B.; Nicolay, N.; Grillet, G.; Hewison, C.; Balasegaram, M. Safety and effectiveness of first line eflornithine for *Trypanosoma brucei gambiense* sleeping sickness in Sudan: Cohort study. *BMJ* **2008**, *336*, 705–708. [CrossRef] [PubMed]

35. Eperon, G.; Balasegaram, M.; Potet, J.; Mowbray, C.; Valverde, O.; Chappuis, F. Treatment options for second-stage gambiense human African trypanosomiasis. *Expert Rev. Antiinfect. Ther.* **2014**, *12*, 1407–1417. [CrossRef] [PubMed]

36. Wang, C.C. A novel suicide inhibitor strategy for antiparasitic drug development. *J. Cell. Biochem.* **1991**, *45*, 49–53. [CrossRef] [PubMed]

37. Li, F.; Hua, S.B.; Wang, C.C.; Gottesdiener, K.M. Trypanosoma brucei brucei: Characterization of an ODC null bloodstream form mutant and the action of alpha-difluoromethylornithine. *Exp. Parasitol.* **1998**, *88*, 255–257. [CrossRef] [PubMed]

38. Xiao, Y.; McCloskey, D.E.; Phillips, M.A. RNA interference-mediated silencing of ornithine decarboxylase and spermidine synthase genes in *Trypanosoma brucei* provides insight into regulation of polyamine biosynthesis. *Eukaryot. Cell.* **2009**, *8*, 747–755. [CrossRef] [PubMed]

39. Iten, M.; Mett, H.; Evans, A.; Enyaru, J.C.; Brun, R.; Kaminsky, R. Alterations in ornithine decarboxylase characteristics account for tolerance of *Trypanosoma brucei rhodesiense* to D,L-α-difluoromethylornithine. *Antimicrob. Agents Chemother.* **1997**, *41*, 1922–1925. [PubMed]

40. Bacchi, C.J.; Nathan, H.C.; Hutner, S.H.; McCann, P.P.; Sjoerdsma, A. Polyamine metabolism: A potential therapeutic target in trypanosomes. *Science* **1980**, *210*, 332–334. [CrossRef] [PubMed]

41. Van Nieuwenhove, S.; Schechter, P.J.; Declercq, J.; Bone, G.; Burke, J.; Sjoerdsma, A. Treatment of gambiense sleeping sickness in the Sudan with oral DFMO (DL-α-difluoromethylornithine), an inhibitor of ornithine decarboxylase; first field trial. *Trans. R. Soc. Trop. Med. Hyg.* **1985**, *79*, 692–698. [CrossRef]

42. Doua, F.; Boa, F.Y.; Schechter, P.J.; Miezan, T.W.; Diai, D.; Sanon, S.R.; De Raadt, P.; Haegele, K.D.; Sjoerdsma, A.; Konian, K. Treatment of human late stage gambiense trypanosomiasis with α-difluoromethylornithine (eflornithine): Efficacy and tolerance in 14 cases in Cote d'Ivoire. *Am. J. Trop. Med. Hyg.* **1987**, *37*, 525–533. [CrossRef] [PubMed]

43. Kuzoe, F.A. Perspectives in research on and control of African trypanosomiasis. *Ann. Trop. Med. Parasitol.* **1991**, *85*, 33–41. [CrossRef] [PubMed]

44. Bacchi, C.J. Progress in Anti-Polyamine Drug Development/Chemotherapy vs. Protozoan-Caused Diseases: The DFMO Story. 2006. Available online: http://wizard.musc.edu/dfmostory.pdf (accessed on 27 December 2017).

45. Yun, O.; Priotto, G.; Tong, J.; Flevaud, L.; Chappuis, F. NECT is next: Implementing the new drug combination therapy for *Trypanosoma brucei gambiense* sleeping sickness. *PLoS Negl. Trop. Dis.* **2010**, *4*, e720. [CrossRef] [PubMed]

46. Gradoni, L.; Iorio, M.A.; Gramiccia, M.; Orsini, S. In Vivo effect of eflornithine (DFMO) and some related compounds on *Leishmania infantum* preliminary communication. *Farmaco* **1989**, *44*, 1157–1166. [PubMed]

47. Mukhopadhyay, R.; Madhubala, R. Effect of a bis(benzyl)polyamine analogue, and DL-α-difluoromethylornithine on parasite suppression and cellular polyamine levels in golden hamster during *Leishmania donovani* infection. *Pharmacol. Res.* **1993**, *28*, 359–365. [CrossRef] [PubMed]

48. Olenyik, T.; Gilroy, C.; Ullman, B. Oral putrescine restores virulence of ornithine decarboxylase-deficient *Leishmania donovani* in mice. *Mol. Biochem. Parasitol.* **2011**, *176*, 109–111. [CrossRef] [PubMed]

49. Boitz, J.M.; Yates, P.A.; Kline, C.; Gaur, U.; Wilson, M.E.; Ullman, B.; Roberts, S.C. Leishmania donovani ornithine decarboxylase is indispensable for parasite survival in the mammalian host. *Infect. Immun.* **2009**, *77*, 756–763. [CrossRef] [PubMed]

50. Wolf, J.E., Jr.; Shander, D.; Huber, F.; Jackson, J.; Lin, C.S.; Mathes, B.M.; Schrode, K.; Eflornithine, H.S.G. Randomized, double-blind clinical evaluation of the efficacy and safety of topical eflornithine HCl 13.9% cream in the treatment of women with facial hair. *Int. J. Dermatol.* **2007**, *46*, 94–98. [CrossRef] [PubMed]

51. Pepin, J.; Milord, F.; Guern, C.; Schechter, P.J. Difluoromethylornithine for arseno-resistant *Trypanosoma brucei gambiense* sleeping sickness. *Lancet* **1987**, *2*, 1431–1433. [CrossRef]

52. Jansen, C.; Mattox, D.E.; Miller, K.D.; Brownell, W.E. An animal model of hearing loss from α-difluoromethylornithine. *Arch. Otolaryngol. Head Neck Surg.* **1989**, *115*, 1234–1237. [CrossRef] [PubMed]

53. Salzer, S.J.; Mattox, D.E.; Brownell, W.E. Cochlear damage and increased threshold in α-difluoromethylornithine (DFMO) treated guinea pigs. *Hear. Res.* **1990**, *46*, 101–112. [CrossRef]

54. Wheeler, D.L.; Ness, K.J.; Oberley, T.D.; Verma, A.K. Inhibition of the development of metastatic squamous cell carcinoma in protein kinase C ε transgenic mice by α-difluoromethylornithine accompanied by marked hair follicle degeneration and hair loss. *Cancer Res.* **2003**, *63*, 3037–3042. [PubMed]

55. Janne, J.; Alhonen, L.; Pietila, M.; Keinanen, T.A. Genetic approaches to the cellular functions of polyamines in mammals. *Eur. J. Biochem.* **2004**, *271*, 877–894. [CrossRef] [PubMed]

56. Pietila, M.; Parkkinen, J.J.; Alhonen, L.; Janne, J. Relation of skin polyamines to the hairless phenotype in transgenic mice overexpressing spermidine/spermine N^1-acetyltransferase. *J. Investig. Dermatol.* **2001**, *116*, 801–805. [CrossRef] [PubMed]

57. Soler, A.P.; Gilliard, G.; Megosh, L.C.; O'Brien, T.G. Modulation of murine hair follicle function by alterations in ornithine decarboxylase activity. *J. Investig. Dermatol.* **1996**, *106*, 1108–1113. [CrossRef] [PubMed]

58. Hamzavi, I.; Tan, E.; Shapiro, J.; Lui, H. A randomized bilateral vehicle-controlled study of eflornithine cream combined with laser treatment versus laser treatment alone for facial hirsutism in women. *J. Am. Acad. Dermatol.* **2007**, *57*, 54–59. [CrossRef] [PubMed]

59. Alberts, D.S.; Dorr, R.T.; Einspahr, J.G.; Aickin, M.; Saboda, K.; Xu, M.J.; Peng, Y.M.; Goldman, R.; Foote, J.A.; Warneke, J.A.; et al. Chemoprevention of human actinic keratoses by topical 2-(difluoromethyl)-DL-ornithine. *Cancer Epidemiol. Biomark. Prev.* **2000**, *9*, 1281–1286.

60. Bartels, P.; Yozwiak, M.; Einspahr, J.; Saboda, K.; Liu, Y.; Brooks, C.; Bartels, H.; Alberts, D.S. Chemopreventive efficacy of topical difluoromethylornithine and/or triamcinolone in the treatment of actinic keratoses analyzed by karyometry. *Anal. Quant. Cytol. Histol.* **2009**, *31*, 355–366. [PubMed]

61. Babbar, N.; Gerner, E.W. Targeting polyamines and inflammation for cancer prevention. *Recent Results Cancer Res.* **2011**, *188*, 49–64. [PubMed]

62. Gerner, E.W.; Meyskens, F.L., Jr. Combination chemoprevention for colon cancer targeting polyamine synthesis and inflammation. *Clin. Cancer Res.* **2009**, *15*, 758–761. [CrossRef] [PubMed]

63. Love, R.R.; Jacoby, R.; Newton, M.A.; Tutsch, K.D.; Simon, K.; Pomplun, M.; Verma, A.K. A randomized, placebo-controlled trial of low-dose α-difluoromethylornithine in individuals at risk for colorectal cancer. *Cancer Epidemiol. Biomark. Prev.* **1998**, *7*, 989–992.

64. Meyskens, F.L., Jr.; Gerner, E.W.; Emerson, S.; Pelot, D.; Durbin, T.; Doyle, K.; Lagerberg, W. Effect of α-difluoromethylornithine on rectal mucosal levels of polyamines in a randomized, double-blinded trial for colon cancer prevention. *J. Natl. Cancer Inst.* **1998**, *90*, 1212–1218. [CrossRef] [PubMed]

65. Rial, N.S.; Meyskens, F.L.; Gerner, E.W. Polyamines as mediators of APC-dependent intestinal carcinogenesis and cancer chemoprevention. *Essays Biochem.* **2009**, *46*, 111–124. [CrossRef] [PubMed]

66. Thompson, P.A.; Wertheim, B.C.; Zell, J.A.; Chen, W.P.; McLaren, C.E.; LaFleur, B.J.; Meyskens, F.L.; Gerner, E.W. Levels of rectal mucosal polyamines and prostaglandin E2 predict ability of DFMO and sulindac to prevent colorectal adenoma. *Gastroenterology* **2010**, *139*, 797–805. [CrossRef] [PubMed]

67. Paz, E.A.; LaFleur, B.; Gerner, E.W. Polyamines are oncometabolites that regulate the LIN28/let-7 pathway in colorectal cancer cells. *Mol. Carcinog.* **2014**, *53*, E96–E106. [CrossRef] [PubMed]

68. Gerner, E.W.; Meyskens, F.L., Jr.; Goldschmid, S.; Lance, P.; Pelot, D. Rationale for, and design of, a clinical trial targeting polyamine metabolism for colon cancer chemoprevention. *Amino Acids* **2007**, *33*, 189–195. [CrossRef] [PubMed]

69. Ignatenko, N.A.; Zhang, H.; Watts, G.S.; Skovan, B.A.; Stringer, D.E.; Gerner, E.W. The chemopreventive agent α-difluoromethylornithine blocks Ki-ras-dependent tumor formation and specific gene expression in Caco-2 cells. *Mol. Carcinog.* **2004**, *39*, 221–233. [CrossRef] [PubMed]

70. LeGendre-McGhee, S.; Rice, P.S.; Wall, R.A.; Sprute, K.J.; Bommireddy, R.; Luttman, A.M.; Nagle, R.B.; Abril, E.R.; Farrell, K.; Hsu, C.H.; et al. Time-serial Assessment of Drug Combination Interventions in a Mouse Model of Colorectal Carcinogenesis Using Optical Coherence Tomography. *Cancer Growth Metastasis* **2015**, *8*, 63–80. [CrossRef] [PubMed]

71. Ibanez, C.; Simo, C.; Valdes, A.; Campone, L.; Piccinelli, A.L.; Garcia-Canas, V.; Cifuentes, A. Metabolomics of adherent mammalian cells by capillary electrophoresis-mass spectrometry: HT-29 cells as case study. *J. Pharm. Biomed. Anal.* **2015**, *110*, 83–92. [CrossRef] [PubMed]

72. Witherspoon, M.; Chen, Q.; Kopelovich, L.; Gross, S.S.; Lipkin, S.M. Unbiased metabolite profiling indicates that a diminished thymidine pool is the underlying mechanism of colon cancer chemoprevention by α-difluoromethylornithine. *Cancer Discov.* **2013**, *3*, 1072–1081. [CrossRef] [PubMed]

73. Meyskens, F.L., Jr.; Emerson, S.S.; Pelot, D.; Meshkinpour, H.; Shassetz, L.R.; Einspahr, J.; Alberts, D.S.; Gerner, E.W. Dose de-escalation chemoprevention trial of α-difluoromethylornithine in patients with colon polyps. *J. Natl. Cancer Inst.* **1994**, *86*, 1122–1130. [CrossRef] [PubMed]

74. Jacoby, R.F.; Cole, C.E.; Tutsch, K.; Newton, M.A.; Kelloff, G.; Hawk, E.T.; Lubet, R.A. Chemopreventive efficacy of combined piroxicam and difluoromethylornithine treatment of APC mutant Min mouse adenomas, and selective toxicity against APC mutant embryos. *Cancer Res.* **2000**, *60*, 1864–1870. [PubMed]

75. Li, H.; Schut, H.A.; Conran, P.; Kramer, P.M.; Lubet, R.A.; Steele, V.E.; Hawk, E.E.; Kelloff, G.J.; Pereira, M.A. Prevention by aspirin and its combination with α-difluoromethylornithine of azoxymethane-induced tumors, aberrant crypt foci and prostaglandin E2 levels in rat colon. *Carcinogenesis* **1999**, *20*, 425–430. [CrossRef] [PubMed]

76. Laukaitis, C.M.; Erdman, S.H.; Gerner, E.W. Chemoprevention in patients with genetic risk of colorectal cancers. *Colorectal Cancer* **2012**, *1*, 225–240. [CrossRef] [PubMed]

77. Raj, K.P.; Zell, J.A.; Rock, C.L.; McLaren, C.E.; Zoumas-Morse, C.; Gerner, E.W.; Meyskens, F.L. Role of dietary polyamines in a phase III clinical trial of difluoromethylornithine (DFMO) and sulindac for prevention of sporadic colorectal adenomas. *Br. J. Cancer* **2013**, *108*, 512–518. [CrossRef] [PubMed]

78. Sporn, M.B.; Hong, W.K. Concomitant DFMO and sulindac chemoprevention of colorectal adenomas: A major clinical advance. *Nat. Clin. Pract. Oncol.* **2008**, *5*, 628–629. [CrossRef] [PubMed]

79. Martinez, M.E.; O'Brien, T.G.; Fultz, K.E.; Babbar, N.; Yerushalmi, H.; Qu, N.; Guo, Y.; Boorman, D.; Einspahr, J.; Alberts, D.S.; et al. Pronounced reduction in adenoma recurrence associated with aspirin use and a polymorphism in the ornithine decarboxylase gene. *Proc. Natl. Acad. Sci. USA* **2003**, *100*, 7859–7864. [CrossRef] [PubMed]

80. Hubner, R.A.; Muir, K.R.; Liu, J.F.; Logan, R.F.; Grainge, M.J.; Houlston, R.S. Members of the, U.C. Ornithine decarboxylase G316A genotype is prognostic for colorectal adenoma recurrence and predicts efficacy of aspirin chemoprevention. *Clin. Cancer Res.* **2008**, *14*, 2303–2309. [CrossRef] [PubMed]

81. Janakiram, N.B.; Mohammed, A.; Bryant, T.; Zhang, Y.; Brewer, M.; Duff, A.; Biddick, L.; Singh, A.; Lightfoot, S.; Steele, V.E.; et al. Potentiating NK cell activity by combination of Rosuvastatin and Difluoromethylornithine for effective chemopreventive efficacy against Colon Cancer. *Sci. Rep.* **2016**, *6*, 37046. [CrossRef] [PubMed]

82. Lao, C.D.; Backoff, P.; Shotland, L.I.; McCarty, D.; Eaton, T.; Ondrey, F.G.; Viner, J.L.; Spechler, S.J.; Hawk, E.T.; Brenner, D.E. Irreversible ototoxicity associated with difluoromethylornithine. *Cancer Epidemiol. Biomark. Prev.* **2004**, *13*, 1250–1252.

83. Pasic, T.R.; Heisey, D.; Love, R.R. α-difluoromethylornithine ototoxicity. Chemoprevention clinical trial results. *Arch. Otolaryngol. Head Neck Surg.* **1997**, *123*, 1281–1286. [CrossRef] [PubMed]

84. Gamble, L.D.; Hogarty, M.D.; Liu, X.; Ziegler, D.S.; Marshall, G.; Norris, M.D.; Haber, M. Polyamine pathway inhibition as a novel therapeutic approach to treating neuroblastoma. *Front. Oncol.* **2012**, *2*, 162. [CrossRef] [PubMed]

85. Saulnier Sholler, G.L.; Gerner, E.W.; Bergendahl, G.; MacArthur, R.B.; VanderWerff, A.; Ashikaga, T.; Bond, J.P.; Ferguson, W.; Roberts, W.; Wada, R.K.; et al. A Phase I Trial of DFMO Targeting Polyamine Addiction in Patients with Relapsed/Refractory Neuroblastoma. *PLoS ONE* **2015**, *10*, e0127246.

86. Louis, C.U.; Shohet, J.M. Neuroblastoma: Molecular pathogenesis and therapy. *Annu. Rev. Med.* **2015**, *66*, 49–63. [CrossRef] [PubMed]

87. Evageliou, N.F.; Hogarty, M.D. Disrupting polyamine homeostasis as a therapeutic strategy for neuroblastoma. *Clin. Cancer Res.* **2009**, *15*, 5956–5961. [CrossRef] [PubMed]

88. Rasmuson, A.; Segerstrom, L.; Nethander, M.; Finnman, J.; Elfman, L.H.; Javanmardi, N.; Nilsson, S.; Johnsen, J.I.; Martinsson, T.; Kogner, P. Tumor development, growth characteristics and spectrum of genetic aberrations in the TH-MYCN mouse model of neuroblastoma. *PLoS ONE* **2012**, *7*, e51297. [CrossRef] [PubMed]

89. Rounbehler, R.J.; Li, W.; Hall, M.A.; Yang, C.; Fallahi, M.; Cleveland, J.L. Targeting ornithine decarboxylase impairs development of MYCN-amplified neuroblastoma. *Cancer Res.* **2009**, *69*, 547–553. [CrossRef] [PubMed]

90. Hogarty, M.D.; Norris, M.D.; Davis, K.; Liu, X.; Evageliou, N.F.; Hayes, C.S.; Pawel, B.; Guo, R.; Zhao, H.; Sekyere, E.; et al. ODC1 is a critical determinant of MYCN oncogenesis and a therapeutic target in neuroblastoma. *Cancer Res.* **2008**, *68*, 9735–9745. [CrossRef] [PubMed]

91. Dang, C.V.; Reddy, E.P.; Shokat, K.M.; Soucek, L. Drugging the 'undruggable' cancer targets. *Nat. Rev. Cancer* **2017**, *17*, 502–508. [CrossRef] [PubMed]

92. Wallick, C.J.; Gamper, I.; Thorne, M.; Feith, D.J.; Takasaki, K.Y.; Wilson, S.M.; Seki, J.A.; Pegg, A.E.; Byus, C.V.; Bachmann, A.S. Key role for p27Kip1, retinoblastoma protein Rb, and MYCN in polyamine inhibitor-induced G1 cell cycle arrest in MYCN-amplified human neuroblastoma cells. *Oncogene* **2005**, *24*, 5606–5618. [CrossRef] [PubMed]

93. Herr, H.W.; Kleinert, E.L.; Relyea, N.M.; Whitmore, W.F., Jr. Potentiation of methylglyoxal-bis-guanylhydrazone by α-difluoromethylornithine in rat prostate cancer. *Cancer* **1984**, *53*, 1294–1298. [CrossRef]

94. Evageliou, N.F.; Haber, M.; Vu, A.; Laetsch, T.W.; Murray, J.; Gamble, L.D.; Cheng, N.C.; Liu, K.; Reese, M.; Corrigan, K.A.; et al. Polyamine Antagonist Therapies Inhibit Neuroblastoma Initiation and Progression. *Clin. Cancer Res.* **2016**, *22*, 4391–4404. [CrossRef] [PubMed]

95. Alhonen-Hongisto, L.; Seppanen, P.; Janne, J. Intracellular putrescine and spermidine deprivation induces increased uptake of the natural polyamines and methylglyoxal bis(guanylhydrazone). *Biochem. J.* **1980**, *192*, 941–945. [CrossRef] [PubMed]

96. Chen, Y.; Weeks, R.S.; Burns, M.R.; Boorman, D.W.; Klein-Szanto, A.; O'Brien, T.G. Combination therapy with 2-difluoromethylornithine and a polyamine transport inhibitor against murine squamous cell carcinoma. *Int. J. Cancer* **2006**, *118*, 2344–2349. [CrossRef] [PubMed]

97. Sunkara, P.S.; Prakash, N.J.; Rosenberger, A.L. An essential role for polyamines in tumor metastases. *FEBS Lett.* **1982**, *150*, 397–399. [CrossRef]

98. Burns, M.R.; Graminski, G.F.; Weeks, R.S.; Chen, Y.; O'Brien, T.G. Lipophilic lysine-spermine conjugates are potent polyamine transport inhibitors for use in combination with a polyamine biosynthesis inhibitor. *J. Med. Chem.* **2009**, *52*, 1983–1993. [CrossRef] [PubMed]

99. Hayes, C.S.; Burns, M.R.; Gilmour, S.K. Polyamine blockade promotes antitumor immunity. *Oncoimmunology* **2014**, *3*, e27360. [CrossRef] [PubMed]

100. Hayes, C.S.; Shicora, A.C.; Keough, M.P.; Snook, A.E.; Burns, M.R.; Gilmour, S.K. Polyamine-blocking therapy reverses immunosuppression in the tumor microenvironment. *Cancer Immunol. Res.* **2014**, *2*, 274–285. [CrossRef] [PubMed]

101. Nowotarski, S.L.; Woster, P.M.; Casero, R.A., Jr. Polyamines and cancer: Implications for chemotherapy and chemoprevention. *Expert Rev. Mol. Med.* **2013**, *15*, e3. [CrossRef] [PubMed]

102. Samal, K.; Zhao, P.; Kendzicky, A.; Yco, L.P.; McClung, H.; Gerner, E.; Burns, M.; Bachmann, A.S.; Sholler, G. AMXT-1501, a novel polyamine transport inhibitor, synergizes with DFMO in inhibiting neuroblastoma cell proliferation by targeting both ornithine decarboxylase and polyamine transport. *Int. J. Cancer* **2013**, *133*, 1323–1333. [CrossRef] [PubMed]

103. Casero, R.A., Jr.; Marton, L.J. Targeting polyamine metabolism and function in cancer and other hyperproliferative diseases. *Nat. Rev. Drug Discov.* **2007**, *6*, 373–390. [CrossRef] [PubMed]

104. Alexander, E.T.; Minton, A.; Peters, M.C.; Phanstiel, O.T.; Gilmour, S.K. A novel polyamine blockade therapy activates an anti-tumor immune response. *Oncotarget* **2017**, *8*, 84140–84152. [CrossRef] [PubMed]

105. Devens, B.H.; Weeks, R.S.; Burns, M.R.; Carlson, C.L.; Brawer, M.K. Polyamine depletion therapy in prostate cancer. *Prostate Cancer Prostatic Dis.* **2000**, *3*, 275–279. [CrossRef] [PubMed]

106. Gitto, S.B.; Pandey, V.; Oyer, J.L.; Copik, A.J.; Hogan, F.C.; Phanstiel, O., 4th; Altomare, D.A. Difluoromethylornithine Combined with a Polyamine Transport Inhibitor Is Effective against Gemcitabine Resistant Pancreatic Cancer. *Mol. Pharm.* **2018**. [CrossRef] [PubMed]

107. Muth, A.; Madan, M.; Archer, J.J.; Ocampo, N.; Rodriguez, L.; Phanstiel, O., 4th. Polyamine transport inhibitors: Design, synthesis, and combination therapies with difluoromethylornithine. *J. Med. Chem.* **2014**, *57*, 348–363. [CrossRef] [PubMed]

108. Massaro, C.; Thomas, J.; Phanstiel Iv, O. Investigation of Polyamine Metabolism and Homeostasis in Pancreatic Cancers. *Med. Sci.* **2017**, *5*, 32. [CrossRef] [PubMed]

109. Phanstiel, O., 4th. An overview of polyamine metabolism in pancreatic ductal adenocarcinoma. *Int. J. Cancer* **2017**. [CrossRef] [PubMed]

110. Mohammed, A.; Janakiram, N.B.; Madka, V.; Ritchie, R.L.; Brewer, M.; Biddick, L.; Patlolla, J.M.; Sadeghi, M.; Lightfoot, S.; Steele, V.E.; et al. Eflornithine (DFMO) prevents progression of pancreatic cancer by modulating ornithine decarboxylase signaling. *Cancer Prev. Res.* **2014**, *7*, 1198–1209. [CrossRef] [PubMed]

111. Bailey, H.H.; Kim, K.; Verma, A.K.; Sielaff, K.; Larson, P.O.; Snow, S.; Lenaghan, T.; Viner, J.L.; Douglas, J.; Dreckschmidt, N.E.; et al. A randomized, double-blind, placebo-controlled phase 3 skin cancer prevention study of α-difluoromethylornithine in subjects with previous history of skin cancer. *Cancer Prev. Res.* **2010**, *3*, 35–47. [CrossRef] [PubMed]

112. Jeter, J.M.; Alberts, D.S. Difluoromethylornithine: The proof is in the polyamines. *Cancer Prev. Res.* **2012**, *5*, 1341–1344. [CrossRef] [PubMed]

113. Kreul, S.M.; Havighurst, T.; Kim, K.; Mendonca, E.A.; Wood, G.S.; Snow, S.; Borich, A.; Verma, A.; Bailey, H.H. A phase III skin cancer chemoprevention study of DFMO: Long-term follow-up of skin cancer events and toxicity. *Cancer Prev. Res.* **2012**, *5*, 1368–1374. [CrossRef] [PubMed]

114. Kim, H.I.; Schultz, C.R.; Buras, A.L.; Friedman, E.; Fedorko, A.; Seamon, L.; Chandramouli, G.V.R.; Maxwell, G.L.; Bachmann, A.S.; Risinger, J.I. Ornithine decarboxylase as a therapeutic target for endometrial cancer. *PLoS ONE* **2017**, *12*, e0189044. [CrossRef] [PubMed]

115. Arisan, E.D.; Obakan, P.; Coker, A.; Palavan-Unsal, N. Inhibition of ornithine decarboxylase alters the roscovitine-induced mitochondrial-mediated apoptosis in MCF-7 breast cancer cells. *Mol. Med. Rep.* **2012**, *5*, 1323–1329. [PubMed]

116. Zhu, Q.; Jin, L.; Casero, R.A.; Davidson, N.E.; Huang, Y. Role of ornithine decarboxylase in regulation of estrogen receptor alpha expression and growth in human breast cancer cells. *Breast Cancer Res. Treat.* **2012**, *136*, 57–66. [CrossRef] [PubMed]

medical sciences

MDPI

Review

Role of Polyamines in Asthma Pathophysiology

Vaibhav Jain [1,2]

1   Centre of Excellence for Translational Research in Asthma & Lung Disease, CSIR-Institute of Genomics and Integrative Biology (CSIR-IGIB), Mall Road, Delhi 110007, India; jain.vaibhav@igib.in or vaibhavj85@outlook.com; Tel.: +91-112-766-2580; Fax: +91-11-2766-7471

2   Academy of Scientific and Innovative Research (AcSIR), Chennai 600113, India

Received: 12 December 2017; Accepted: 2 January 2018; Published: 6 January 2018

Abstract: Asthma is a complex disease of airways, where the interactions of immune and structural cells result in disease outcomes with airway remodeling and airway hyper-responsiveness. Polyamines, which are small-sized, natural super-cations, interact with negatively charged intracellular macromolecules, and altered levels of polyamines and their interactions have been associated with different pathological conditions including asthma. Elevated levels of polyamines have been reported in the circulation of asthmatic patients as well as in the lungs of a murine model of asthma. In various studies, polyamines were found to potentiate the pathogenic potential of inflammatory cells, such as mast cells and granulocytes (eosinophils and neutrophils), by either inducing the release of their pro-inflammatory mediators or prolonging their life span. Additionally, polyamines were crucial in the differentiation and alternative activation of macrophages, which play an important role in asthma pathology. Importantly, polyamines cause airway smooth muscle contraction and thus airway hyper-responsiveness, which is the key feature in asthma pathophysiology. High levels of polyamines in asthma and their active cellular and macromolecular interactions indicate the importance of the polyamine pathway in asthma pathogenesis; therefore, modulation of polyamine levels could be a suitable approach in acute and severe asthma management. This review summarizes the possible roles of polyamines in different pathophysiological features of asthma.

Keywords: spermine; spermidine; putrescine; polyamine metabolism; mast cells; eosinophils; neutrophils; M2 macrophages; airway smooth muscle cells

---

1. Introduction

Polyamines were discovered late in the 16th century as crystals in human semen by Antonie van Leeuwenhoek [1]. Now, polyamines are defined as positively charged aliphatic alkylamines which are natural supercations carrying two, three and four positive charges on putrescine $(H_3N^+(CH_2)_4{}^+NH_3)$, spermidine $(H_3N^+(CH_2)_4{}^+NH_2(CH_2)_3{}^+NH_3)$ and spermine $(H_3N^+(CH_2)_3{}^+NH_2 (CH_2)_4{}^+NH_2(CH_2)_3{}^+NH_3)$, respectively, at physiological pH [2]. These positively charged polyamines interact with negatively charged molecules in a cell such as DNA, RNA, proteins, and proteoglycans; hence, polyamines are involved in a myriad of biological processes [3,4]. Broadly, they are involved in the processes that play important roles in cell survival or death pathways, such as proliferation, differentiation, autophagy, apoptosis and translation [5–8]. Interestingly, depending on the concentration of polyamines and cell type, a cell may undergo proliferation, differentiation, apoptosis or quiescence. Generally, polyamines are thought to be required for cell proliferation and growth and high levels of polyamines have been observed in multiple cancerous conditions [9]. In contrast, their levels decline with aging, where replicative senescence of cells plays an important role [6,10]. However, the effects of polyamines are contextual and depending on their concentrations and cell type, their effects may vary. Cells with high proliferation rates may need higher levels of polyamines than their optimal levels.

However, in cells that are not directed for proliferation, supra-optimal levels of polyamines may lead to apoptosis, while sub-optimal levels of polyamines may lead to a quiescent state [8].

Asthma is a complex disease that is defined by chronic airway inflammation, remodeling and hyperresponsiveness. The pathogenesis of asthma is multifactorial and includes complex interactions between genetic and environmental factors leading to multiple phenotypes that differ in their clinical and physiological features as well as their response to therapy [11]. For example, two phenotypes of asthma include patients differing in their responsiveness to corticosteroid therapy. Unlike in corticosteroid-responsive asthmatics, where the airway inflammation is majorly eosinophilic, patients with severe asthma predominantly show neutrophilic airway inflammation and are poor responders to corticosteroid treatment [12]. This indicates that there is a need to discover alternative treatments for asthma, since existing treatments are not very effective for all severities of asthma. High levels of polyamines in the circulation of asthmatics with active symptoms and in the lungs of a murine model of asthma [13,14] indicates that polyamines could play an important role in respiratory diseases. Additionally, the interaction of polyamines with the immune and structural cells of asthma pathophysiology indicates that the polyamine pathway plays an active role in the asthmatic response.

This review summarizes the studies highlighting the importance of polyamines in asthma pathophysiology. The outcomes from various studies indicate that the polyamine pathway may play a decisive role and could be an important therapeutic target in different phenotypes of asthma such as acute and severe asthma.

2. Homeostasis of Polyamines

The levels of polyamines are tightly regulated by their finely tuned metabolism and transport (Figure 1) [15,16]. Polyamine metabolism is a very specific and well-studied process in comparison to the poorly explored polyamine transport mechanism. Of note, there are pharmacological modulators available for the key enzymes of polyamine metabolic pathway. The modulators of polyamine metabolism are used either to alter the levels of polyamines or to study the importance of polyamine metabolic enzymes in a cell. Of the modulators of polyamine metabolism, the most widely used compounds are: 2-difluoromethylornithine (DFMO) and bis(ethyl)norspermine (BENSPM) (Figure 1). DFMO is an irreversible inhibitor of the polyamine anabolic enzyme, ornithine decarboxylase (ODC). DFMO reduces the levels of putrescine and spermidine [17]; however, DFMO is generally ineffective in bringing down the levels of spermine due to other compensatory mechanisms of polyamine uptake and incomplete inhibition of ODC by DFMO [18,19]. BENSPM is a spermine analogue that suppresses the polyamine biosynthetic enzyme ODC, while it potently induces the polyamine catabolic enzymes spermidine/spermine N1-acetyltransferase (SAT1) and spermine oxidase (SMOX) and depletes the natural polyamines pools, i.e., putrescine, spermidine and spermine. BENSPM also competes for the import of polyamines [20,21].

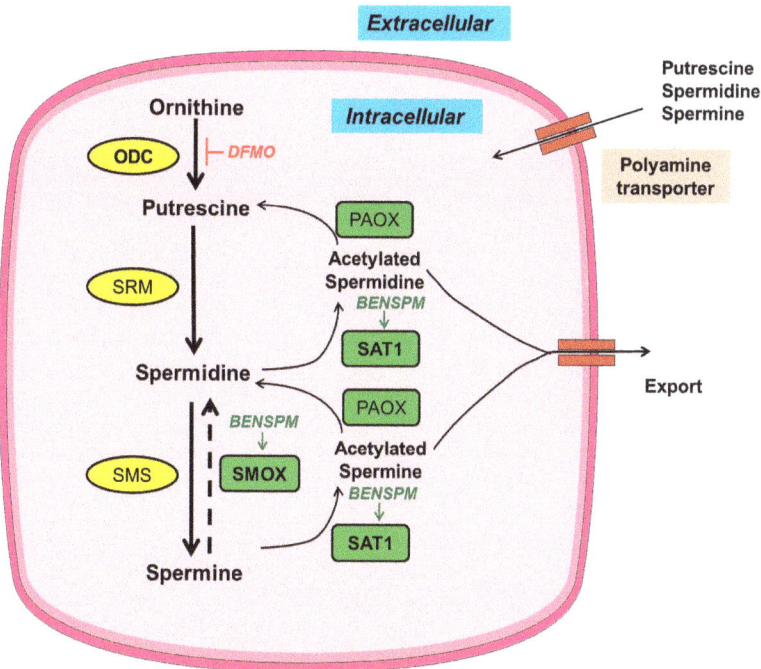

Figure 1. Schematic depiction of polyamine metabolism and transport. Anabolic and catabolic enzymes in yellow circles or green rectangles, respectively, with key metabolic enzymes in bold. Pharmacological inducer and inhibitor are italicized in green and red, respectively. Ornithine decarboxylase (ODC) synthesizes putrescine from ornithine; putrescine further gets converted into spermidine by spermidine synthase (SRM) and further to the largest polyamine spermine by spermine synthase (SMS). During catabolism, spermine can be back converted into smaller polyamines, i.e., spermidine and putrescine, by the concerted actions of spermidine/spermine N1-acetyltransferase (SAT1) and polyamine oxidase (PAOX) with an intermediate acetylation step mediated by SAT1; acetylated polyamines are either exported out of the cells or catabolized by PAOX. Spermine oxidase (SMOX), another enzyme of polyamine catabolism, directly converts spermine into spermidine without an intermediate acetylation step. In addition to de novo synthesis, polyamines can also be transported. BENSPM: bis(ethyl)norspermine; DMFO: 2-difluoromethylornithine.

3. Asthma Pathophysiology

Asthma is a heterogeneous airway disease that is associated with chronic airway inflammation. Asthmatic patients show a history of wheezing, shortness of breath, chest tightness and cough, all with variations in intensity over time. Airway inflammation in asthma involves a complex interplay of structural and immune cells [22]. The immune component includes mainly leukocytes, both granulocytes and lymphocytes. Dendritic cells process and present antigens to T lymphocytes, resulting in the development of type-2 T helper cell (Th2) responses [23]. This Th2 immune response plays a dominant role in asthma pathophysiology. Th2 cells produce cytokines like interleukin (IL)-4, IL-5 and IL-13. These cytokines and Th2 cells together enhance allergic response by inducing several pathways [24]. IL-4 promotes immunoglobulin class switching from immunoglobulin G (IgG) to IgE isotype in B cells [25]. Additionally, IL-4 induces the expression of IgE receptors on mast cells and basophils, which bind IgE and subsequently release preformed mediators that are important in asthma pathology [26]. IL-13 is another key cytokine of asthma; it shows pleiotropic effects on multiple cells. IL-13 is involved in the class switching of immunoglobulin to IgE isotype and promotes the

Th2 response, along with airway hyper-responsiveness and airway remodeling [27]. Interestingly, administration or overexpression of either the IL-4 or IL-13 cytokine alone, in naïve mice lungs, can generate key features of allergic airway disease [28,29]. IL-5 is an important cytokine for the survival, growth and activation of eosinophils, which promote innate and adaptive immune response during allergic airway inflammation [30]. In addition, macrophages play an active role in asthma. In asthma, monocytes differentiate into the M2 or alternatively activated (M2) macrophage (AAM) phenotype in a Th2-dominant microenvironment [31]. It has been shown that the abundance of these AAMs negatively correlates with the lung function of asthma patients [32].

In asthma, in addition to the immune cells, the active involvement of structural cells, i.e., of airway smooth muscle cells (ASMCs), fibroblasts and airway epithelial cells, have been explored. Remodeling of lung architecture in response to a Th2 environment increases the thickness of the sub-epithelial basement membrane in lung bronchus due to the proliferation of ASMCs and fibroblasts. Further, ASMCs and fibroblasts secrete collagen and other extracellular matrix molecules contributing to the thickness of the sub-epithelial region. Airway remodeling has been thought to contribute to airway hyper-responsiveness by increasing the stiffness of airways [33]. In addition to ASMCs and fibroblasts, current literature suggests that airway epithelium could play an active and causative role in asthma pathogenesis [34]. Airway epithelium is the first line of defense between lung and external environment. In asthma, airway epithelium has been reported to be damaged and is sensitive to apoptosis [35]. Since airway epithelium is exposed to the external environment, it encounters various environmental insults, allergens and antigens, which bring airway epithelial cells under stress. Injured or stressed airway epithelial cells secrete many factors such as cytokines. These cytokines, such as thymic stromal lymphopoietin (TSLP), IL-25 and IL-33, together called alarmins, are released by stressed airway epithelial cells. These alarmins can govern Th2 immune responses and other key features of asthma, such as airway hyper-responsiveness and airway remodeling [36]. Moreover, airway epithelial cells undergo mucus metaplasia and epithelial-to-mesenchymal transition, in which airway epithelial cells change their phenotype to mucus-producing goblet cells and a fibroblast phenotype, respectively, that contributes to airway wall remodeling [37]. Taken together, in asthma, immune and structural cells interact with each other, resulting in the disease pathophysiology; however, the cause and effect relationship between immune and structural component of asthma pathogenesis remains elusive.

4. Alteration of Polyamines in Asthma

The first report of polyamines in asthma was in the blood of asthmatic patients; the levels of polyamines were reported to be high in the blood of asthmatic patients, particularly during asthmatic attacks [13]. Additionally, elevated polyamine levels were reported in the lungs of a murine model of asthma [38]. In addition to the high levels in blood, the polyamine spermine was found to be significantly high in the bronchoalveolar lavage fluids of asthmatic patients, which is a direct reflection of lung microenvironment [14].

5. Polyamines Affect Immune Cells in Asthma

Eosinophils, neutrophils, basophils, mast cells and macrophages are all inflammatory cells involved in asthma pathology. The inflammatory response in allergic asthma is eosinophilic. However, along with eosinophilia, severe asthmatic patients show the involvement of a neutrophilic response [39]. Eosinophils and neutrophils are inflammatory granulocytes; they release several preformed inflammatory mediators stored in their secretory granules upon their activation in an allergic environment of the lungs. These mediators, such as histamines, leukotrienes, reactive oxygen species (ROS), and toxic granule proteins, induce asthmatic exacerbations and inflammation in the lungs by causing bronchoconstriction, tissue damage, airway inflammation and remodeling. Histamines are biogenic amines derived from the amino acid histidine and are present in mast cell granules; they induce bronchoconstriction, vasodilation and mucus secretion [40]. The generation of ROS by granulocytes in asthma is accompanied with tissue damage during airway inflammation [41]. Several oxidative products, like H_2O_2 and superoxide anions, are released that themselves cause damage and may further react with other molecules to generate damaging oxidative products such as leukotrienes, reactive nitrogen species, etc. Granulocytes also release leukotrienes, which are oxidative products of lipids and powerful bronchoconstrictors [42,43]. These mediators released by granulocytes are potential therapeutic targets for asthma. Polyamines affect immune cells during asthma in several ways (Figure 2), as briefly discussed below.

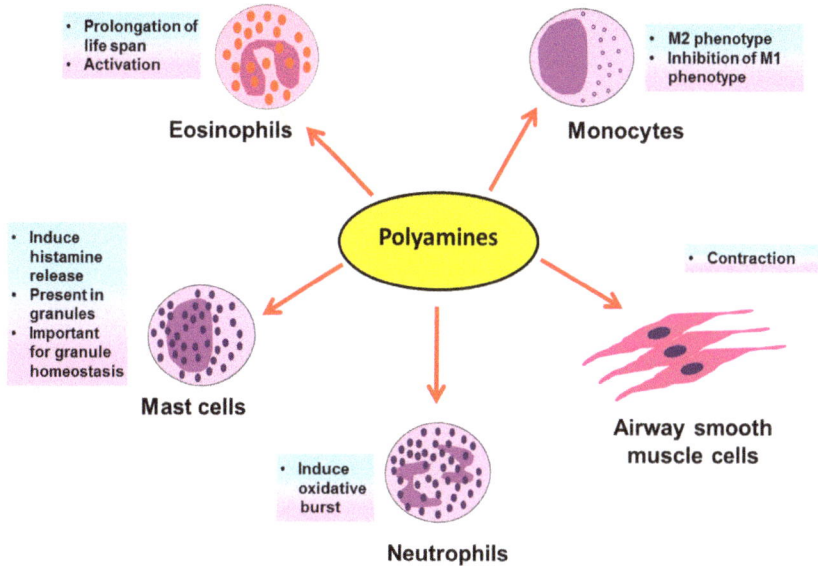

Figure 2. Possible effects of polyamines on immune and structural cells in asthma.

5.1. Polyamines and Mast Cells

The association of high levels of polyamines with the functional response of immune cells was uncovered by Kurosawa et al., who demonstrated the effect of polyamines on mast cells. Polyamines, specifically spermine and spermidine at higher concentrations (1 mM) were demonstrated to induce the rapid release of histamine, a bronchoconstrictor, from mast cells (Figure 2). Additionally, at lower concentrations (0.1 mM) polyamines alone were ineffective; however, they did potentiate IgE-induced histamine release [44]. The mechanism behind this induction was speculated to be the ability of spermine and spermidine to activate the phosphatidylinositol kinase enzyme in the secretory granules, which play a crucial role in the release of histamine from mast cells [45]. Another report for mast cells came from a well-known interaction of polyamines and proteoglycans. The proteoglycan serglycin is very important for maintaining secretory granule structure and the packaging of secretory compounds like histamine and serotonin in mast cell granules. Interestingly, polyamines, specifically spermine and spermidine, were found to be present in the mast cell secretory granules and the concerted interactions of both, polyamines and serglycin, was found to be crucial for maintaining secretory compounds in granules. By using DFMO, polyamines were reduced in mast cell granules and it was found that the ultrastructure of mast cell granules was disrupted along with a disturbance in the homeostasis of secretory compounds; the morphology observed with DFMO treatment was similar to the morphology seen after depletion of serglycin, showing the importance of both polyamines and serglycin in granule morphology and function [46]. Recently a transporter, vesicular polyamine transporter (VPAT) was identified in the mast cell. VPAT was found to be responsible for the active storage and release of polyamines from the secretory granules of mast cells [47].

5.2. Polyamines and Eosinophils

In context of another important granulocyte in asthma pathology, the eosinophil, an interesting study demonstrated that polyamines, especially spermine (median effective concentration of 15 μmol/L), prolong normal life span of eosinophils [48] (Figure 2). In this study, spermine inhibited the normal apoptosis of eosinophils from normal subjects in addition to asthmatic subjects, thereby prolonging their life span in ex vivo culture. Spermine was the most active polyamine followed by spermidine; however, putrescine was ineffective in inducing survival of eosinophils. In this report, spermine was found to inhibit apoptosis linked to mitochondrial permeability transition (mPT), along with the activities of effector apoptotic caspases 3/7 and the initiator caspases 8 and 9. Further, spermine induced the expression of the adhesion molecule integrin, CD11b, on eosinophils. Though increased expression of CD11b on eosinophils is related with the capacity of their super oxide generation [49], spermine alone was not sufficient; however, it potentiated the oxidative burst induced by a bacterial chemotactic peptide, N-formyl-methionyl-leucine-phenylalanine (FMLP) peptide.

5.3. Polyamines and Neutrophils

In severe asthma, neutrophils play a crucial role. To study the role of neutrophils in asthma, neutrophils from the blood of asthmatics are stimulated with FMLP or phorbol myristate acetate (PMA) to induce an oxidative burst [50], followed by the addition of different chemicals to be tested for their effects on neutrophils. In one study, FMLP- and PMA-stimulated neutrophils from mild and severe asthmatics were tested for their steroid responsiveness. In contrast to mild asthmatics, neutrophils from severe asthmatics were unresponsive to treatment with the steroid prednisolone, in terms of reducing oxidative burst [51]. Interestingly, in a separate study on human neutrophils, the polyamines putrescine, spermidine and spermine (10–100 μM) induced respiratory burst (Figure 2) generated by FMLP, possibly by enhancing of the availability of exogenous Ca^{2+} [52].

5.4. Polyamines and Monocytes/Macrophages

Macrophages are important antigen-presenting cells in asthma and constitute the most abundant immune cells in the lung (approximately 70% of the total immune cells) [32]. In the asthmatic patient's lung environment, monocytes majorly differentiate into AAMs [32,53]. These M2 macrophages are very important for the generation of the Th2 immune response in asthma. M2 macrophages are characterized by the increased expression of IL-4 and IL-13, which are very crucial cytokines in asthma pathogenesis. Interestingly, theses cytokines also promote the proliferation and differentiation of M2 macrophages in an autocrine manner [31]. Further, depletion of M2 macrophages in a murine model of allergic asthma was shown to attenuate inflammation and remodeling of airways [54]. A pivotal evidence of the interaction of polyamines and the Th2 response in asthma is evident from a study done in macrophages, where IL-4, a classical cytokine of allergic asthma, induced polyamine synthesis in macrophages and this production of polyamines orchestrated them into AAMs, while inhibiting the production of classically activated pro-inflammatory M1 macrophages [55] (Figure 2). Interestingly, the depletion of all polyamines after inducing polyamine catabolism using the drug BENSPM markedly reduced the expression levels of several genes or markers of AAMs, confirming a crucial role of polyamines in the polarization of AAMs. However, as the use of DFMO was not effective in reducing the polyamine spermine, it was used to uncover a set of genes modulated by the levels of putrescine and spermidine. The polyamines were needed as cofactors to induce the expression of several AAM genes induced by IL-4. Out of a total of 15 markers of AAM tested in this study, 9 markers were affected by the depletion of polyamines; the remaining unaltered markers were termed as polyamine-independent markers of AAMs. By using DFMO and BENSPM, this study exclusively determined the set of genes specifically induced by the individual polyamines putrescine, spermidine and spermine.

6. Polyamines and Structural Airway Smooth Muscle Cells

Along with inflammation, airway remodeling is a key feature of asthma pathology. Airway smooth muscle cells (ASMCs) are the major contributors in airway remodeling. A reversible contractility of the ASMCs leads to airflow limitation in asthmatic patients [56]. ASMC proliferation rate was found to be higher in asthmatics in comparison to ASMCs from normal individuals which leads to increase the numbers of ASMCs in airways, thereby thickening the airway wall [57]. ASMCs express smooth muscle-specific contractile proteins like α-smooth muscle actin, smooth muscle-specific heavy chain and calponin, myosin light chain kinase and others [58]. These proteins impart smooth muscles a contractile phenotype. The mechanism of ASMCs contraction includes the interactions of actin and myosin light chain proteins [59] and requires an increase in intracellular Ca^{++}. The main source of this calcium is intracellular sarcoplasmic reticulum (SR) stores. The release of Ca^{++} from the SR involves the entry of inositol 1,4,5-trisphosphate (IP3) into the SR. The formation of IP3 starts from membrane-bound phosphatidylinositol 4-phosphate (PI4P), which after phosphorylation by PIP5K1γ gets converted into phosphatidylinositol 4,5-bisphosphate (PIP2). PIP2 is the precursor of soluble IP3 and this reaction is catalyzed by the enzyme phospholipase C (PLC), which generates membrane-bound residual di-acyl glycerol (DAG) and soluble IP3 from PIP2 [60].

Polyamines are well known for their effects on cell proliferation and viability. Airway remodeling in asthma requires ASMCs proliferation [57]. During proliferation, ASMCs change from the contractile to the synthetic phenotype. The role of polyamines in this context was speculated and the expression of ODC was found to be responsible for this transition in arterial smooth muscle cells, since inhibiting ODC using DFMO prevented this transition. In murine ASMCs, polyamines, specifically spermine, were shown to induce acetylcholine-mediated contraction in ex vivo cultures of ASMCs [61] (Figure 2). A possible effect of spermine in this contraction could be due to the reduction in the availability of nitric oxide (NO) for the relaxation of smooth muscles, since L-arginine is a common precursor for the synthesis of both polyamines as well as NO. Additionally, polyamines, mainly spermine, along with Mg^{2+} ions are potent inducers of PIP5K1γ in ASMCs, which is a determinant enzyme for the formation of the IP3 molecule, an inducer of the release of sarcoplasmic Ca^{2+} [62]. An interesting role of the polyamine catabolic enzyme SAT1 in smooth muscle contraction was published by Chen et al., who showed the intracellular interaction of SAT1 with a transmembrane protein α9β1 integrin is important for the smooth muscle relaxation [63]. This interaction was important for the degradation of larger polyamines, spermine and spermidine, to the smallest polyamine putrescine, which does not induce PIP5K1γ and thereby leads to the relaxation of smooth muscle contraction. In this study, treatment of isolated murine tracheal rings with BENSPM induced polyamine catabolism and led to the relaxation of methacholine (a bronchoconstrictor)-induced ASMCs; however, induction of SAT1 was not effective when α9β1 integrin was knocked out. In addition to in vitro findings in tracheal ASMCs, spermine nebulization to naïve mice was shown to induce airway hyperresponsiveness [14].

7. Discussion

Though polyamines are needed for cell survival and proliferation, if not maintained optimally, their altered levels may dispose a cell to various kinds of stress. Asthma is a complex disease which can be defined by chronic airway inflammation, remodeling and hyperresponsiveness; it is interesting to note that polyamines have been linked with all of these features. Polyamines have been found to be high in the circulation of asthmatics with active symptoms and in the lungs of murine model of experimental asthma [14,44]. Since polyamines have been found to be high in the circulation as well as in the lungs, it is pertinent that they may be involved in various molecular or cellular events, locally as well as systemically. Different well-designed studies have reported their roles in these scenarios.

In immune cells such as mast cells and neutrophils, polyamines increase the pathogenic potential by inducing an active state; in the case of mast cells, by imparting structural support to secretory granules and also by inducing the release of preformed inflammatory mediators; whereas in neutrophils, by enhancing the generation of ROS. While one study on mast cells, extracellular addition of polyamines enhanced histamine release [44], another report demonstrated that polyamines themselves were present in the secretory granules of mast cells [46]. Interestingly, polyamine-containing granules were lacking histamines, but depletion of polyamines after using DFMO resulted in an increased secretion of histamine, indicating a definite but incompletely explored role of polyamines in histamine storage and secretion from mast cells. The discovery of a mast cell polyamine transporter, VPAT [47], could be tested in an in vivo model of asthma or patient samples. Its potential to play an important role in asthma pathology could implicate it as an important therapeutic target.

With another granulocyte, the neutrophil, polyamines were found to potentiate the ROS generation mediated by FMLP [52]. However, this observation was not further explored in full detail and it was speculated that polyamines either increase the interaction of the bacterial peptide with neutrophils or enhance the entry of exogenous Ca^{2+}. Moreover, this finding could be tested in severe asthmatic conditions, where a neutrophilic response is dominant and patients are resistant to existing steroid therapy.

Eosinophils from normal and asthmatic patients, when exposed to spermine-rich conditions, survive longer than their natural life span and increase the expression of the activation marker CD11b on their surface. Spermine was demonstrated to inhibit mPT-induced apoptosis of eosinophils along with caspase cascade activation [48]. This interesting observation could be tested further in the context of other granulocytes in asthma where polyamines may also show survival-promoting effects. Hence, a polyamine-rich environment may induce the active repertoire of inflammatory immune cells in asthma, which may further amplify this response with their active interactions with other immune cells.

In addition to enhancing the inflammatory potential of granulocytes, polyamines play a crucial role in the differentiation of monocytes towards the M2 phenotype in the presence of a Th2 cytokine of asthma [55]. This study highlighted the induction of several polyamine-dependent genes important for the M2 phenotype, whereas genes important for the M1 phenotype were repressed in the presence of polyamines. Earlier reports had shown the importance of polyamines in the modulation of gene expression by affecting the acetylation of histones, ultimately affecting the access of transcription factors to their binding sites on DNA [64,65]—a mechanism speculated for the polyamine-mediated alternative activation of macrophages. Interestingly, during *Helicobacter pylori* and *Citrobacter rodentium* infections, ODC was found to inhibit M1, while ODC inhibition promoted the M1 phenotype. Further, myeloid-specific deletion of ODC promoted the M1 phenotype, with the secretion of M1-specific cytokines and chemokines that resulted in the resolution of bacterial infection in mice [66]. However, this deletion did not promote the M2 phenotype in contrast to the previous reports, demonstrating that the presence or absence of polyamines can switch the macrophage phenotype to M1 or M2, respectively. Moreover, during *H. pylori* and *C. rodentium* infection, spermine and spermidine were not very effective in suppressing the M1 phenotype, however, putrescine played a decisive role in chromatin remodeling (for euchromatin transformation) and in inhibition of M1 differentiation.

These findings together suggest that the presence of polyamines is a determinant for differentiation towards the M2 phenotype in an environment-dependent manner. In the presence of M2-promoting factors like IL-4, polyamines act as essential cofactors for M2 polarization and simultaneously inhibit the gene expression of the M1 phenotype. However, another study demonstrated that during bacterial infection, depletion of polyamines is required for M1 polarization, but it does not promote the M2 phenotype in the absence of other required factors for M2 polarization. In conclusion, polyamines are cofactors for the differentiation towards M2 phenotype in the presence of a Th2-dominant environment, and the loss of either the polyamine anabolic enzyme ODC or polyamines, specifically putrescine, orchestrates the M1 response.

ASMCs are the crucial component of asthma-associated airway hyper-responsiveness and airway remodeling. In asthma, spermine could play an important role in the induction of ASMCs contraction and thus airway hyper-responsiveness, by either reducing the availability of NO for ASMCs relaxation [14] or by activating the PIP5K1γ enzyme in ASMCs [63]. While the interaction between the polyamine catabolic enzyme SAT1 and intracellular domain of α9β1 integrin was found to be important for the relaxation of ASMCs, SAT1 alone was not effective in reducing the local concentration of polyamines in the absence of integrin [63]. Hence, induction of SAT1 and restoration of reduced levels of α9β1 integrin, both could be effective in relaxation of ASMCs contractions in asthma. However, the role of SAT1 in asthma needs to be explored further.

The dysregulation of polyamine anabolism in asthma was thought to be an important driver for the high levels of polyamine found, and an attempt was made to bring down these high levels by inhibiting ODC using DFMO in an allergic acute model of asthma. Though DFMO could not bring down the high levels of spermine and the histological features of asthma, it could successfully reduce airway hyper-responsiveness in a murine asthma model [14]. In addition to their effects on airway mechanics, polyamines induce the expression of transforming growth factor beta (TGF-β) in intestinal epithelial cells [67]. TGF-β is an important cytokine for airway remodeling during asthma. It has multiple effects on different immune and structural cells of the lung and has an active role in pulmonary fibrosis [68]. Notably, spermine levels were reported high in the sputum of cystic fibrosis (CF) patients and were associated with pulmonary exacerbations. Moreover, treating CF patients with antibiotics improved bronchoconstriction with a reduction in the levels of spermine [61]. Additionally, spermine treatment to mouse bronchial rings enhanced acetylcholine-induced constriction. These findings in structural cells indicate a pathophysiological role of polyamines in asthma. In addition to polyamine metabolism, polyamine transport may also play a crucial role in asthma pathogenesis, but due to a lack of in-depth knowledge of polyamine transport in mammalian systems, this area requires research to reach the level of therapeutics.

The polyamine pathway seems to be a very promising target in asthma due to its involvement in the important processes which play crucial roles in multiple pathways of asthma pathogenesis. Since asthma is a complex disease, finding a common node involving polyamines which regulates multiple pathways of asthma could be beneficial.

8. Conclusions and Future Perspective

To the best of my knowledge, this review summarizes the key research reports to date, relevant to asthma and polyamines. The polyamine pathway seems a central node, which can regulate multiple arms of asthma pathogenesis. Though the effects of polyamines on immune and structural cells have been reported, in depth knowledge of polyamine involvement in asthma pathogenesis still requires further investigation. Different pharmacological modulators of the polyamine pathway could be tested in asthmatic conditions. Hence, positive outcomes from findings of the polyamine pathway in asthma provide an impetus to investigate the polyamine pathway in more detail in asthma.

Acknowledgments: I apologize to authors whose works could not be cited due to space limitations. V.J. received Senior Research Fellowship from The Department of Biotechnology (DBT), New Delhi, India. The author acknowledges Balaram Ghosh, Anurag Agrawal and Ulganathan Mabalirajan for the critical review of this manuscript.

Conflicts of Interest: The author declares no competing interest.

References

1. Dobell, C. *Antony van Leeuwenhoek and His "Little Animals"; Being Some Account of the Father of Protozoology and Bateriology and His Multifarious Discoveries in These Disciplines*; John Bale, Sons & Danielson Ltd.: London, UK, 1932.
2. Woster, P.M. *Polyamine Cell Signaling*; Humana Press: New York, NY, USA, 2006; pp. 3–24.
3. Miller-Fleming, L.; Olin-Sandoval, V.; Campbell, K.; Ralser, M. Remaining Mysteries of Molecular Biology: The Role of Polyamines in the Cell. *J. Mol. Biol.* **2015**, *427*, 3389–3406. [CrossRef] [PubMed]
4. Tabor, C.W.; Tabor, H. Polyamines. *Annu. Rev. Biochem.* **1984**, *53*, 749–790. [CrossRef] [PubMed]
5. Eisenberg, T.; Knauer, H.; Schauer, A.; Büttner, S.; Ruckenstuhl, C.; Carmona-Gutierrez, D.; Ring, J.; Schroeder, S.; Magnes, C.; Antonacci, L.; et al. Induction of autophagy by spermidine promotes longevity. *Nat. Cell Biol.* **2009**, *11*, 1305–1314. [CrossRef] [PubMed]

6.   Minois, N.; Carmona-Gutierrez, D.; Madeo, F. Polyamines in aging and disease. *Aging* **2011**, *3*, 716–732. [CrossRef] [PubMed]

7.   Pegg, A.E. Functions of Polyamines in Mammals. *J. Biol. Chem.* **2016**, *291*, 14904–14912. [CrossRef] [PubMed]

8.   Seiler, N.; Raul, F. Polyamines and apoptosis. *J. Cell. Mol. Med.* **2005**, *9*, 623–642. [CrossRef] [PubMed]

9.   Nowotarski, S.L.; Woster, P.M.; Casero, R.A., Jr. Polyamines and cancer: Implications for chemotherapy and chemoprevention. *Expert Rev. Mol. Med.* **2013**, *15*, e3. [CrossRef] [PubMed]

10.   Campisi, J. Replicative senescence: An old lives' tale? *Cell* **1996**, *84*, 497–500. [CrossRef]

11.   Wenzel, S.E. Asthma phenotypes: The evolution from clinical to molecular approaches. *Nat. Med.* **2012**, *18*, 716–725. [CrossRef] [PubMed]

12.   Wood, L.G.; Baines, K.J.; Fu, J.; Scott, H.A.; Gibson, P.G. The Neutrophilic Inflammatory Phenotype Is Associated With Systemic Inflammation in Asthma. *Chest* **2012**, *142*, 86–93. [CrossRef] [PubMed]

13.   Kurosawa, M.; Shimizu, Y.; Tsukagoshi, H.; Ueki, M. Elevated levels of peripheral-blood, naturally occurring aliphatic polyamines in bronchial asthmatic patients with active symptoms. *Allergy* **1992**, *47*, 638–643. [CrossRef] [PubMed]

14.   North, M.L.; Grasemann, H.; Khanna, N.; Inman, M.D.; Gauvreau, G.M.; Scott, J.A. Increased ornithine-derived polyamines cause airway hyperresponsiveness in a mouse model of asthma. *Am. J. Respir. Cell Mol. Biol.* **2013**, *48*, 694–702. [CrossRef] [PubMed]

15.   Wallace, H.M.; Fraser, A.V.; Hughes, A. A perspective of polyamine metabolism. *Biochem. J.* **2003**, *376*, 1–14. [CrossRef] [PubMed]

16.   Seiler, N.; Delcros, J.G.; Moulinoux, J.P. Polyamine transport in mammalian cells. *Int. J. Biochem. Cell Biol.* **1996**, *28*, 843–861. [CrossRef]

17.   Babbar, N.; Gerner, E.W. Targeting polyamines and inflammation for cancer prevention. *Clin. Cancer Prev.* **2011**, *188*, 49–64. [CrossRef]

18.   Muth, A.; Madan, M.; Archer, J.J.; Ocampo, N.; Rodriguez, L.; Phanstiel, O. Polyamine transport inhibitors: Design, synthesis, and combination therapies with difluoromethylornithine. *J. Med. Chem.* **2014**, *57*, 348–363. [CrossRef] [PubMed]

19.   Pegg, A.E.; McCann, P.P. Polyamine metabolism and function. *Am. J. Physiol.* **1982**, *243*, C212–C221. [CrossRef] [PubMed]

20.   Pegg, A.E. Spermidine/spermine-N^1-acetyltransferase: A key metabolic regulator. *Am. J. Physiol. Endocrinol. Metab.* **2008**, *294*, E995–E1010. [CrossRef] [PubMed]

21.   Porter, C.W.; Bernacki, R.J.; Miller, J.; Bergeron, R.J. Antitumor activity of N^1,N^{11}-bis(ethyl)norspermine against human melanoma xenografts and possible biochemical correlates of drug action. *Cancer Res.* **1993**, *53*, 581–586. [PubMed]

22.   Ramakrishna, L.; de Vries, V.C.; Curotto de Lafaille, M.A. Cross-roads in the lung: Immune cells and tissue interactions as determinants of allergic asthma. *Immunol. Res.* **2012**, *53*, 213–228. [CrossRef] [PubMed]

23.   Nakano, H.; Free, M.E.; Whitehead, G.S.; Maruoka, S.; Wilson, R.H.; Nakano, K.; Cook, D.N. Pulmonary CD103+ dendritic cells prime Th2 responses to inhaled allergens. *Mucosal Immunol.* **2012**, *5*, 53–65. [CrossRef] [PubMed]

24.   Holgate, S.T. Innate and adaptive immune responses in asthma. *Nat. Med.* **2012**, *18*, 673–683. [CrossRef] [PubMed]

25.   Lebman, D.A.; Coffman, R.L. Interleukin 4 causes isotype switching to IgE in T cell-stimulated clonal B cell cultures. *J. Exp. Med.* **1988**, *168*, 853–862. [CrossRef] [PubMed]

26.   Pawankar, R.; Okuda, M.; Yssel, H.; Okumura, K.; Ra, C. Nasal mast cells in perennial allergic rhinitics exhibit increased expression of the Fc epsilonRI, CD40L, IL-4, and IL-13, and can induce IgE synthesis in B cells. *J. Clin. Investig.* **1997**, *99*, 1492–1499. [CrossRef] [PubMed]

27.   Wills-Karp, M. Interleukin-13 in asthma pathogenesis. *Curr. Allergy Asthma Rep.* **2004**, *4*, 123–131. [CrossRef] [PubMed]

28.   Zhu, Z.; Homer, R.J.; Wang, Z.; Chen, Q.; Geba, G.P.; Wang, J.; Zhang, Y.; Elias, J.A. Pulmonary expression of interleukin-13 causes inflammation, mucus hypersecretion, subepithelial fibrosis, physiologic abnormalities, and eotaxin production. *J. Clin. Investig.* **1999**, *103*, 779–788. [CrossRef] [PubMed]

29.   Perkins, C.; Wills-Karp, M.; Finkelman, F.D. IL-4 induces IL-13-independent allergic airway inflammation. *J. Allergy Clin. Immunol.* **2006**, *118*, 410–419. [CrossRef] [PubMed]

30. Varricchi, G.; Canonica, G.W. The role of interleukin 5 in asthma. *Expert Rev. Clin. Immunol.* **2016**, *12*, 903–905. [CrossRef] [PubMed]

31. Martinez, F.O.; Helming, L.; Gordon, S. Alternative Activation of Macrophages: An Immunologic Functional Perspective. *Annu. Rev. Immunol.* **2009**, *27*, 451–483. [CrossRef] [PubMed]

32. Girodet, P.-O.; Nguyen, D.; Mancini, J.D.; Hundal, M.; Zhou, X.; Israel, E.; Cernadas, M. Alternative Macrophage Activation Is Increased in Asthma. *Am. J. Respir. Cell Mol. Biol.* **2016**, *55*, 467–475. [CrossRef] [PubMed]

33. Siddiqui, S.; Martin, J.G. Structural aspects of airway remodeling in asthma. *Curr. Allergy Asthma Rep.* **2008**, *8*, 540–547. [CrossRef] [PubMed]

34. Holgate, S.T. Epithelium dysfunction in asthma. *J. Allergy Clin. Immunol.* **2007**, *120*, 1233–1244. [CrossRef] [PubMed]

35. Zhou, C.; Yin, G.; Liu, J.; Liu, X.; Zhao, S. Epithelial apoptosis and loss in airways of children with asthma. *J. Asthma Off. J. Assoc. Care Asthma* **2011**, *48*, 358–365. [CrossRef] [PubMed]

36. Bartemes, K.R.; Kita, H. Dynamic role of epithelium-derived cytokines in asthma. *Clin. Immunol.* **2012**, *143*, 222–235. [CrossRef] [PubMed]

37. Avila, P.C. Plasticity of airway epithelial cells. *J. Allergy Clin. Immunol.* **2011**, *128*, 1225–1226. [CrossRef] [PubMed]

38. Zimmermann, N.; King, N.E.; Laporte, J.; Yang, M.; Mishra, A.; Pope, S.M.; Muntel, E.E.; Witte, D.P.; Pegg, A.A.; Foster, P.S.; et al. Dissection of experimental asthma with DNA microarray analysis identifies arginase in asthma pathogenesis. *J. Clin. Investig.* **2003**, *111*, 1863–1874. [CrossRef] [PubMed]

39. Fahy, J.V. Eosinophilic and Neutrophilic Inflammation in Asthma: Insights from Clinical Studies. *Proc. Am. Thorac. Soc.* **2009**, *6*, 256–259. [CrossRef] [PubMed]

40. White, M.V. The role of histamine in allergic diseases. *J. Allergy Clin. Immunol.* **1990**, *86*, 599–605. [CrossRef]

41. Vargas, L.; Patiño, P.J.; Montoya, F.; Vanegas, A.C.; Echavarría, A.; García de Olarte, D. A study of granulocyte respiratory burst in patients with allergic bronchial asthma. *Inflammation* **1998**, *22*, 45–54. [CrossRef] [PubMed]

42. Sampson, A.P. The role of eosinophils and neutrophils in inflammation. *Clin. Exp. Allergy J. Br. Soc. Allergy Clin. Immunol.* **2000**, *30* (Suppl. 1), 22–27. [CrossRef]

43. Reuter, S.; Stassen, M.; Taube, C. Mast cells in allergic asthma and beyond. *Yonsei Med. J.* **2010**, *51*, 797–807. [CrossRef] [PubMed]

44. Kurosawa, M.; Uno, D.; Kobayashi, S. Naturally occurring aliphatic polyamines-induced histamine release from rat peritoneal mast cells. *Allergy* **1991**, *46*, 349–354. [CrossRef] [PubMed]

45. Kurosawa, M.; Uno, D.; Hanawa, K.; Kobayashi, S. Polyamines stimulate the phosphorylation of phosphatidylinositol in rat mast cell granules. *Allergy* **1990**, *45*, 262–267. [CrossRef] [PubMed]

46. García-Faroldi, G.; Rodríguez, C.E.; Urdiales, J.L.; Pérez-Pomares, J.M.; Dávila, J.C.; Pejler, G.; Sánchez-Jiménez, F.; Fajardo, I. Polyamines Are Present in Mast Cell Secretory Granules and Are Important for Granule Homeostasis. *PLoS ONE* **2010**, *5*, e15071. [CrossRef] [PubMed]

47. Takeuchi, T.; Harada, Y.; Moriyama, S.; Furuta, K.; Tanaka, S.; Miyaji, T.; Omote, H.; Moriyama, Y.; Hiasa, M. Vesicular Polyamine Transporter Mediates Vesicular Storage and Release of Polyamine from Mast Cells. *J. Biol. Chem.* **2017**, *292*, 3909–3918. [CrossRef] [PubMed]

48. Ilmarinen, P.; Moilanen, E.; Erjefält, J.S.; Kankaanranta, H. The polyamine spermine promotes survival and activation of human eosinophils. *J. Allergy Clin. Immunol.* **2015**, *136*, 482–484. [CrossRef] [PubMed]

49. Walker, C.; Rihs, S.; Braun, R.K.; Betz, S.; Bruijnzeel, P.L. Increased expression of CD11b and functional changes in eosinophils after migration across endothelial cell monolayers. *J. Immunol.* **1993**, *150*, 4061–4071. [PubMed]

50. Combadière, C.; el Benna, J.; Pedruzzi, E.; Hakim, J.; Périanin, A. Stimulation of the human neutrophil respiratory burst by formyl peptides is primed by a protein kinase inhibitor, staurosporine. *Blood* **1993**, *82*, 2890–2898. [PubMed]

51. Mann, B.S.; Chung, K.F. Blood neutrophil activation markers in severe asthma: Lack of inhibition by prednisolone therapy. *Respir. Res.* **2006**, *7*, 59. [CrossRef] [PubMed]

52. Guarnieri, C.; Georgountzos, A.; Caldarera, I.; Flamigni, F.; Ligabue, A. Polyamines stimulate superoxide production in human neutrophils activated by N-fMet-Leu-Phe but not by phorbol myristate acetate. *Biochim. Biophys. Acta* **1987**, *930*, 135–139. [CrossRef]

53. Melgert, B.N.; ten Hacken, N.H.; Rutgers, B.; Timens, W.; Postma, D.S.; Hylkema, M.N. More alternative activation of macrophages in lungs of asthmatic patients. *J. Allergy Clin. Immunol.* **2011**, *127*, 831–833. [CrossRef] [PubMed]

54. Lee, Y.G.; Jeong, J.J.; Nyenhuis, S.; Berdyshev, E.; Chung, S.; Ranjan, R.; Karpurapu, M.; Deng, J.; Qian, F.; Kelly, E.A.B.; et al. Recruited Alveolar Macrophages, in Response to Airway Epithelial–Derived Monocyte Chemoattractant Protein 1/CCL2, Regulate Airway Inflammation and Remodeling in Allergic Asthma. *Am. J. Respir. Cell Mol. Biol.* **2015**, *52*, 772–784. [CrossRef] [PubMed]

55. Van den Bossche, J.; Lamers, W.H.; Koehler, E.S.; Geuns, J.M.C.; Alhonen, L.; Uimari, A.; Pirnes-Karhu, S.; Van Overmeire, E.; Morias, Y.; Brys, L.; et al. Pivotal Advance: Arginase-1-independent polyamine production stimulates the expression of IL-4-induced alternatively activated macrophage markers while inhibiting LPS-induced expression of inflammatory genes. *J. Leukoc. Biol.* **2012**, *91*, 685–699. [CrossRef] [PubMed]

56. Noble, P.B.; Pascoe, C.D.; Lan, B.; Ito, S.; Kistemaker, L.E.M.; Tatler, A.L.; Pera, T.; Brook, B.S.; Gosens, R.; West, A.R. Airway smooth muscle in asthma: Linking contraction and mechanotransduction to disease pathogenesis and remodelling. *Pulm. Pharmacol. Ther.* **2014**, *29*, 96–107. [CrossRef] [PubMed]

57. Johnson, P.R.A.; Roth, M.; Tamm, M.; Hughes, M.; Ge, Q.; King, G.; Burgess, J.K.; Black, J.L. Airway Smooth Muscle Cell Proliferation Is Increased in Asthma. *Am. J. Respir. Crit. Care Med.* **2001**, *164*, 474–477. [CrossRef] [PubMed]

58. Halayko, A.J.; Salari, H.; MA, X.; Stephens, N.L. Markers of airway smooth muscle cell phenotype. *Am. J. Physiol.* **1996**, *270*, L1040–L1051. [CrossRef] [PubMed]

59. Gunst, S.J.; Tang, D.D. The contractile apparatus and mechanical properties of airway smooth muscle. *Eur. Respir. J.* **2000**, *15*, 600–616. [CrossRef] [PubMed]

60. Berridge, M.J. Inositol trisphosphate and calcium signalling mechanisms. *Biochim. Biophys. Acta Mol. Cell Res.* **2009**, *1793*, 933–940. [CrossRef] [PubMed]

61. Grasemann, H.; Shehnaz, D.; Enomoto, M.; Leadley, M.; Belik, J.; Ratjen, F. L-ornithine derived polyamines in cystic fibrosis airways. *PLoS ONE* **2012**, *7*, e46618. [CrossRef] [PubMed]

62. Chen, H.; Baron, C.B.; Griffiths, T.; Greeley, P.; Coburn, R.F. Effects of polyamines and calcium and sodium ions on smooth muscle cytoskeleton-associated phosphatidylinositol (4)-phosphate 5-kinase. *J. Cell. Physiol.* **1998**, *177*, 161–173. [CrossRef]

63. Chen, C.; Kudo, M.; Rutaganira, F.; Takano, H.; Lee, C.; Atakilit, A.; Robinett, K.S.; Uede, T.; Wolters, P.J.; Shokat, K.M.; et al. Integrin α9β1 in airway smooth muscle suppresses exaggerated airway narrowing. *J. Clin. Investig.* **2012**, *122*, 2916–2927. [CrossRef] [PubMed]

64. Childs, A.C.; Mehta, D.J.; Gerner, E.W. Polyamine-dependent gene expression. *Cell. Mol. Life Sci.* **2003**, *60*, 1394–1406. [CrossRef] [PubMed]

65. Hobbs, C.A.; Gilmour, S.K. High levels of intracellular polyamines promote histone acetyltransferase activity resulting in chromatin hyperacetylation. *J. Cell. Biochem.* **2000**, *77*, 345–360. [CrossRef]

66. Hardbower, D.M.; Asim, M.; Luis, P.B.; Singh, K.; Barry, D.P.; Yang, C.; Steeves, M.A.; Cleveland, J.L.; Schneider, C.; Piazuelo, M.B.; et al. Ornithine decarboxylase regulates M1 macrophage activation and mucosal inflammation via histone modifications. *Proc. Natl. Acad. Sci. USA* **2017**, *114*, E751–E760. [CrossRef] [PubMed]

67. Wang, J.Y.; Viar, M.J.; Li, J.; Shi, H.J.; McCormack, S.A.; Johnson, L.R. Polyamines are necessary for normal expression of the transforming growth factor-beta gene during cell migration. *Am. J. Physiol.* **1997**, *272*, G713–G720. [CrossRef] [PubMed]

68. Das, S.; Kumar, M.; Negi, V.; Pattnaik, B.; Prakash, Y.S.; Agrawal, A.; Ghosh, B. MicroRNA-326 regulates profibrotic functions of transforming growth factor-β in pulmonary fibrosis. *Am. J. Respir. Cell Mol. Biol.* **2014**, *50*, 882–892. [CrossRef] [PubMed]

medical sciences

MDPI

Review

Regulation of Polyamine Metabolism by Curcumin for Cancer Prevention and Therapy

Tracy Murray-Stewart and Robert A. Casero Jr. *

Johns Hopkins University, Sidney Kimmel Comprehensive Cancer Center, Baltimore, MD 21287, USA; tmurray2@jhmi.edu
* Correspondence: rcasero@jhmi.edu; Tel.: +1-410-955-8580

Received: 17 November 2017; Accepted: 14 December 2017; Published: 18 December 2017

Abstract: Curcumin (diferuloylmethane), the natural polyphenol responsible for the characteristic yellow pigment of the spice turmeric (*Curcuma longa*), is traditionally known for its antioxidant, anti-inflammatory, and anticarcinogenic properties. Capable of affecting the initiation, promotion, and progression of carcinogenesis through multiple mechanisms, curcumin has potential utility for both chemoprevention and chemotherapy. In human cancer cell lines, curcumin has been shown to decrease ornithine decarboxylase (ODC) activity, a rate-limiting enzyme in polyamine biosynthesis that is frequently upregulated in cancer and other rapidly proliferating tissues. Numerous studies have demonstrated that pretreatment with curcumin can abrogate carcinogen-induced ODC activity and tumor development in rodent tumorigenesis models targeting various organs. This review summarizes the results of curcumin exposure with regard to the modulation of polyamine metabolism and discusses the potential utility of this natural compound in conjunction with the exploitation of dysregulated polyamine metabolism in chemopreventive and chemotherapeutic settings.

Keywords: curcumin; diferuloylmethane; ornithine decarboxylase; polyamine; NF-κB; chemoprevention; carcinogenesis; polyphenol

1. Introduction

Chemoprevention entails the long-term use of synthetic or natural agents by healthy individuals, particularly those with a predisposing cancer risk, to delay disease onset. As such, potential side effects and off-target effects must be absolutely minimal. Natural products derived from foods are therefore at an advantage due to their accessibility and history of safe consumption. Epithelial cancers are often age-related cancers: through its long-term, direct interaction with environmental and dietary factors, the epithelium has the greatest potential for interactions that might prevent or modulate the course of tumorigenesis. Naturally, gastrointestinal (GI) cancers have one of the greatest potentials for dietary factor influence. As approximately 20% of cancers worldwide are associated with infection or inflammation [1,2], the anti-inflammatory and antioxidant properties associated with many natural products might be of particular value. Nutritional components also have the potential to participate in therapeutic strategies, and elucidating the molecular mechanisms of these agents, including traditional medicines, is providing clues as to how they might best be incorporated into treatment regimens.

2. Polyamines and Cancer

Increases in polyamine biosynthesis and intracellular polyamine content are some of the most consistent biochemical alterations observed in cancer cells of all types, indicating their importance in tumorigenesis [3,4]. The mammalian polyamines include spermine, spermidine, and putrescine, which are essential polycations with pleiotropic roles in cellular proliferation and survival (Figure 1) [3]. Due to their positive charge at physiological pH, many of the essential functions of polyamines stem from their interactions with negatively charged cellular components, including DNA, RNA, certain proteins, and ion channels [5–8].

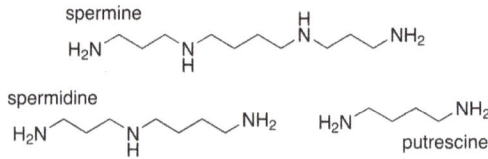

Figure 1. Chemical structures of the primary mammalian polyamines.

In neoplastic cells, loss of polyamine homeostasis occurs and is accompanied by dysregulated proliferation involving upregulated biosynthesis, downregulated catabolism, and increased uptake (Figure 2) [4,9–11]. The activity of ornithine decarboxylase (ODC), the initial rate-limiting step in polyamine biosynthesis, has been directly correlated with the rates of DNA synthesis and cellular proliferation in multiple tissue types [3].

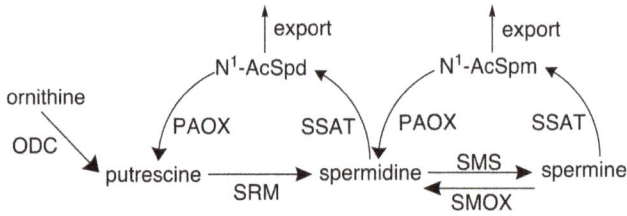

Figure 2. The mammalian polyamine pathway. Polyamines are derived from the amino acid ornithine, which is decarboxylated by ornithine decarboxylase (ODC) to form the diamine putrescine. Putrescine undergoes the sequential addition of 2 aminopropyl groups to form spermidine followed by spermine. These reactions are catalyzed by the spermidine and spermine synthases (SRM and SMS, respectively), using decarboxylated S-adenosylmethionine as the aminopropyl donor. Catabolism of spermine back to spermidine can occur through direct oxidation via spermine oxidase (SMOX) or by acetylation at the N^1 position by spermidine/spermine N^1-acetyltransferase (SSAT), followed by oxidation by the acetylpolyamine oxidase (PAOX). This latter two-step mechanism also back-converts spermidine to putrescine via an N^1-acetylspermidine (N^1-AcSpd) intermediate. Alternatively, acetylated spermine and spermidine can be readily exported from the cell.

The requirement for polyamines increases over the course of tumorigenesis, and studies in multiple human cancer types have demonstrated an elevation of ODC activity and/or polyamines in neoplastic or tumor tissue relative to adjacent normal tissue [12–15]. In a cohort of 50 primary breast tumors, the level of ODC activity demonstrated a strong negative correlation with both disease-free and overall survival, indicating ODC as a poor prognostic factor [16]. Polyamines have been implicated in oncogenic and viral transformation, and ODC activity is rapidly induced upon exposure to oncogenic or growth-promoting stimuli [10,17–19]. In particular, polyamine biosynthesis is upregulated at multiple steps by the *c-MYC* oncogene [20–22], and the activation of *k-RAS* rapidly induces ODC activity to promote malignant transformation and oncogenesis [23–25]. Colorectal carcinoma biopsies with activating *k-RAS* mutations were shown to have enhanced polyamine biosynthesis compared to those with wild-type *k-RAS* [26]. The expression level of ODC has been shown to directly correlate with the potential to promote tumorigenesis in both lymphomas and in solid tumors [27,28]. Furthermore, ODC activity rapidly increases with exposure to chemical carcinogens or tumor promoters, and this elevated expression is often utilized as a biomarker of tumor promotion in carcinogenesis models [3,29–31].

2.1. Targeting Polyamine Metabolism for Cancer Prevention

The potential targeting of polyamine biosynthesis as an antiproliferative strategy came with the recognition that elevated polyamine biosynthesis was a general requirement for the survival of cancer cells [32–34]. In fact, the ability of an agent to inhibit ODC activity is commonly considered a predictor of chemopreventive activity [14]. The most widely studied and successful inhibitor of polyamine biosynthesis, α-difluoromethylornithine (DFMO), or eflornithine, is enzyme-activated and irreversibly inhibits ODC through covalently binding with its active site [35]. DFMO typically elicits cytostatic effects in cell culture models through the depletion of putrescine and spermidine, with variable effects on spermine, and it is capable of preventing tumor formation in numerous animal models [36]. Early clinical trials investigating the prevention of colorectal cancer using low doses of DFMO have established the safety of its administration as well as its efficacy in reducing polyamine levels in colorectal mucosa [37].

Subsequent studies in colorectal cancer models have described enhanced antitumor benefits when combining DFMO with common non-steroidal anti-inflammatory drugs (NSAIDs), including sulindac and celecoxib. In addition to DFMO inhibiting ODC activity, the addition of an NSAID further decreased intracellular polyamine content by stimulating polyamine catabolism and export through activation of spermidine/spermine-N^1-acetyltransferase (SSAT), resulting in an additive reduction in the formation of colon tumors [38–40]. The combination of DFMO and sulindac has since been clinically investigated with impressive outcomes including a 70% reduction in the number of total metachronous colorectal adenomas [41,42], and additional studies are ongoing. Clinical trials have also been conducted investigating the efficacy of DFMO as a chemopreventive agent in individuals with a history of non-melanoma skin cancer, actinic keratosis (a squamous cell carcinoma precursor), Barrett's esophagus, and prostate cancer (reviewed in [43]). Although mostly in early clinical phases, DFMO was safely administered in each of these studies, and the results warranted further trials. Importantly, these studies have provided strong evidence supporting the targeting of polyamine metabolism as a valid strategy for the prevention of tumorigenesis, particularly in those susceptible to colorectal cancer.

It should be noted that the long-term use of DFMO is not without minor, but unwanted, side effects: a dose-related, but generally reversible, ototoxicity has frequently occurred in patients on long-term treatment. However, a reduced daily oral dose of DFMO has been shown to be sufficient in reducing polyamine levels in the colorectal mucosa while minimizing these side effects [31]. Furthermore, combination therapies, such as that with DFMO and NSAIDs, allow for lower doses of the individual agents, thereby lessening the risk for side effects. This is also important for the long-term use of NSAIDs as chemoprevention, as the most common side effects are gastrointestinal mucosal injury and renal toxicity, with cardiovascular, central nervous system (CNS) and platelet side effects occurring less frequently [44].

2.2. Targeting Polyamine Metabolism for Cancer Treatment

Although DFMO effectively inhibits cellular proliferation in a variety of cancer cell types in vitro and in vivo, its efficacy as a monotherapy against established tumors in clinical trials has been mostly unsuccessful due to compensatory uptake of polyamines released into the intestinal lumen through the turnover of gut mucosal cells as well as the microbiome and diet (reviewed in [45,46]). Polyamine analogues have thus been developed with the ability to downregulate polyamine biosynthesis through negative feedback mechanisms, inhibit uptake through the polyamine transporter, and induce the catabolism of the natural polyamines [47–49]. These structural mimetics are capable of competing with natural polyamines for binding sites, but are unable to substitute for their growth-sustaining functions. Treatment of a wide variety of cancer cell lines both in vitro and in xenograft mouse models with members of the symmetrically substituted bis(ethyl)polyamine analogues has resulted in tumor-specific cytotoxicity that is associated with depletion of the natural polyamines and, in some cases, generation of reactive oxygen species (ROS) [33,34,47,48]. Early clinical trials with a second-generation analogue, PG-11047 [50], have shown it to be well tolerated as both a single agent and in combination with common chemotherapeutic agents (clinicaltrials.gov #NCT00293488, NCT00705653, NCT00705874). Most recently, two of these analogues, bis(ethyl)norspermine (BENSpm) and PG-11047, were used to generate polycationic polymers capable of targeting polyamine catabolism while simultaneously acting as nanocarriers for the delivery of therapeutic nucleic acids, including microRNA (miRNA) and small interfering RNA (siRNA) [51,52].

3. Dietary Factors with Potential to Moderate Carcinogenesis through the Polyamine Pathway

Although drug development targeting the polyamine pathway is progressing, evidence also suggests that many naturally occurring compounds already present in our diet can affect polyamine metabolism, with ultimate effects on cancer prevention or treatment. The interaction with dietary components is greatest in the gastrointestinal tract, and it is here that the anticancer potential of dietary or environmental factors might be most advantageous. As with DFMO and NSAIDs, colorectal cancer (CRC) model systems have historically been used to study the efficacy of both synthetic and naturally occurring dietary agents, including plant polyphenols, phytoestrogens, and probiotics, with regard to chemopreventive and chemotherapeutic potential.

3.1. Plant Polyphenols

Naturally occurring polyphenols include several subclasses of structurally related plant substances long recognized for their health benefits in traditional medicine. They can be classified as phenolic acids (highly concentrated in coffee, teas, pomegranate, and berries), stilbenes (including resveratrol; found in grapes, wine, and blueberries), tannins (grapes, tea, coffee, lentil, and walnuts), diferuloylmethanes (turmeric), and flavonoids, which constitute the largest subclass of phenolic compounds and are the major source in the average diet. Many flavonoids are produced by plants as a means of protection against parasites, oxidative injury and harsh environmental stress conditions, and can be further classified into groups including, but not limited to, anthocyanins (blue- or purple-pigmented fruits), flavanols (including catechins found in teas, dark chocolate, and cocoa), flavanones (citrus fruits) and isoflavones (phytoestrogens in soy products, such as genistein). Representative compounds of nearly all of these groups have been demonstrated to affect the polyamine metabolic pathway, primarily through an inhibition of ODC activity, resulting in decreased tumorigenesis. The impact of certain flavonoids, including resveratrol, genistein, and green tea (-)-epigallocatechin-3-gallate (EGCG) on polyamine metabolism and colorectal carcinogenesis, as well as in other carcinogenesis models, is evident, and these data have been comprehensively reviewed by Russo and colleagues [53,54]. The remainder of the current review will therefore focus only on curcumin and its potential in the regulation of polyamine metabolism.

3.2. Curcumin

Curcumin, or 1,7-bis(4-hydroxy 3-methoxy phenyl)-1,6-heptadiene-3,5-dione, is a naturally occurring polyphenol that has been the focus of many studies in a variety of medical fields (Figure 3). A phenolic compound, this yellowish orange pigment's only source is the rhizomatous turmeric plant (*Curcuma longa*), which is cultivated mostly in India and Southeast Asia. Curcuminoids constitute only approximately 5% of turmeric root powder, and exist in 3 forms: curcumin, also referred to as diferuloylmethane (60–70% of crude extract), desmethoxycurcumin (20–27%), and bisdesmethoxycurcumin (10–15%) [55].

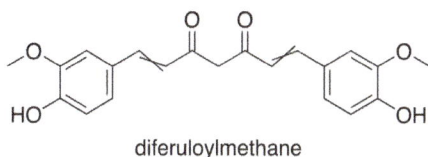

diferuloylmethane

Figure 3. Chemical structure of curcumin, the principle active curcuminoid component of turmeric

Commonly used as a spice and food coloring as well as in skin care products and textile dyes, curcumin has been used for centuries in traditional Chinese medicine and Indian Ayurvedic medicine for its health-promoting properties. The pathologies believed to benefit from curcumin are diverse, and likely stem from its potential to regulate key molecular processes involved in the pathology of many diseases. Of particular interest in cancer etiology and prevention are the antioxidant, anti-inflammatory, and antiproliferative benefits attributed to dietary curcumin. As an antioxidant, curcumin can act as a free radical scavenger, inhibit the generation of free radicals and subsequent oxidative damage, and induce the activity of antioxidant molecules and enzymes involved in detoxification processes, such as glutathione-S-transferase (GST) and nuclear factor E2-related factor (NRF-2) [56,57].

The role of curcumin in the anti-inflammatory response is associated with its ability to downregulate certain transcription factors that promote the production of inflammatory gene products. Perhaps most significantly, curcumin inhibits nuclear factor-kappa B (NF-κB) activation, preventing its translocation into the nucleus where it could directly induce the transcription of genes associated with cell survival and inflammation [58,59]. Affected genes include the free-radical-producing enzymes cyclooxygenase-2 (COX2), lipoxygenase (LOX), and inducible nitric oxide synthase (iNOS) as well as pro-inflammatory cytokines, such as interferon gamma (IFNγ), tumor necrosis factor alpha (TNFα), interleukin (IL)-1, IL-6, and others. In addition, curcumin invokes growth inhibitory effects through other NF-κB target gene products, including cyclin D1 and c-MYC, induces apoptosis in cancer cells, and demonstrates anti-angiogenic activity. Overall, the importance of curcumin in cancer chemoprevention and treatment may originate from its inhibitory effect on molecules linking inflammation and cancer. As several genes encoding polyamine metabolic enzymes are also regulated by many of the above-mentioned molecules, the potential exists for transcriptional modulation of the polyamine pathway via curcumin.

4. Investigations into the Antitumor Potential of Curcumin through Modulating Polyamine Metabolism

4.1. Evidence of the Chemopreventive Activity of Curcumin in Carcinogenesis Models

Animal models of carcinogenesis involve the administration of carcinogens and/or toxicants that act as tumor initiators or promoters. Studies with DFMO established a critical role for ODC induction by tumor promoters such as phorbol esters in the early stages of tumor development [60]. Subsequently, ODC has been used as an indicator of tumor promotion induced by a variety of agents in multiple carcinogenesis model systems [31].

4.1.1. Topical Application of Curcumin in Animal Models of Skin Cancer

The effects of curcumin on polyamine metabolism were first investigated in the CD-1 skin carcinogenesis mouse model [61]. The tumor promoter 12-O-tetradecanoylphorbol-13-acetate (TPA) is well established to rapidly induce ODC activity [30], and this ODC induction contributes to tumorigenesis [3]. Topical application of curcumin to the epidermis concurrently with TPA potently inhibited the induction of epidermal ODC activity in a dose-dependent manner. Similarly, TPA-induced DNA synthesis was progressively inhibited by increasing doses of curcumin. Ultimately, in a two-stage initiation-promotion model using 7,12-dimethylbenz[a]anthracene (DMBA) followed by TPA, curcumin potently reduced TPA-induced tumor promotion, resulting in a 98% decrease in the number of skin tumors observed [61]. In a similar strategy, both topical and intraperitoneal (i.p.) administrations of curcumin were investigated for their effects on TPA-induced ODC mRNA and activity in mouse epidermis. Both routes of administration were capable of inhibiting the induction of ODC mRNA and activity in a near-parallel, dose-dependent manner, indicating that modulation of ODC by curcumin occurs primarily at the mRNA transcript level [62].

The results of these studies were further verified in vitro using ME308 mouse keratinocytes established from DMBA-initiated mouse skin [63]. Lee and Pezzuto [64] used this system to investigate an extensive panel of potential chemopreventive agents, including curcumin, for their ability to inhibit TPA-induced ODC activity. Their data revealed that co-incubation of curcumin with TPA provided one of the most potent inhibitory effects on ODC activity, with a half maximal inhibitory concentration (IC$_{50}$) of 4 μM. In comparison, the irreversible ODC inhibitor DFMO prevented TPA-induced ODC activity with an IC$_{50}$ of 20 μM. A similar inhibitory effect for curcumin compared to DFMO was obtained in a subsequent screen using the rat 2C5 tracheal epithelial cell line to investigate TPA-induced ODC activity [65]. Of note, curcumin was not toxic to this non-tumorigenic immortalized cell line.

Studies into the mechanism of tumor promotion by TPA revealed a critical role for protein kinase C (PKC) signaling in mediating many TPA-induced tumorigenic effects, including ODC activity. Topical application of curcumin to the dorsal skin of Swiss bare mice prevented TPA-induced PKC translocation, resulting in effects analogous to those observed using a known selective inhibitor of PKC. These effects of PKC inhibition ultimately included the repression of ODC induction, ROS generation, apoptosis, and hyperplasia, which were associated with alterations in TPA-induced kinases and transcription factors [66].

Irradiation of CD-1 mice with ultraviolet A (UVA) has been shown to enhance the tumor-promoting effects of TPA on the epidermis beyond that observed with TPA alone. These effects include increased ODC activity and dermatitis, as evidenced by dermal infiltrating inflammatory cells, and topical pretreatment with curcumin significantly prevented these increases as well as those observed with TPA alone [67]. In the same model, it was subsequently determined that although ODC mRNA was induced by TPA, as previously reported [62], it was not significantly enhanced by addition of UVA; ODC activity was, however, induced by UVA in addition to TPA, suggesting post-transcriptional regulation of ODC by UVA. Importantly, pretreatment with curcumin could block the UVA-TPA-stimulated induction of both ODC mRNA and protein [68].

4.1.2. Dietary Curcumin in Rodent Models of Carcinogenesis

Following the initial finding that curcumin prevented TPA-induced ODC activity and tumorigenesis when applied topically, studies were conducted investigating the effects of dietary administration of curcumin. At the time of these studies, the chemopreventive effect of NSAIDs on colon tumorigenesis was becoming apparent. As curcumin has traditionally been used in the treatment of a variety of inflammatory conditions, its potential for inhibiting colon tumorigenesis was investigated. F344 rats were fed diets containing 2000 p.p.m. curcumin for two weeks prior to subcutaneous injections of azoxymethane (AOM), a carcinogen that specifically induces distal colon tumors in rodents with a pathology mimicking that of sporadic human colon cancers [69]. In the group receiving the curcumin-supplemented diet, AOM-induced ODC activity was significantly decreased in the colonic mucosa as well as the liver, where AOM is metabolically activated. Additionally, animals receiving dietary curcumin prior to AOM exposure demonstrated a greater than 50% reduction in the number of AOM-induced aberrant crypt foci (ACF), which are early preneoplastic lesions in the colon [70].

The F344 rat strain was also used to investigate the protective effect of curcumin on 4-nitroquinoline 1-oxide (NQO)-induced oral carcinogenesis. 4-NQO is easily administered to rats in the drinking water and produces tongue lesions including squamous cell neoplasms that are comparable to those in human oral carcinogenesis [71]. To study the efficacy of curcumin in preventing the initiation of tumorigenesis, rats were fed a diet containing 500 p.p.m. curcumin starting one week prior to and throughout 4-NQO exposure for 8 weeks. A second group received no curcumin until one week after the 8-week 4-NQO exposure (post-initiation), and curcumin remained in the diet of this group until the conclusion of the study 22 weeks later. Rats in both curcumin groups demonstrated impressive reductions in the frequency of tongue neoplasms: 4-NQO-induced carcinomas were reduced from 54% of the animals to 5% when curcumin was added at the initiation phase and to 15% when added post-initiation. The number of animals with preneoplastic squamous cell dysplasias was similarly decreased. Analyses of the polyamine content of normal-appearing sections of tongue tissue at the end of the study revealed that 4-NQO exposure significantly elevated the levels of spermidine and spermine, as well as total polyamine content, above that of untreated animals, consistent with an induction of ODC activity. Importantly, these elevated levels were prevented when curcumin was also administered, with initiation phase exposure providing the greatest benefit and maintaining polyamine pools that did not significantly differ from those of rats not receiving carcinogen. These reduced polyamine levels and carcinoma numbers were also accompanied by reductions in proliferation biomarkers in the tongue epithelium [72]. Of note is a previous study conducted by the same investigators that analyzed the effects of the ODC inhibitor DFMO on NQO-induced oral carcinogenesis with very similar results [73].

In a third model system, male ddY mice were pretreated with 1% dietary curcumin for 4 weeks prior to receiving i.p. injections of the renal carcinogen ferric nitrilotriacetate (Fe-NTA). Twelve hours after administration of Fe-NTA, ODC activity levels in the mouse kidney were increased approximately 4.4-fold, while pretreatment with curcumin inhibited this increase by 63%, with no effect on basal enzyme activity. Concurrently, Fe-NTA-generated oxidative stress, a mechanism associated with its tumor-promoting abilities, and nephrotoxicity were also alleviated by curcumin pretreatment [74].

4.2. Investigations into Polyamine-Associated Effects of Curcumin on Established Tumors—Potential for Cancer Treatment

Unlike non-tumorigenic cells, cancer cells typically respond to curcumin exposure through inducing apoptosis and cell death [75]. Several characteristics of cancer cells are responsible for this differential sensitivity, including increased curcumin uptake and ROS generation, lower glutathione levels, and the constitutive activation of NF-κB that often mediates the survival of cancer cells [75–78]. In addition to being a c-MYC-regulated gene, evidence exists for the regulation of ODC by NF-κB [79], and the SSAT catabolic enzyme is also an NF-κB target [80]. Studies have also suggested that NF-κB can be activated by the elevated levels of polyamines present in tumor cells [81]. Thus, the inhibition of polyamine biosynthesis by curcumin might indirectly inhibit NF-κB activation. In spite of these potential mechanistic links in terms of regulation by transcription factors, relatively few studies have investigated the effect of curcumin exposure on polyamine metabolism in cell lines. Mehta et al. [82] first analyzed the effect of curcumin on a panel of 8 breast cancer cell lines representing multidrug-resistant (MDR), estrogen-dependent, and estrogen-independent breast cancers. Impressive growth inhibitory effects were observed in all of the cell lines (1–26% viability following 1 μg/mL treatment for 72 h), including an Adriamycin-resistant MCF-7 line (~15% viability). This antiproliferative activity was time- and dose-dependently correlated with curcumin-induced inhibition of ODC activity. Interestingly, in MCF-7 cells, there was no apparent evidence of apoptosis and the apoptosis-related genes examined remained unchanged [82]. Flow cytometric studies in MDA-231 cells indicated temporary growth arrest at the G2/M checkpoint after 24 h of curcumin exposure; however, cells appeared to re-enter the cell cycle with longer exposure times. The results of this study were the first to suggest the potential use of curcumin as an antiproliferative agent against breast cancer cells.

In a more recent study using SK-BR-3 breast cancer cells, which overexpress Human Epidermal Growth Factor (HER)-2, curcumin inhibited growth, and a flow cytometric assay measuring bromolated deoxyuridine triphosphate (BrdU) incorporation indicated the induction of apoptosis even at the lowest concentration examined (2.5 µM after 72 h). Analysis of intracellular polyamine pools following curcumin treatment revealed substantial decreases in spermidine and spermine levels (85% and 50%, respectively), suggesting that the loss of intracellular polyamines might be important in the antiproliferative mechanism of curcumin [83].

Curcumin inhibits the activation and nuclear translocation of NF-κB, thereby preventing stimulation of NF-κB target genes [58]. The involvement of NF-κB signaling in curcumin-mediated changes in polyamine metabolism was recently investigated in human breast cancer [84]. Pretreatment of MCF-7 breast cancer cells with an inhibitor of NF-κB prior to curcumin treatment suggested that curcumin-mediated alterations in c-MYC, ODC, SSAT, and PAOX protein levels occurred in part through this signaling pathway [84]. Interestingly, this response pattern to curcumin was altered by overexpression of the B-cell lymphoma-2 (*BCL2*) gene, an alteration associated with chemo- and radioresistance [85]. Although the MCF-7/BCL2 cells were less sensitive to the effects of curcumin than the parental strain, significant inhibition of colony formation remained evident. Of note, as BCL2 overexpression in itself upregulates NF-κB [86], the basal expression levels of c-MYC, ODC, and SSAT were also substantially increased in this cell line [84], providing further evidence for regulation of these enzymes by NF-κB.

The ability to stimulate apoptosis in cancer cells of various origins is a key anti-carcinogenic property of curcumin. In the HL-60 promyelocytic leukemia cell line, Liao et al. [87] provided comprehensive evidence indicating a role for ODC in the mechanism of curcumin-induced apoptosis. Treatment of these cells with curcumin quickly inhibited ODC enzymatic activity in a time- and dose-dependent manner that correlated with growth inhibition; furthermore, overexpression of ODC or pretreatment of wildtype cells with a caspase inhibitor increased survival in the presence of curcumin. Relative to the parental cell line, ODC-overexpressing HL-60 cells were protected from the apoptotic hallmarks of curcumin treatment and presented little evidence of DNA fragmentation, ROS production, loss of mitochondrial membrane potential or cytochrome *c* release, cleavage of procaspases 3 and 9, downregulation of BCL2, or apoptosis-related morphological changes. Importantly, the loss of curcumin-induced DNA fragmentation observed in the ODC-overexpressing cells was restored by DFMO treatment or siRNA targeting ODC [87].

5. Translational Potential, Clinical Trials, and Limitations

In addition to the anecdotal evidence accompanying centuries of traditional medicine, curcumin has been safely administered to humans in many registered clinical trials, with nearly 70 trials completed that targeted a variety of conditions (clinicaltrials.gov). One patient population with potential to benefit from curcumin supplementation includes individuals with familial adenomatous polyposis (FAP), a hereditary form of colorectal cancer resulting from a germ-line mutation of the *adenomatous polyposis coli* (*APC*) gene. ODC activity is elevated in normal-appearing colonic mucosa as well as polyps of patients with FAP [88]; furthermore, pre-symptomatic FAP patients contain elevated colorectal mucosa levels of putrescine [89]. Studies in the Min/+ mouse, a model of FAP with one mutant and one wild type copy of the *Apc* gene, demonstrated a 64% reduction in adenoma formation following daily dietary curcumin intake [90]. In a clinical study of FAP patients, the combination of curcumin and a second polyphenol, quercetin, effectively reduced adenoma polyp number and size; however, treatment arms with the individual agents were not conducted [91]. A recently completed randomized, placebo-controlled phase 2 trial (clinicaltrials.gov identifier #NCT00641147) specifically investigated the effect of daily dietary curcumin supplementation on the regression of adenomas in FAP patients over the course of one year. According to the reported results, no significant benefit was observed for the treatment group in terms of polyp number or size, nor were changes observed in the levels of polyamines, suggesting a lack of drug availability.

The poor solubility, bioavailability, and stability of curcumin are common impediments to its clinical utility, particularly when given orally. However, its stability is increased in acidic environments such as the stomach, and the requirement for systemic bioavailability is lessened with the potential for direct contact. Strategies improving this bioavailability are a current area of research and include such approaches as the use of adjuvants that interfere with the metabolism of curcumin, structural analogues of curcumin, and curcumin-containing nanoparticles [57,92]. The structure of curcumin has been widely modified, with particular focus on changes in the β-diketone structure and aryl substitution pattern of the molecule. Of these structural analogues, the incorporation of a 3,5-dibenzylidenepiperidin-4-one framework elicits enhanced antioxidant and antiproliferative actions relative to curcumin, potentially offering an improved pharmacokinetic profile [93–95].

As curcumin has potential therapeutic value against multiple human conditions, enhancing its bioavailability and ascertaining its efficacy in clinical trials could significantly impact the treatment and health of many individuals around the world.

6. Conclusions

The ability of curcumin to specifically alter the signaling pathways required for cancer cell survival strongly suggests its potential in chemopreventive and chemotherapeutic strategies, particularly in inflammation- or ROS-associated carcinogenesis. Modulation of polyamine pathway enzymes and the levels of intracellular polyamines appear to contribute to the anticancer potential of curcumin both in terms of carcinogenesis and in the treatment of established tumors. Therefore, establishing the molecular mechanisms underlying the regulation of polyamines by curcumin will potentially add to our understanding of how to most effectively target and prevent tumor cell proliferation and might provide insight on how to best supplement or substitute current more toxic therapies with curcumin.

Acknowledgments: Support for this review was provided by National Institutes of Health grant R01-CA204345 (RAC). No funds were received for covering the costs to publish in open access.

Author Contributions: T.M.S. conceptualized, researched, illustrated, wrote, and edited the paper. R.A.C. researched, reviewed, and edited the report.

Conflicts of Interest: The authors declare no conflict of interest. The founding sponsors had no role in the design of the study; in the collection, analyses, or interpretation of data; in the writing of the manuscript, and in the decision to publish the results.

References

1. Wu, Y.; Antony, S.; Meitzler, J.L.; Doroshow, J.H. Molecular mechanisms underlying chronic inflammation-associated cancers. *Cancer Lett.* **2014**, *345*, 164–173. [CrossRef] [PubMed]
2. Okada, F. Inflammation-related carcinogenesis: Current findings in epidemiological trends, causes and mechanisms. *Yonago Acta Med.* **2014**, *57*, 65–72. [PubMed]
3. Pegg, A.E. Polyamine metabolism and its importance in neoplastic growth and a target for chemotherapy. *Cancer Res.* **1988**, *48*, 759–774. [PubMed]
4. Gerner, E.W.; Meyskens, F.L., Jr. Polyamines and cancer: Old molecules, new understanding. *Nat. Rev. Cancer* **2004**, *4*, 781–792. [CrossRef] [PubMed]
5. Thomas, T.J.; Tajmir-Riahi, H.A.; Thomas, T. Polyamine-DNA interactions and development of gene delivery vehicles. *Amino Acids* **2016**, *48*, 2423–2431. [CrossRef] [PubMed]
6. Lentini, A.; Abbruzzese, A.; Caraglia, M.; Marra, M.; Beninati, S. Protein-polyamine conjugation by transglutaminase in cancer cell differentiation: Review article. *Amino Acids* **2004**, *26*, 331–337. [CrossRef] [PubMed]
7. Baronas, V.A.; Kurata, H.T. Inward rectifiers and their regulation by endogenous polyamines. *Front. Physiol.* **2014**, *5*, 325. [CrossRef] [PubMed]
8. Williams, K. Modulation and block of ion channels: A new biology of polyamines. *Cell. Signal.* **1997**, *9*, 1–13. [CrossRef]
9. Thomas, T.; Thomas, T.J. Polyamine metabolism and cancer. *J. Cell. Mol. Med.* **2003**, *7*, 113–126. [CrossRef] [PubMed]
10. Shantz, L.M.; Levin, V.A. Regulation of ornithine decarboxylase during oncogenic transformation: Mechanisms and therapeutic potential. *Amino Acids* **2007**, *33*, 213–223. [CrossRef] [PubMed]

11. Casero, R.A.; Pegg, A.E. Polyamine catabolism and disease. *Biochem. J.* **2009**, *421*, 323–338. [CrossRef] [PubMed]

12. Mohan, R.R.; Challa, A.; Gupta, S.; Bostwick, D.G.; Ahmad, N.; Agarwal, R.; Marengo, S.R.; Amini, S.B.; Paras, F.; MacLennan, G.T.; et al. Overexpression of ornithine decarboxylase in prostate cancer and prostatic fluid in humans. *Clin. Cancer Res.* **1999**, *5*, 143–147. [PubMed]

13. Wallace, H.M.; Caslake, R. Polyamines and colon cancer. *Eur. J. Gastroenterol. Hepatol.* **2001**, *13*, 1033–1039. [CrossRef] [PubMed]

14. Thomas, T.; Thomas, T.J. Polyamines in cell growth and cell death: Molecular mechanisms and therapeutic applications. *Cell. Mol. Life Sci.* **2001**, *58*, 244–258. [CrossRef] [PubMed]

15. Gilmour, S.K. Polyamines and nonmelanoma skin cancer. *Toxicol. Appl. Pharmacol.* **2007**, *224*, 249–256. [CrossRef] [PubMed]

16. Manni, A.; Mauger, D.; Gimotty, P.; Badger, B. Prognostic influence on survival of increased ornithine decarboxylase activity in human breast cancer. *Clin. Cancer Res.* **1996**, *2*, 1901–1906. [PubMed]

17. Hibshoosh, H.; Johnson, M.; Weinstein, I.B. Effects of overexpression of ornithine decarboxylase (ODC) on growth control and oncogene-induced cell transformation. *Oncogene* **1991**, *6*, 739–743. [PubMed]

18. Auvinen, M.; Paasinen, A.; Andersson, L.C.; Holtta, E. Ornithine decarboxylase activity is critical for cell transformation. *Nature* **1992**, *360*, 355–358. [CrossRef] [PubMed]

19. Smith, M.K.; Trempus, C.S.; Gilmour, S.K. Co-operation between follicular ornithine decarboxylase and v-Ha-ras induces spontaneous papillomas and malignant conversion in transgenic skin. *Carcinogenesis* **1998**, *19*, 1409–1415. [CrossRef] [PubMed]

20. Wagner, A.J.; Meyers, C.; Laimins, L.A.; Hay, N. c-Myc induces the expression and activity of ornithine decarboxylase. *Cell. Growth Differ.* **1993**, *4*, 879–883. [PubMed]

21. Bello-Fernandez, C.; Packham, G.; Cleveland, J.L. The ornithine decarboxylase gene is a transcriptional target of c-Myc. *Proc. Natl Acad. Sci. USA* **1993**, *90*, 7804–7808. [CrossRef] [PubMed]

22. Forshell, T.P.; Rimpi, S.; Nilsson, J.A. Chemoprevention of B-cell lymphomas by inhibition of the Myc target spermidine synthase. *Cancer Prev. Res. (Phila)* **2010**, *3*, 140–147. [CrossRef] [PubMed]

23. Holtta, E.; Sistonen, L.; Alitalo, K. The mechanisms of ornithine decarboxylase deregulation in c-Ha-ras oncogene-transformed NIH 3T3 cells. *J. Biol. Chem.* **1988**, *263*, 4500–4507. [PubMed]

24. Shantz, L.M. Transcriptional and translational control of ornithine decarboxylase during Ras transformation. *Biochem. J.* **2004**, *377*, 257–264. [CrossRef] [PubMed]

25. Shantz, L.M.; Pegg, A.E. Ornithine decarboxylase induction in transformation by H-Ras and RhoA. *Cancer Res.* **1998**, *58*, 2748–2753. [PubMed]

26. Linsalata, M.; Notarnicola, M.; Caruso, M.G.; Di Leo, A.; Guerra, V.; Russo, F. Polyamine biosynthesis in relation to K-ras and p-53 mutations in colorectal carcinoma. *Scand. J. Gastroenterol.* **2004**, *39*, 470–477. [CrossRef] [PubMed]

27. Nilsson, J.A.; Keller, U.B.; Baudino, T.A.; Yang, C.; Norton, S.; Old, J.A.; Nilsson, L.M.; Neale, G.; Kramer, D.L.; Porter, C.W.; et al. Targeting ornithine decarboxylase in Myc-induced lymphomagenesis prevents tumor formation. *Cancer Cell.* **2005**, *7*, 433–444. [CrossRef] [PubMed]

28. O'Brien, T.G.; Megosh, L.C.; Gilliard, G.; Soler, A.P. Ornithine decarboxylase overexpression is a sufficient condition for tumor promotion in mouse skin. *Cancer Res.* **1997**, *57*, 2630–2637. [PubMed]

29. Gilmour, S.K.; Robertson, F.M.; Megosh, L.; O'Connell, S.M.; Mitchell, J.; O'Brien, T.G. Induction of ornithine decarboxylase in specific subpopulations of murine epidermal cells following multiple exposures to 12-O-tetradecanoylphorbol-13-acetate, mezerein and ethyl phenylpropriolate. *Carcinogenesis* **1992**, *13*, 51–56. [CrossRef] [PubMed]

30. O'Brien, T.G.; Simsiman, R.C.; Boutwell, R.K. Induction of the polyamine-biosynthetic enzymes in mouse epidermis by tumor-promoting agents. *Cancer Res.* **1975**, *35*, 1662–1670. [PubMed]

31. Meyskens, F.L., Jr.; Gerner, E.W. Development of difluoromethylornithine (DFMO) as a chemoprevention agent. *Clin. Cancer Res.* **1999**, *5*, 945–951. [PubMed]

32. Russell, D.; Snyder, S.H. Amine synthesis in rapidly growing tissues: Ornithine decarboxylase activity in regenerating rat liver, chick embryo, and various tumors. *Proc. Natl. Acad. Sci. USA* **1968**, *60*, 1420–1427. [CrossRef] [PubMed]

33. Casero, R.A., Jr.; Marton, L.J. Targeting polyamine metabolism and function in cancer and other hyperproliferative diseases. *Nat. Rev. Drug Discov.* **2007**, *6*, 373–390. [CrossRef] [PubMed]

34. Murray-Stewart, T.R.; Woster, P.M.; Casero, R.A., Jr. Targeting polyamine metabolism for cancer therapy and prevention. *Biochem. J.* **2016**, *473*, 2937–2953. [CrossRef] [PubMed]

35. Metcalf, B.W.; Bey, P.; Danzin, C.; Jung, M.J.; Casara, P.; Vevert, J.P. Catalytic irreversible inhibition of mammalian ornithine decarboxylase (E.C.4.1.1.17) by substrate and product analogs. *J. Am. Chem. Soc.* **1978**, *100*, 2551–2553. [CrossRef]

36. McCann, P.P.; Pegg, A.E. Ornithine decarboxylase as an enzyme target for therapy. *Pharmacol. Ther.* **1992**, *54*, 195–215. [CrossRef]

37. Meyskens, F.L., Jr.; Gerner, E.W.; Emerson, S.; Pelot, D.; Durbin, T.; Doyle, K.; Lagerberg, W. Effect of α-difluoromethylornithine on rectal mucosal levels of polyamines in a randomized, double-blinded trial for colon cancer prevention. *J. Natl. Cancer Inst.* **1998**, *90*, 1212–1218. [CrossRef] [PubMed]

38. Gerner, E.W.; Meyskens, F.L., Jr. Combination chemoprevention for colon cancer targeting polyamine synthesis and inflammation. *Clin. Cancer Res.* **2009**, *15*, 758–761. [CrossRef] [PubMed]

39. Ignatenko, N.A.; Besselsen, D.G.; Stringer, D.E.; Blohm-Mangone, K.A.; Cui, H.; Gerner, E.W. Combination chemoprevention of intestinal carcinogenesis in a murine model of familial adenomatous polyposis. *Nutr. Cancer* **2008**, *60*, 30–35. [CrossRef] [PubMed]

40. Jacoby, R.F.; Cole, C.E.; Tutsch, K.; Newton, M.A.; Kelloff, G.; Hawk, E.T.; Lubet, R.A. Chemopreventive efficacy of combined piroxicam and difluoromethylornithine treatment of Apc mutant Min mouse adenomas, and selective toxicity against Apc mutant embryos. *Cancer Res.* **2000**, *60*, 1864–1870. [PubMed]

41. Meyskens, F.L., Jr.; McLaren, C.E.; Pelot, D.; Fujikawa-Brooks, S.; Carpenter, P.M.; Hawk, E.; Kelloff, G.; Lawson, M.J.; Kidao, J.; McCracken, J.; et al. Difluoromethylornithine plus sulindac for the prevention of sporadic colorectal adenomas: A randomized placebo-controlled, double-blind trial. *Cancer Prev. Res. (Phila)* **2008**, *1*, 32–38. [CrossRef] [PubMed]

42. Zell, J.A.; Pelot, D.; Chen, W.P.; McLaren, C.E.; Gerner, E.W.; Meyskens, F.L. Risk of cardiovascular events in a randomized placebo-controlled, double-blind trial of difluoromethylornithine plus sulindac for the prevention of sporadic colorectal adenomas. *Cancer Prev. Res. (Phila)* **2009**, *2*, 209–212. [CrossRef] [PubMed]

43. Nowotarski, S.L.; Woster, P.M.; Casero, R.A., Jr. Polyamines and cancer: Implications for chemotherapy and chemoprevention. *Expert Rev. Mol. Med.* **2013**, *15*, e3. [CrossRef] [PubMed]

44. Tsioulias, G.J.; Go, M.F.; Rigas, B. NSAIDs and colorectal cancer control: Promise and challenges. *Curr. Pharmacol. Rep.* **2015**, *1*, 295–301. [CrossRef] [PubMed]

45. Schechter, P.J.; Barlow, J.L.R.; Sjoerdsma, A. Clinical aspects of inhibition of ornithine decarboxylase with emphasis on therapeutic trials of eflornithine (DFMO) in cancer and protozoan diseases. In *Inhibition of Polyamine Metabolism. Biological Significance and Basis for New Therapies*; McCann, P.P., Pegg, A.E., Sjoerdsma, A., Eds.; Academic Press: Orlando, FL, USA, 1987; pp. 345–364.

46. Seiler, N. Thirty years of polyamine-related approaches to cancer therapy. Retrospect and prospect. Part 1. Selective enzyme inhibitors. *Curr. Drug Targets* **2003**, *4*, 537–564. [CrossRef] [PubMed]

47. Battaglia, V.; DeStefano Shields, C.; Murray-Stewart, T.; Casero, R.A., Jr. Polyamine catabolism in carcinogenesis: Potential targets for chemotherapy and chemoprevention. *Amino Acids* **2014**, *46*, 511–519. [CrossRef] [PubMed]

48. Casero, R.A., Jr.; Wang, Y.; Stewart, T.M.; Devereux, W.; Hacker, A.; Wang, Y.; Smith, R.; Woster, P.M. The role of polyamine catabolism in anti-tumour drug response. *Biochem. Soc. Trans.* **2003**, *31*, 361–365. [CrossRef] [PubMed]

49. Murray-Stewart, T.; Casero, R., Jr. Mammalian polyamine catabolism. In *Polyamines*; Kusano, T., Suzuki, H., Eds.; Springer: Tokyo, Japan, 2015; pp. 61–75.

50. Reddy, V.K.; Valasinas, A.; Sarkar, A.; Basu, H.S.; Marton, L.J.; Frydman, B. Conformationally restricted analogues of $^{1}N,^{12}N$-bisethylspermine: Synthesis and growth inhibitory effects on human tumor cell lines. *J. Med. Chem.* **1998**, *41*, 4723–4732. [CrossRef] [PubMed]

51. Xie, Y.; Murray-Stewart, T.; Wang, Y.; Yu, F.; Li, J.; Marton, L.J.; Casero, R.A., Jr.; Oupicky, D. Self-immolative nanoparticles for simultaneous delivery of microRNA and targeting of polyamine metabolism in combination cancer therapy. *J. Control. Release* **2017**, *246*, 110–119. [CrossRef] [PubMed]

52. Murray-Stewart, T.; Ferrari, E.; Xie, Y.; Yu, F.; Marton, L.J.; Oupicky, D.; Casero, R.A., Jr. Biochemical evaluation of the anticancer potential of the polyamine-based nanocarrier Nano11047. *PLoS ONE* **2017**, *12*, e0175917. [CrossRef]

53. Linsalata, M.; Orlando, A.; Russo, F. Pharmacological and dietary agents for colorectal cancer chemoprevention: Effects on polyamine metabolism (review). *Int. J. Oncol.* **2014**, *45*, 1802–1812. [CrossRef] [PubMed]

54. Linsalata, M.; Russo, F. Nutritional factors and polyamine metabolism in colorectal cancer. *Nutrition* **2008**, *24*, 382–389. [CrossRef] [PubMed]

55. Nelson, K.M.; Dahlin, J.L.; Bisson, J.; Graham, J.; Pauli, G.F.; Walters, M.A. The essential medicinal chemistry of curcumin. *J. Med. Chem.* **2017**, *60*, 1620–1637. [CrossRef] [PubMed]

56. Park, W.; Amin, A.R.; Chen, Z.G.; Shin, D.M. New perspectives of curcumin in cancer prevention. *Cancer Prev. Res. (Phila)* **2013**, *6*, 387–400. [CrossRef] [PubMed]

57. Kumar, G.; Mittal, S.; Sak, K.; Tuli, H.S. Molecular mechanisms underlying chemopreventive potential of curcumin: Current challenges and future perspectives. *Life Sci.* **2016**, *148*, 313–328. [CrossRef] [PubMed]

58. Singh, S.; Aggarwal, B.B. Activation of transcription factor NF-κB is suppressed by curcumin (diferuloylmethane) [corrected]. *J. Biol. Chem.* **1995**, *270*, 24995–25000. [CrossRef] [PubMed]

59. Jobin, C.; Bradham, C.A.; Russo, M.P.; Juma, B.; Narula, A.S.; Brenner, D.A.; Sartor, R.B. Curcumin blocks cytokine-mediated NF-κB activation and proinflammatory gene expression by inhibiting inhibitory factor I-κB kinase activity. *J. Immunol.* **1999**, *163*, 3474–3483. [PubMed]

60. O'Brien, T.G. The induction of ornithine decarboxylase as an early, possibly obligatory, event in mouse skin carcinogenesis. *Cancer Res.* **1976**, *36*, 2644–2653. [PubMed]

61. Huang, M.T.; Smart, R.C.; Wong, C.Q.; Conney, A.H. Inhibitory effect of curcumin, chlorogenic acid, caffeic acid, and ferulic acid on tumor promotion in mouse skin by 12-O-tetradecanoylphorbol-13-acetate. *Cancer Res.* **1988**, *48*, 5941–5946. [PubMed]

62. Lu, Y.P.; Chang, R.L.; Huang, M.T.; Conney, A.H. Inhibitory effect of curcumin on 12-O-tetradecanoylphorbol-13-acetate-induced increase in ornithine decarboxylase mRNA in mouse epidermis. *Carcinogenesis* **1993**, *14*, 293–297. [CrossRef] [PubMed]

63. Strickland, J.E.; Greenhalgh, D.A.; Koceva-Chyla, A.; Hennings, H.; Restrepo, C.; Balaschak, M.; Yuspa, S.H. Development of murine epidermal cell lines which contain an activated *ras*^Ha oncogene and form papillomas in skin grafts on athymic nude mouse hosts. *Cancer Res.* **1988**, *48*, 165–169. [PubMed]

64. Lee, S.K.; Pezzuto, J.M. Evaluation of the potential of cancer chemopreventive activity mediated by inhibition of 12-O-tetradecanoyl phorbol 13-acetate-induced ornithine decarboxylase activity. *Arch. Pharm. Res.* **1999**, *22*, 559–564. [CrossRef] [PubMed]

65. White, E.L.; Ross, L.J.; Schmid, S.M.; Kelloff, G.J.; Steele, V.E.; Hill, D.L. Screening of potential cancer-preventing chemicals for inhibition of induction of ornithine decarboxylase in epithelial cells from rat trachea. *Oncol. Rep.* **1998**, *5*, 717–722. [CrossRef] [PubMed]

66. Garg, R.; Ramchandani, A.G.; Maru, G.B. Curcumin decreases 12-O-tetradecanoylphorbol-13-acetate-induced protein kinase C translocation to modulate downstream targets in mouse skin. *Carcinogenesis* **2008**, *29*, 1249–1257. [CrossRef] [PubMed]

67. Ishizaki, C.; Oguro, T.; Yoshida, T.; Wen, C.Q.; Sueki, H.; Iijima, M. Enhancing effect of ultraviolet A on ornithine decarboxylase induction and dermatitis evoked by 12-O-tetradecanoylphorbol-13-acetate and its inhibition by curcumin in mouse skin. *Dermatology* **1996**, *193*, 311–317. [CrossRef] [PubMed]

68. Oguro, T.; Yoshida, T. Effect of ultraviolet A on ornithine decarboxylase and metallothionein gene expression in mouse skin. *Photodermatol. Photoimmunol. Photomed.* **2001**, *17*, 71–78. [CrossRef] [PubMed]

69. Chen, J.; Huang, X.F. The signal pathways in azoxymethane-induced colon cancer and preventive implications. *Cancer Biol. Ther.* **2009**, *8*, 1313–1317. [CrossRef] [PubMed]

70. Rao, C.V.; Simi, B.; Reddy, B.S. Inhibition by dietary curcumin of azoxymethane-induced ornithine decarboxylase, tyrosine protein kinase, arachidonic acid metabolism and aberrant crypt foci formation in the rat colon. *Carcinogenesis* **1993**, *14*, 2219–2225. [CrossRef] [PubMed]

71. Tanaka, T.; Ishigamori, R. Understanding carcinogenesis for fighting oral cancer. *J. Oncol.* **2011**, *2011*, 603740. [CrossRef] [PubMed]

72. Tanaka, T.; Makita, H.; Ohnishi, M.; Hirose, Y.; Wang, A.; Mori, H.; Satoh, K.; Hara, A.; Ogawa, H. Chemoprevention of 4-nitroquinoline 1-oxide-induced oral carcinogenesis by dietary curcumin and hesperidin: Comparison with the protective effect of β-carotene. *Cancer Res.* **1994**, *54*, 4653–4659. [PubMed]

73. Tanaka, T.; Kojima, T.; Hara, A.; Sawada, H.; Mori, H. Chemoprevention of oral carcinogenesis by DL-α-difluoromethylornithine, an ornithine decarboxylase inhibitor: Dose-dependent reduction in 4-nitroquinoline 1-oxide-induced tongue neoplasms in rats. *Cancer Res.* **1993**, *53*, 772–776. [PubMed]

74. Okazaki, Y.; Iqbal, M.; Okada, S. Suppressive effects of dietary curcumin on the increased activity of renal ornithine decarboxylase in mice treated with a renal carcinogen, ferric nitrilotriacetate. *Biochim. Biophys. Acta* **2005**, *1740*, 357–366. [CrossRef] [PubMed]

75. Syng-Ai, C.; Kumari, A.L.; Khar, A. Effect of curcumin on normal and tumor cells: Role of glutathione and Bcl-2. *Mol. Cancer Ther.* **2004**, *3*, 1101–1108. [PubMed]

76. Fang, J.; Lu, J.; Holmgren, A. Thioredoxin reductase is irreversibly modified by curcumin: A novel molecular mechanism for its anticancer activity. *J. Biol. Chem.* **2005**, *280*, 25284–25290. [CrossRef] [PubMed]

77. Kunwar, A.; Barik, A.; Mishra, B.; Rathinasamy, K.; Pandey, R.; Priyadarsini, K.I. Quantitative cellular uptake, localization and cytotoxicity of curcumin in normal and tumor cells. *Biochim. Biophys. Acta* **2008**, *1780*, 673–679. [CrossRef] [PubMed]

78. Karin, M.; Cao, Y.; Greten, F.R.; Li, Z.W. NF-κB in cancer: From innocent bystander to major culprit. *Nat. Rev. Cancer* **2002**, *2*, 301–310. [CrossRef] [PubMed]

79. Tacchini, L.; De Ponti, C.; Matteucci, E.; Follis, R.; Desiderio, M.A. Hepatocyte growth factor-activated NF-κB regulates HIF-1 activity and ODC expression, implicated in survival, differently in different carcinoma cell lines. *Carcinogenesis* **2004**, *25*, 2089–2100. [CrossRef] [PubMed]

80. Babbar, N.; Gerner, E.W.; Casero, R.A., Jr. Induction of spermidine/spermine N^1-acetyltransferase (SSAT) by aspirin in Caco-2 colon cancer cells. *Biochem. J.* **2006**, *394*, 317–324. [CrossRef] [PubMed]

81. Shah, N.; Thomas, T.; Shirahata, A.; Sigal, L.H.; Thomas, T.J. Activation of nuclear factor κB by polyamines in breast cancer cells. *Biochemistry* **1999**, *38*, 14763–14774. [CrossRef] [PubMed]

82. Mehta, K.; Pantazis, P.; McQueen, T.; Aggarwal, B.B. Antiproliferative effect of curcumin (diferuloylmethane) against human breast tumor cell lines. *Anticancer Drugs* **1997**, *8*, 470–481. [CrossRef] [PubMed]

83. Thomas, T.J.; Santhakumaran, L.M.; Parikh, M.S.; Thomas, T. A possible mechanism for the growth inhibitory action of curcumin on HER-2 over expressing SK-BR-3 breast cancer cells involves the polyamine pathway. *Cancer Res.* **2004**, *64*, 168–169.

84. Berrak, O.; Akkoc, Y.; Arisan, E.D.; Coker-Gurkan, A.; Obakan-Yerlikaya, P.; Palavan-Unsal, N. The inhibition of PI3K and NFκB promoted curcumin-induced cell cycle arrest at G2/M via altering polyamine metabolism in Bcl-2 overexpressing MCF-7 breast cancer cells. *Biomed. Pharmacother.* **2016**, *77*, 150–160. [CrossRef] [PubMed]

85. Thomas, S.; Quinn, B.A.; Das, S.K.; Dash, R.; Emdad, L.; Dasgupta, S.; Wang, X.Y.; Dent, P.; Reed, J.C.; Pellecchia, M.; et al. Targeting the Bcl-2 family for cancer therapy. *Expert Opin. Ther. Targets* **2013**, *17*, 61–75. [CrossRef] [PubMed]

86. Ricca, A.; Biroccio, A.; Del Bufalo, D.; Mackay, A.R.; Santoni, A.; Cippitelli, M. Bcl-2 over-expression enhances NF-κB activity and induces MMP-9 transcription in human MCF7[ADR] breast-cancer cells. *Int. J. Cancer* **2000**, *86*, 188–196. [CrossRef]

87. Liao, Y.F.; Hung, H.C.; Hour, T.C.; Hsu, P.C.; Kao, M.C.; Tsay, G.J.; Liu, G.Y. Curcumin induces apoptosis through an ornithine decarboxylase-dependent pathway in human promyelocytic leukemia HL-60 cells. *Life Sci.* **2008**, *82*, 367–375. [CrossRef] [PubMed]

88. Luk, G.D.; Baylin, S.B. Ornithine decarboxylase as a biologic marker in familial colonic polyposis. *N. Engl. J. Med.* **1984**, *311*, 80–83. [CrossRef] [PubMed]

89. Giardiello, F.M.; Hamilton, S.R.; Hylind, L.M.; Yang, V.W.; Tamez, P.; Casero, R.A., Jr. Ornithine decarboxylase and polyamines in familial adenomatous polyposis. *Cancer Res.* **1997**, *57*, 199–201. [PubMed]

90. Perkins, S.; Verschoyle, R.D.; Hill, K.; Parveen, I.; Threadgill, M.D.; Sharma, R.A.; Williams, M.L.; Steward, W.P.; Gescher, A.J. Chemopreventive efficacy and pharmacokinetics of curcumin in the min/+ mouse, a model of familial adenomatous polyposis. *Cancer Epidemiol. Biomark. Prev.* **2002**, *11*, 535–540.

91. Cruz-Correa, M.; Shoskes, D.A.; Sanchez, P.; Zhao, R.; Hylind, L.M.; Wexner, S.D.; Giardiello, F.M. Combination treatment with curcumin and quercetin of adenomas in familial adenomatous polyposis. *Clin. Gastroenterol. Hepatol.* **2006**, *4*, 1035–1038. [CrossRef] [PubMed]

92. Adiwidjaja, J.; McLachlan, A.J.; Boddy, A.V. Curcumin as a clinically-promising anti-cancer agent: Pharmacokinetics and drug interactions. *Expert Opin. Drug Metab. Toxicol.* **2017**, *13*, 953–972. [CrossRef] [PubMed]

93. Padhye, S.; Chavan, D.; Pandey, S.; Deshpande, J.; Swamy, K.V.; Sarkar, F.H. Perspectives on chemopreventive and therapeutic potential of curcumin analogs in medicinal chemistry. *Mini Rev. Med. Chem.* **2010**, *10*, 372–387. [CrossRef] [PubMed]

94. Pati, H.N.; Das, U.; Quail, J.W.; Kawase, M.; Sakagami, H.; Dimmock, J.R. Cytotoxic 3,5-bis(benzylidene)piperidin-4-ones and *N*-acyl analogs displaying selective toxicity for malignant cells. *Eur. J. Med. Chem.* **2008**, *43*, 1–7. [CrossRef] [PubMed]

95. Youssef, K.M.; El-Sherbeny, M.A.; El-Shafie, F.S.; Farag, H.A.; Al-Deeb, O.A.; Awadalla, S.A. Synthesis of curcumin analogues as potential antioxidant, cancer chemopreventive agents. *Arch. Pharm. (Weinheim)* **2004**, *337*, 42–54. [CrossRef] [PubMed]

MDPI

St. Alban-Anlage 66

4052 Basel

Switzerland

Tel. +41 61 683 77 34

Fax +41 61 302 89 18

www.mdpi.com

Medical Sciences Editorial Office

E-mail: medsci@mdpi.com

www.mdpi.com/journal/medsci

www.ingramcontent.com/pod-product-compliance
Lightning Source LLC
Chambersburg PA
CBHW051729210326
41597CB00032B/5657